Eric Oberheim

Feyerabend's Philosophy

Quellen und Studien zur Philosophie

Herausgegeben von
Jens Halfwassen, Jürgen Mittelstraß,
Dominik Perler

Band 73

Walter de Gruyter · Berlin · New York

Feyerabend's Philosophy

by

Eric Oberheim

Walter de Gruyter · Berlin · New York

♾ Printed on acid-free paper which falls within
the guidelines of the ANSI to ensure permanence and durability.

ISBN-13: 978-3-11-018907-0
ISBN-10: 3-11-018907-0

ISSN 0344-8142

Bibliographic information published by the Deutsche Nationalbibliothek

The Deutsche Nationalbibliothek lists this publication in the Deutsche Nationalbibliografie; detailed bibliographic data are available in the Internet at http://dnb.d-nb.de.

Cover design: Christopher Schneider, Berlin
Printed in Germany

for Christopher

Preface

The aim and scope of this investigation is to reconstruct Feyerabend's philosophy, as it developed from the logical positivist climate of his university studies in Vienna in the late 1940s, through to his *opus Magnus*, *Against Method* (1975, 1988, 1993). Given the received view that Feyerabend simply had no coherent overall philosophy, the task set may seem to be impossible. On this popular conception, Feyerabend was an epistemological anarchist. He was the anti-philosophy philosopher. Nevertheless, I will argue that a detailed inspection of Feyerabend's ideas reveals a common thread that binds his diverse considerations together. A careful examination of his arguments exposes a coherent, positive conception of philosophy that he developed in the 1950s, and practiced through to the early 1990s. Feyerabend was a philosophical pluralist who challenged many dogmatic presuppositions prevalent in philosophy and the sciences for over half a century.

My interests in Paul Feyerabend's ideas began while I was working toward my M. Phil. at the University of Cambridge. I had been reading Kuhn, and had become fascinated by conceptual change and the notion of incommensurability. Since then I have become convinced that a chief danger of professional philosophy is getting stuck in some one point of view. As my Ph.D. supervisor Paul Hoyningen-Huene once remarked, "If you join any one philosophical school, you loose ten IQ points'. No matter how fruitful some system of thought may appear to be, its merits can only be appreciated through contrast to alternatives. Improving understanding requires comparing the merits of radically different competing points of view. Incommensurability does not imply incomparability. To the contrary, the best way of testing the merits of some theory is by comparing it to incommensurable competitors because they offer the most fruitful critical comparisons. Allegiance to a single point of view creates the myths that hamper progress.

The investigation to follow is often based on a close scrutiny of Feyerabend's original texts. I hope that some of the material that I have chosen to present reflects Feyerabend's sense of humor – he was very funny. I hope that the following investigation will be useful to anyone interested in the ideas he discussed. I would like to take this opportunity to thank some of the individuals who contributed to the completion of this project.

My friends Daniel Sirtes and Vitali Filipov inspired many stimulating discussions that contributed directly to the ideas here developed. Marcel Weber made many helpful suggestions on an earlier draft of chapter 5, and has been the source of some very good advice. I have also had many stimulating discussions on the topics in this dissertation with colleagues such as Hanne Andersen, Insuk Ko, Helmut Heit, Tatjana Tarkian, Gonzalo Munévar and Peter Schaber. Matteo Collodel, Torbjørn Gunderden, Straun Jacobs, Ilko Kabiershi, and Dabiel Kuby also contributed helpful suggestions. Howard Sankey has been a source of clear and insightful thinking that helped me sort things out, especially concerning scientific realism. I would like to thank Dr. Brigitte Parakenings for her continued support from the Feyerabend Archive at the University of Konstanz, and Grazia Borrini-Feyerabend for her hospitality and for allowing me access to some of Paul Feyerabend's books and papers. I would also like to thank Peter Lipton for pointing me in the right direction. I am pleased to express my appreciation to Nicola Teubner, who bore the brunt of my frustrations and shared my enthusiasm. Her systematic thinking helped me keep my feet on the ground. Lastly, I wish to acknowledge the guiding role that Paul Hoyningen-Huene has played in my intellectual development. It was Paul Hoyningen, not Paul Feyerabend, who first taught me the merits of a pluralistic approach to philosophy, aimed at progress through constructive criticism.

Berlin, August 2006 Eric Oberheim

Contents

Analytic Table Of Contents

Introduction

Paul Feyerabend was one of the most dynamic philosophers of the second half of the twentieth century. The introduction provides an overview of his remarkable development; from his lectures on Nazi ideology that he gave as an officer in the German army during the Second World War, to his last major lecture series on the role of philosophy in Western society. The overview serves three purposes. It introduces the major themes and topics discussed in the book. It explains why attempting to reconstruct Feyerabend's philosophy is an interesting and important project, and it explains how and why the book is structured as it is.

The book has three parts. Part I examines and assesses the three main influences on Feyerabend formative years. They were Ludwig Wittgenstein, Karl Popper and Felix Ehrenhaft. Part II reconstructs Feyerabend's critiques. It focuses on the central idea motivating Feyerabend's philosophy: incommensurability. The basic idea of incommensurability is that theoretical change in the course of scientific advance causes meaning changes that can lead to new conceptions of reality. Feyerabend repeatedly used the notion of incommensurability to challenge various forms of conceptual conservativism or dogmatism. A thesis is conceptual conservative if it unjustifiably favors existing concepts over potential improvements. Part III attempts to synthesize Feyerabend's diverse ideas into a form of philosophical pluralism. Feyerabend used a pluralistic philosophical method to develop and defend his pluralistic conception of knowledge.

Part I. Feyerabend's Philosophical Development

Part I examines Feyerabend's philosophical development in relation to three influential figures. The main aims are to locate Feyerabend's ideas into a wider historical setting; to clear away some of the misleading preconceptions of Feyerabend's ideas as solely belonging to one or another of these specific philosophical schools; and to understand and appreciate the ideas and events that most significantly influenced Feyerabend's formative years.

Chapter 1. Facing Feyerabend
Some preliminary problems

Chapter one discusses some of the potential obstacles that a critical reconstruction of Feyerabend's philosophy must confront, and it identifies and justifies the various strategies that will be used to cope with these difficulties. There are three types of potential obstacles.

1.1. The wide range of interpretations

Feyerabend collected many diverse labels, from logical positivist to postmodernist, from realist to anti-realist, from rationalist to relativist, and any number of things in between or even beyond. He has been understood and developed as a Popperian and as a Wittgensteinian, as well as a neo-Kantian and a neo-Hegelian. There has been an equally wide range of assessments of Feyerabend's ideas. To some, he was little more than the court jester of philosophy of science. To others, he was one of the most important philosophical figures of the second half of the twentieth century. In the course of the book, I identify and challenge many misunderstandings of Feyerabend's ideas, and I offer a new interpretation of Feyerabend based on a much more detailed and thorough consideration of his texts.

1.2. Rhetoric, polemic and provocation

Feyerabend often complained bitterly about being misread, but it was at least partly his own fault. His texts are filled with rhetoric, polemic and intentional provocations. It is often difficult to recognize and distinguish serious convictions from sarcasm and jokes. This section offers a short survey of some of Feyerabend's more amusing polemics, and it explains how they have contributed to the misunderstandings of his views. I attempt to overcome this obstacle by very closely analyzing Feyerabend's texts (published and unpublished), and by comparing the many different presentations of the same basic points.

1.3. Referencing and reliability

The task of reconstructing Feyerabend's philosophy is complicated by two factors: Feyerabend's relationship to his own past, and by his conscious pursuit of the unconventional. Feyerabend was not always forthcoming about his sources. As a scholar, he was downright sloppy. This can make it difficult to identify the traditions to which the ideas that he developed rightly belong. Second, the reprints of Feyerabend's papers often differ from the originals. For example, he systematically removed his acknowledgements to Popper from the two volumes of collected papers published in 1981, sometimes replacing them with acknowledgments to Duhem or Mill. To diffuse these kinds of potential difficulties, I always cite from original publications, and then compare them to the later reprints, commenting on significant changes. I also use published and unpublished letters, as well as other achieve material. For example, in order to identify the sources he used in developing his ideas, I have examined his personal copies of various books and the notes that he made in their margins.

Chapter 2. Ludwig Wittgenstein
Meaning and Ontology

John Preston has influentially interpreted Feyerabend's semantics, especially those ideas underpinning the incommensurability thesis, as owing much to Wittgenstein's later philosophy. This chapter challenges such interpretations, and attempts to sort out the complicated issue of Wittgenstein's influence on Feyerabend's philosophical development.

2.1. Feyerabend's distinction between meaning and use

This section argues that far from endorsing Wittgenstein's conception of meaning as use (as some would have it), Feyerabend actually explicitly rejected it. It begins with a detailed analysis of Feyerabend's semantic views. The semantic theory underlying his incommensurability thesis (a contextual theory of meaning according to which the meaning of a term is determined by the theory to which it belongs) actually relies on a sharp distinction between use and meaning.

2.2. Feyerabend's critique of Wittgensteinianism

Furthermore, interpreting Feyerabend as a Wittgensteinian must confront the fact that Feyerabend was actually extremely critical of many aspects of Wittgensteinianism, especially Wittgenstein's conception of philosophy itself. This section examines and assesses Feyerabend's various critiques of Wittgenstein's later philosophy and of the linguistic philosophy that it spawned.

2.3. Feyerabend's adoption of nominalism

Wittgenstein's influence on Feyerabend concerns ontology, not semantics. Wittgenstein's arguments solidified Feyerabend's nominalist convictions with respect to natural kinds. How we carve up the world conceptually is based on our decisions. It is not forced upon us by some natural order. Feyerabend first adopted such a nominalism in the early to mid-1950s. It is one of very few factors that remained relatively consistent through Feyerabend's otherwise pluralistic philosophical development.

2.4. Conclusions

This section summarizes the main points of chapter two, and foreshadows how they will bear on Parts II and III.

Chapter 3. Karl Popper
Using and abusing critical rationalism

The most influential figure in Feyerabend's philosophical development is undoubtedly Karl Popper. Many commentators see Feyerabend as a member of the Popperian school until at least the late 1960s. Then in a dramatic reversal, Feyerabend became one of Popper's staunchest critics. This chapter examines Feyerabend's relation to Popper, from its beginnings in Alpbach in the late 1940s, through to the increasingly harsh criticisms in the 1970s and 80s. The main argument is that Feyerabend was actually much less of a Popperian than is commonly thought. He rejected the main tenets of Popper's philosophy in the early 1960s (at the latest), and he had been keen to distance himself from Popper's school already in the early 1950s.

3.1. Falling for falsificationism

The story begins by recounting how the two first met, and what initially drew Feyerabend to Popper and his ideas. The focus then turns to the Duhemian argument that Popper used to convince Feyerabend of the relative merits of deductivism over inductivism.

3.2. Critical rationalist elements in Feyerabend's philosophy

There are three main aspects of Feyerabend's philosophy that can be correctly called Popperian. These are falsificationism, conventionalism concerning methodology, and the essential role of criticism for promoting progress. These ideas are discussed and developed. Only the last plays a fundamental and lasting role in Feyerabend's philosophy.

3.3. From critical rationalism to incommensurability

Even though Feyerabend sometimes endorsed certain elements of Popper's critical rationalism, contrary to received opinion, Feyerabend should not be lumped with the Popperians – not even in the 1950s and early 1960s. There are three main justifications for this controversial claim. The first two concern the content of Feyerabend's ideas. The last is based on a sociological point. To begin with, the central idea of Feyerabend's philosophy (incommensurability) does not fit within critical rationalism. By 1962, when Feyerabend developed his pluralistic test-model according to which theories are not simply tested against the evidence, but compete with each other, Feyerabend had explicitly rejected Popper's falsificationist philosophy. Second, Feyerabend adopted and practiced forms of philosophy beyond the confines of critical rationalism, even very early on. For example, in the early 1960s, he practiced philosophy of science with both normative and descriptive approaches. Lastly, Feyerabend was keenly aware of the dangers of conformism. He loathed ideologies and 'schools' of any kind. He valued his intellectual independence to the extent that when Popper offered him an assistance ship in 1952, he turned the offer down because he felt that he was becoming too tightly identified with Popper and his views. For these reasons Feyerabend should not be considered to have been a Popperian.

3.4. Casting off critical rationalism

Feyerabend's relationship to Popper continued to deteriorate, and Feyerabend began to criticize Popper's ideas (and sometimes even Popper personally). Drawing heavily on Feyerabend's correspondences, this section concludes by explaining how and why Feyerabend finally broke with Popper.

Chapter 4. Felix Ehrenhaft
The impotence of experiment

The decisive episode shaping Feyerabend's philosophical development has gone largely untold in the secondary literature. It happened while Feyerabend was still at university in Vienna. The events that unfolded taught Feyerabend a lesson about experience that grounded his entire philosophical outlook.

4.1. Ehrenhaft, oil drops and experimental effects

Prior to his return to Vienna, Felix Ehrenhaft had challenged Millikan's empirical evidence for an elementary electric charge based on the oil drop experiments. Although Ehrenhaft's experimental method was more 'scientific' than Millikan's, Ehrenhaft was defeated in this infamous dispute, and at considerable lose to his reputation. Feyerabend wondered how it could be possible to be a physicist, a professor of physics, and a charlatan at the same time. And if such a combination was possible, did this not reveal a very fascinating aspect of physics?

Ehrenhaft used many simple and straightforward experiments to challenge various well-entrenched physical theories. Most notably, Ehrenhaft claimed to have experimentally demonstrated the existence of magnetic monopoles, which are impossible according to orthodox theory. Feyerabend was especially familiar with these experiments, having worked with Ehrenhaft, even writing up protocols and selling them on campus.

Ehrenhaft demonstrated many of his astounding effects in Alpbach. Feyerabend witnessed the subsequent, often heated, discussions among leading scientists and philosophers, such as Dirac and Popper, about these inexplicable effects. Although the effects appear to clash directly with some basic tenets of contemporary physics, Ehrenhaft failed even to sway those theoretical physicists who were inclined toward positivism.

4.2. The impotence of Ehrenhaft's experiments

Back in Vienna, Feyerabend formed a student club with the explicit goal of explaining Ehrenhaft's experiments and the unusual phenomena that they produced according to orthodox theory. Although they too failed, they were not converted either. They simply believed that a better and more sophisticated approach would do the trick.

In the positivist climate prevalent in Vienna in the late 1940s, Feyerabend was convinced that scientific knowledge is firmly based on the observable facts. However, this episode taught him that recalcitrant experimental evidence, by itself, lacks the power necessary to undermine the authority of a well-entrenched theory. Ehrenhaft's inability to convince the theoretical physicists exposed "an iron curtain surrounding Maxwell's equations . . . an iron curtain of precisely the kind that had protected Galileo's opponents."

4.3. Conclusions

Although it would take more than two decades for those lessons to mature into *Against Method*, his views on the limited role of observation in theory testing, his pluralistic test-model and his arguments for pluralism are best understood in the context of the lessons Feyerabend drew first-hand from the impotence of Ehrenhaft's experiments.

Part II. Feyerabend's Assault on Conceptual Conservativism

As in Part I, Part II sets systematic considerations into an historical narrative. It focuses on the notion of incommensurability, which lies at the heart of Feyerabend's philosophy. Chapter five presents the idea of incommensurability, and investigates how Feyerabend developed and used it. Chapter six is concerned with incommensurability and its relation to scientific realism.

Chapter 5. Incommensurability as attack on conceptual conservativism

Chapter five begins by analyzing the idea of incommensurability as Feyerabend developed it in his landmark 'Explanation, Reduction and Empiricism' (1962). The chapter then traces the idea of incommensurability back through many of Feyerabend's earlier publications to his unpublished Ph.D. thesis. In each case, it identifies and examines the main influences on Feyerabend's developing ideas. The penultimate section reconstructs Feyerabend's uses of the notion of incommensurability after the 1962 introduction of the term.

The main conclusions are that Pierre Duhem, on of the 20th century's greatest historian of science, introduced an incommensurability thesis into the philosophy of science, and that it was Duhem (not Thomas Kuhn, Ludwig Fleck or Michael Polanyi, as has been variously claimed) who was the original source of Feyerabend's notion of incommensurability; and that Feyerabend's philosophy is best understood as an assault on *conceptual conservativism*, based on the idea of that theory change can cause meaning changes that can result in new conceptions, even of the very nature of reality itself.

5.1. Preliminaries

The chapter begins with a brief etymological account of the term 'incommensurable', and then it investigates the historical origins of Thomas Kuhn's notion of incommensurability. Feyerabend developed the notion of incommensurability *independently* of Kuhn, who developed his version of the idea based on insights drawn from Ludwig Fleck.

5.2. Formal accounts of reduction and explanation
(1962)

Feyerabend first used the term 'incommensurable' in 1962. Basically the idea is that in the course of scientific advance, theoretical change causes meaning changes that can lead to new conceptions of reality. He used the idea to attack logical empiricist accounts of reduction and explanation for being conceptually conservative. He claimed that such accounts unjustifiably favor existing concepts over potential improvements. If they would be enforced, they would hinder scientific progress.

5.3. Niels Bohr and the principle of complementarity
(1958)

Niels Bohr made some remarks to Feyerabend about the development of Quantum Theory that directly shaped Feyerabend's developing notion of incommensurability. Feyerabend then turned the idea of incommensurability against Bohr. He used it to attack an alleged form of conceptual conservativism underlying the principle of complementarity.

5.4. Physics and ontology
(1954)

The notion of incommensurability first appeared in Feyerabend's published work in a relatively unknown German paper called *Physik und Ontologie* (1954). Feyerabend attacked analytic accounts of ontology on the grounds that there have been meaning changes due to theoretical advance that have altered our very conception of the nature of reality. Elizabeth Anscombe and Wittgenstein's later philosophy directly contributed to the development of the ontological aspect of Feyerabend's incommensurability thesis.

5.5. Werner Heisenberg and cumulative progress
(1951)

Feyerabend delineated the notion of incommensurability for the first time in his unpublished Ph.D. thesis (Vienna, 1951). He used the idea to challenge the implicit conceptual conservativism in Werner Heisenberg's *cumu-*

lative account of scientific progress, according to which established scientific results must be presupposed by all further research.

5.6. Logical Positivism and protocol sentences (1949-1958)

Feyerabend first began to develop his notion of incommensurability in 1949 in discussions of the logical positivist's account of the role of observation sentences, or 'protocol sentences', in theory testing. In his Ph.D. thesis, and in the 1958 article based on it, Feyerabend used the notion of incommensurability to attack the conceptual conservativism inherit in such accounts of testing theories. He argued that protocol sentences do not provide a neutral means with which to test theories, because the meanings of their terms must be interpreted according to the theories being tested. Moreover, he argued that the meanings of these observational terms may change in the course of scientific advance. Feyerabend based this perceptual component of his incommensurability thesis on work done by Wolfgang Köhler, and on advances in the psychology of perception due to the rise of Gestalt psychology.

5.7. Pierre Duhem and theoretical transitions (1906)

Feyerabend originally developed his notion of incommensurability from Pierre Duhem, who had already argued that meaning change undermines accounts of testing theories directly against observable facts. Duhem had already pointed out that the meanings of the terms used to test theories must be interpreted according to the theories being tested; that theoretical transitions cause changes in the meanings of these terms; and that such meaning changes have serious negative implications for convergent accounts of the growth of knowledge. The section ends with an assessment of the main similarities and differences between Duhem and Feyerabend's notions of incommensurability.

5.8. Conclusions

Although Feyerabend's philosophy discusses a wide range of topics, there is a common strand that ties his diverse ideas together. In every major exposition, from his 1951 unpublished doctoral thesis through to *Against Method* (1975, 1986, 1993), Feyerabend developed and used the notion of incommensurability in order to attack some form of conceptual conservativism. Each of the conceptually conservative accounts Feyerabend attacked purportedly fails to recognize the true significance and consequences of meaning change in the course of scientific advance. Feyerabend's philosophy is best understood as an assault on various forms of conceptual conservativism that he believed to be rampant in philosophical accounts of knowledge and understanding.

A thorough examination of the historical development of Feyerabend's notion of incommensurability reveals that Feyerabend first encountered the notion in Duhem's writings, which had been recommended to him by Popper in Alpbach in 1948. He refined the idea primarily in light of Köhler's contributions to Gestalt psychology, Wittgenstein's insights about language and reality, and Bohr's reflections on the nature of theoretical advance.

Chapter 6. Incommensurability and scientific realism

Chapter six focuses on Feyerabend's views on scientific realism. The widely publicized interpretation of Feyerabend as originally a scientific realist who slowly retreated from that position is incorrect. Originally, Feyerabend was a positivist. Then in the 1950s, he adopted a normative realism, but he consistently argued against scientific realism. Moreover, he developed the notion of incommensurability based on an anti-realist Neo-Kantian metaphysical position.

6.1. Scientific realism vs. normative realism

Feyerabend is often understood to have been a proponent of scientific realism who believed that scientific theories tell us about a world which exists independently of measurement and observation. As his philosophy developed he supposedly retreated from this position. The distinction between scientific realism and normative realism belies this widespread interpretation of Feyerabend's philosophical development.

6.2. Feyerabend's normative realism

In the late 1940s, Feyerabend was a logical positivist. Then in the early 1950s, due to an argument from Walter Hollitscher, he adopted a form of normative realism according to which scientific realism is unjustifiable as a description of actual scientific theories, but acceptable as an ideal on the grounds that it contributes to scientific advance.

6.3. Feyerabend's anti-realism

Feyerabend developed the notion of incommensurability within a neo-Kantian metaphysical position, according to which theories co-constitute the phenomena we experience. This is a very similar metaphysical picture to the one Feyerabend found in Einstein's philosophical remarks. Although Feyerabend developed a Kantian ontology, like Einstein, he repeatedly and adamantly rejected any form of transcendental necessity.

6.4. Conclusions

Feyerabend did not 'retreat' from scientific realism, as Preston would have it. He often adopted an anti-scientific realist, neo-Kantian metaphysical stance, and he repeatedly challenged many aspects of scientific realism, even in his very early publications

Part III. Feyerabend's Philosophical Pluralism

The principal aim of Part III is to reconstruct Feyerabend's pluralistic conception of knowledge and to show how it complements his pluralistic conception of philosophy. Feyerabend was a pluralist in two closely related respects. First, he supported the use of a plurality of competing alternative views in the pursuit of knowledge. He argued that developing a wide range of divergent points of views better encourages and preserves scientific advance than sticking to just one theoretical perspective – no matter how successful it may seem. He was a *theoretical pluralist* with respect to knowledge. Moreover, as a philosopher, Feyerabend was never tied to any single point of view. Instead, he adopted and developed a plurality of different ideas and approaches within philosophy. He often switched between mutually exclusive philosophical perspectives in order better to make an informed assessment of their relative merits. He was a *methodological pluralist* with respect to philosophy. Thus, Feyerabend practiced methodological pluralism in philosophy to support and complement theoretical pluralism with respect to knowledge.

Chapter 7. Feyerabend's methods

Feyerabend's early ideas have been reconstructed as an attempt to develop a single, new methodological position. This chapter argues that Feyerabend did not have a single, coherent philosophical position, and that he did not attempt to develop a new methodology. Feyerabend's work is a series of *immanent* constructive critiques. Consequently, appreciating Feyerabend's ideas requires a *piecemeal approach.*

7.1. A new methodology?

On the standard interpretation, Feyerabend attempted to develop a novel 'rational model for the acquisition of knowledge'. Such a model specifies an aim for science, as well as a methodology for attaining that aim. Such a unificationist reconstruction of Feyerabend's philosophy is unsubstantiated. Feyerabend did not try to develop a novel rational model for the acquisition of knowledge. Trying to squeeze Feyerabend's diverse considerations into a single methodological position leads to untenable inconsistencies in those ideas.

Feyerabend did sometimes temporarily accepted some particular methodological principles, but not for the purpose of developing his own new methodological system. He accepted certain methodological principles for the sake of launching immanent criticisms of certain well-known methodological views.

7.2. Constructive critiques

Feyerabend's publications are a series of constructive criticisms. A comparison of Feyerabend's major papers exposes a revealing pattern that directly bears on his critical style and pluralistic mode of philosophy. Every major publication between the 1950s and the 1970s is explicitly a critique of some existing position. This fits with Feyerabend conception of philosophy, according to which the chief task of philosophy is to promote progress by identifying and challenging the assumptions hidden in authoritative views.

7.3. *Ad hominem* arguments

Feyerabend practiced his pluralistic philosophical method with a particular argumentative tool: the use of *ad hominem* or immanent critique. Such critiques accept certain components of some position in order to undermine that position. This has been a primary cause of confusion surrounding Feyerabend's ideas. His use of some particular idea or distinction for the sake of an efficacious argument did not commit him to the correctness of that idea or distinction.

Whereas in *Against Method*, he was like an undercover agent playing the game of reason in order to undercut the authority of reason, in the

1960s, he was like an undercover agent playing the game of empiricism, in order to undercut the authority of experience.

Chapter 8. The role of alternatives in promoting progress

This chapter develops Feyerabend's pluralistic conception of knowledge and the arguments for pluralism that he used to support it. It sets Feyerabend's pluralistic conception of knowledge into the context of his pluralistic philosophical method. It begins by developing Feyerabend's pluralistic model for how to test scientific theories – not only against the available empirical evidence, but also against each other. Then it investigates how Feyerabend used the notion of incommensurability as an argument for pluralism. Feyerabend used the notion of incommensurable concepts as an argument for promoting pluralism, not to argue that the content of theories cannot be compared. This lays the ground work for a detailed reconstruction of Feyerabend's methodological argument for pluralism, and the historical episode of a crucial experiment between incommensurable theories that he used to support it: Einstein's prediction of the statistical behavior of Brownian particles. The chapter concludes by probing Feyerabend's pluralistic conception of knowledge, and by contrasting it with more standard views.

8.1. Feyerabend's pluralistic test-model

Feyerabend explicitly rejected methodological reconstructions of theory testing based on simple models of testing a theory against the empirical evidence. Both the Popperian and logical empiricist models of theory testing are based on the idea that a single theory is tested against the given facts. Feyerabend argued that any such test-model is an oversimplification that errs in two closely related respects. Facts are not simply 'given', and theories are not merely compared to the facts. Theories are compared against each other. They compete. Feyerabend developed a pluralistic test-model according to which competition between two theories increases the empirical content of both.

8.2. Incommensurability as an argument for pluralism

Incommensurability is often thought to imply incomparability. However, Feyerabend did not argue that the content of incommensurable theories cannot be compared. Quite to the contrary, he argued that theoretical pluralism, based on the use of incommensurable alternatives, is preferable to theoretical monism on the grounds that it increases the testability and empirical content of our theories. Thus, Feyerabend used the notion of incommensurability as part of an argument for promoting theoretical pluralism. He argued that scientific advance is best promoted by encouraging completion between incommensurable theoretical alternatives.

8.3. Feyerabend's methodological argument for pluralism

Feyerabend's pluralistic test-model supports a general methodological argument for pluralism. Theoretical pluralism is methodologically preferable to theoretical monism because it provides more severe means for testing our beliefs than simply testing them against the available empirical evidence. Theoretical pluralism thus becomes an essential feature of scientific advance. This is the methodological basis for Feyerabend's controversial claim that sound scientific method promotes the invention and elaboration of theories that are inconsistent with the accepted points of view, even if those points of view happen to be highly confirmed and generally accepted.

8.4. Entropy and anomaly: The case of Brownian Motion

Feyerabend refined his general argument for pluralism and illustrated it with a specific historical case: Einstein's prediction (and Perrin's subsequent experimental confirmation) of the statistical behavior of Brownian particles. According to Feyerabend, this is a crucial experiment between incommensurable theories. On the basis of this case, he argued that the invention of an alternative theory is sometimes *necessary* in order to discover that some specific phenomenon actually refutes an existing theory. For example, a well-known phenomenon, Brownian Motion, could be proven to refute Phenomenological Thermodynamics only *after* the invention of an incommensurable alternative to it.

8.5. Feyerabend's pluralistic conception of knowledge

Feyerabend did not share the received conception of scientific progress as convergent on an ideal point of view. Rather, he conceived of scientific progress as essentially a divergent process. Scientific progress is conceived of as an ever-increasing ocean of alternative theoretical perspectives. The aim of science is not to discover the truth. The aim of science is to develop our mental capacities through a process of promoting a plurality of competing points of views that force each other into greater articulation.

According to Feyerabend, a pluralism that promotes the constructive criticism of diverse points of view is always preferable to a monism that only allows for a single theory to be tested against the purported facts. Pluralism is preferable because it better promotes progress.

Chapter 9. Feyerabend's philosophical pluralism (1950s-1990s)

Feyerabend was a philosophical pluralist, who often temporarily adopted mutually exclusive points of view, in order best to appreciate their comparative strengths and weaknesses. His arguments for pluralism were part and parcel of the pluralistic philosophical method, complemented and sanctioned by his pluralistic conception of knowledge.

9.1. Commitments and perspectives

Feyerabend was an open-minded, self-critical thinker, who often explicitly changed his opinions, and sometimes even his whole philosophical outlook. He believed that philosophical theses should be informed: They should take the latest scientific developments, as well as to the best available historical evidence, into account. However, he was also acutely aware of the highly *speculative* nature of the epistemological and metaphysical issues prevalent in the philosophy of science, and accordingly, he was not particularly tied to any specific positions. To Feyerabend, the speculative nature of the issues at stake did not recommend caution. He was a champion of bold conjectures, creative connections, and the big picture.

9.2. Pluralism and opportunism

Feyerabend was like a chameleon, adapting to the philosophical landscape as it changed dramatic over four decades. That Feyerabend practiced such a pluralistic philosophical method after 1975 is uncontroversial. However, Feyerabend practiced a pluralistic philosophical method already in the early 1960s. As the 1970s progressed, Feyerabend began to defend ever more radical ideas. His defenses of such ideas are best understood in the context of his views about the necessary role of alternative views for promoting the kind of critical discussion that safeguards progress, not as serious expressions of his own convictions.

9.3. From philosophical pluralism to epistemological anarchism

Feyerabend's philosophy is often divided into two periods: early and late. The transition is said to have occurred with *Against Method* in 1975. However, such interpretations of Feyerabend's philosophy are unsuccessful because Feyerabend's pluralistic conception of knowledge, as well as his corresponding pluralistic philosophical method, did not substantively change from the early 1960s through the his more famous 1970s publications. He merely made more polemic and provocative presentations of his pluralistic philosophical ideas. After all, *Against Method* is merely a collage of his earlier work.

Conclusions: Feyerabend's Philosophy? Pluralism for progress

Feyerabend argued against many particular views on the grounds that that they would constrain the freedom necessary to develop alternative points of view, and thereby hinder scientific advance. For Feyerabend this was not merely some abstract danger. He repeatedly argued about cases from contemporary research, and was directly connected to actual developments in science and philosophy. In Feyerabend's view, the continuing entrenchment of successful theoretical traditions makes fundamental criticism of them extremely difficult. Feyerabend argued that even in the face of observations that conflict with a reigning theoretical tradition (e.g. Ehrenhaft's effects, the perihelion of Mercury, Brownian Motion), scientists tend to trust the established theory even over their own observations. They simply explain the recalcitrant observations away. However, they may be taken more seriously if there is an alternative theoretical perspective that purports to explain them.

Feyerabend believed that his role as a philosopher of science in the 1950s and 1960s was to help contribute to scientific advance by attempting to prevent science from becoming dogma in the name of empiricism. This was the motivation for Feyerabend's pluralistic view of knowledge and the pluralistic philosophical method that he both preached and practiced. Throughout his carrier, Feyerabend's use and improvement of a plurality of different positions and ideas cohered with his pluralistic conception of knowledge. His pluralistic philosophical method both promoted and was conducive to the development of many diverse issues in philosophy. The goal of Feyerabend's pluralistic philosophy is to challenge presumptions taken for granted by established or popular points of view – to use criticism to open the way for improvement.

On this conception, knowledge is not a series of self-consistent theories that converges toward an ideal view, and the advancement of knowledge is not a gradual approach to the truth, and this holds both for science and philosophy. Such knowledge is an increasing set of competing alternative theoretical perspectives that force each other to improve, and thereby contribute to the development of our mental faculties.

Introduction

If the chief task of philosophy is to challenge the prevalent presuppositions of some particular period in our intellectual development, then Paul Feyerabend was surely one of the greats. He is widely recognized as an influential and controversial figure in twentieth century philosophy – one of the big four philosophers of science, together with Karl Popper, Thomas Kuhn, and Imre Lakatos – who is credited with having reshaped the philosophical landscape. His extraordinarily eventful life was as unconventional as it was controversial, and the remarkable events that transpired contributed directly to how he repeatedly redeveloped his ideas. He was a master of seizing the day, managing to be smack at the center of many of the major developments. He addressed a wide range of topics, and his ideas persistently proved to be a source of both inspiration as well as detestation. By effectively navigating the tumultuous philosophical environment from the Vienna Circle to Wittgenstein's later philosophy, from Popper's falsificationism to Kuhn's historicism, from Berkeley in the late 1960s to the post-modernist 1990s, he became something of a living legend, always right where things were hottest.

In many circles, the name 'Feyerabend' has come to be associated with irrationality, with anarchism, or with a form of 'toothless' relativism according to which everyone is entitled to their own perspective because one perspective cannot be judged to be better or worse than another. After all, his slogan was 'Anything goes!'. This kind of misunderstanding of Feyerabend's philosophy appears to be especially rampant among some sociologists, who take the notion of incommensurability to be an explanation of the inevitability of extensive miscommunication between people with different agendas. The name 'Feyerabend' is invoked to smooth over conflicts. It is used as a call for the tolerance of other points of view that is necessary in this pluralistic age. Unfortunately, such calls for pluralism and tolerance are often used to smoother criticism, and this is exactly not what Feyerabend's philosophy was about. Feyerabend was never a 'toothless' relativist, who uncritically accepted an idea or perspective. Quite to the contrary, Feyerabend supported pluralism to encourage the kind of *criticism* of different points of view that leads to their improvement.

As Feyerabend is widely recognized as an influential proponent of pluralism and tolerance, there is quite some irony in the little known fact that during the war, Feyerabend held a lecture that drew on some of the most despicable aspects of Nazi ideology. While this may not seem directly relevant to his later academic career, his wartime experiences may have been a crucial factor in shaping his subsequent philosophical development, especially his sensitivity to the dangers of conformism. Throughout his career, he vehemently rejected all forms of dogmatism in favor of his trademark pleas for *pluralism* and *tolerance* of the unusual in all matters philosophical. A brief overview of this remarkable development will serve well as an introduction to the subsequent investigation in Part I.

Like many young Austrian men, after the *Anschluss* Feyerabend was drafted into work service (*Arbeitsdienst*) and at the end of 1942 he joined the army. He even considered joining the SS, because he thought that he might look good in the uniform. Eventually he volunteered for officers' school to try to avoid the front-line. In his autobiography, *Killing Time* (1995a), he recounts how in 1943 he was tying to devise a simple and convincing diagram of the Lorentz transformations, when a fellow soldier informed him of his mother's suicide. He reports having felt nothing. In November of that year, he was finally sent into battle, and as with most things he took a hand to, he proved rather successful.

His interests in the arts matched his early interests in the sciences, and so he proposed staging a performance with his troops. War heroics, it would seem, were just not enough. Unfortunately, his theater ambitions were frustrated by his commanding officer who would not allow him to stage a performance. He was, however, allowed to give a lecture propagating Nazi ideology. With the burning desire of a youth keen to be at the center of attention, Feyerabend studied Alfred Bäumler's justification of the war. Bäumler (1887-1968) was a leading academic proponent of National Socialism in Nazi Germany. He was appointed Professor of Political Education at Berlin University in 1933 and emerged as the chief liaison between the universities and Rosenberg's office of ideological education. As is well known, Bäumler tried to transform Nietzsche's thought to serve Nazi ideology. Feyerabend's lectures drew heavily on Baumler's work. His lecture notes can be found in his Wartime diary in the Philosophical Archive in the University of Konstanz, Germany.

Having won the Iron Cross for acts of bravery in combat on the Eastern Front, Feyerabend was quickly promoted to lance corporal, sergeant, and then by the end 1944 to lieutenant. Prone to taking theatrical risks, eventually he was shot in the face, hand, and spine while directing traffic in the middle of a firefight. He was paralyzed from the waist down for the rest of his life.

After the war, he was granted the opportunity to attend university in Vienna, as he had never officially joined the Nazi party. And so he entered university in a wheel chair, with a talent for provocation and an ego to match it. He busied himself nurturing his interes'ts in theatre, music, philosophy, history, sociology, mathematics, astronomy and physics. And he developed the habit of being well read in a wide range of areas. (Like his handicap, this habit appears to have remained with him to the end of his days.) He did not just study. He sang and he argued. He sang Schummann with the director of the Burtheater, Hanns Eisler, who then introduced him to Bertold Brecht. Feyerabend turned down Brecht's offer to become his assistant – a decision he later ruefully regretted as one of the biggest mistakes he had ever made. Toward the end of his life he reevaluated that decision: "I suspect I would have detested the collective pressure of the partly fearful, partly dedicated and certainly pushy and closely knit group that surrounded Brecht" (1995a, p.73). As we shall see, this reevaluation mirrors the experience he had with Karl Popper and the Popperians. Feyerabend highly valued his intellectual independence, and that shaped many of the key elements of his philosophical development.

As a student in the positive climate of the 1940s in Vienna, one of his favorite pastimes was attending seminars in the social sciences, listening politely, and then declaring the entire discussion meaningless on the grounds that the claims were not empirically testable. Originally, he wanted to turn in a thesis in theoretical physics, but his project was leading nowhere, and so he concentrated on philosophy of science, and developed a doctoral thesis under the supervision of Viktor Kraft. He co-founded a student group (the 'Kraft Circle') modeled on the Vienna Circle – arguably the most influential movement of twentieth century Western philosophy. Invited speakers included Elizabeth Anscombe and Ludwig Wittgenstein. At meetings, Feyerabend presented what would later become his doctoral thesis - *Zur Theorie der Basissätze* - a treatise on the roles of language and perception in science and its development that served as the basis for the subsequent development of his ideas, especially two of his major early articles 'An attempt at a realistic interpretation of experience' (1958a), and 'Explanation, reduction and empiricism' (1962a).

In 1948 while still working toward his Ph.D. thesis, he began as secretary for the Austrian College Society's Alpbach summer schools a main hub of scientific and political activity in the late 1940s and early 1950s that had been organized by some prominent members of the Austrian resistance. The existing secondary literature on Feyerabend makes little mention of some crucial episodes that took place in these formative years, in particular Feyerabend's relationship to the physicist Felix Ehrenhaft. The epistemological and methodological lessons Feyerabend drew from the impotence of Ehrenhaft's experiments are indispensable for understanding Feyerabend's subsequent critiques of empiricism.

What is well known is that at this world center of intellectual, artistic, economic and political exchange, Feyerabend first met Karl Popper who was chairing the philosophy discussions. This was to be the start of a long and complicated relationship, which is discussed at length in chapter 3. Having completing his Ph.D. in 1951, Feyerabend had planned to join Wittgenstein at Cambridge on British Council scholarship, but Wittgenstein died, and so, Feyerabend decided to go to Popper in London where he worked on the philosophy of Quantum Physics in addition to his attempts to come to terms with Wittgenstein's influential ideas. His work on foundational problems of Quantum Physics led him to attend a lecture and seminar with Neils Bohr, who would become both an inspiration and a target of much criticism. The inspiration concerns a theme common to both Wittgenstein and Bohr's work that impressed a lesson on Feyerabend that grounded his entire ontological outlook: How we experience the world depends on the concepts that we bring to it, and a variety of different approaches may be need to reveal it in all of its abundance. But Bohr would also become the target of Feyerabend's pointed criticism. Feyerabend repeatedly attacked dogmatic tendencies in Bohr's interpretation of Quantum Mechanics. Decades later, in a typical reversal, Feyerabend explicitly retracted his earlier criticisms, and defended some of Bohr's ideas.

In the early 1960s, Feyerabend took a job at the University of California at Berkeley, where he met Thomas Kuhn and became an influential proponent of what was to become the next big movement - the historical turn in the philosophy of science. Kuhn and Feyerabend are often understood as having shared many of the same ideas. For example, they supposedly independently introduced the notion of incommensurability in 1962 – an idea commonly misunderstood as implying the rational incomparability of scientific theories. Actually, the two disagreed fervently in many respects, most notably about Kuhn's views concerning the positive functional role of dogma in scientific advance (see 1995b, 1970c and 2007). According to Paul Hoyningen-Huene, Kuhn was so displeased with Feyerabend's charge of dogmatism that sometime in the early 1960s their relationship cooled down significantly for many years.

One of Feyerabend's closest friends and colleges was a student of Popper's, Imre Lakatos, who was busy trying to save Popper's falsificationist methodology (we propose, nature disposes) from Kuhn's historicist critiques. Lakatos convinced Feyerabend to present his ideas in the form of a book. Although Feyerabend was already professionally well established, he had never published a single book before 1975.[1] The two agreed

1 Feyerabend had written a book length manuscript *Einführung in die Naturphilosophie* (*An Introduction to the Philosophy of Nature*, my translation) in 1971, but it was never published. Apparently, Feyerabend was never quite satisfied with anything that he wrote. He reworked

to co-author *For and Against Method.* Lakatos was to defend rationalism, and Feyerabend was to criticize it. After Lakatos's unexpected death, Feyerabend went ahead and published his half. It is a rhetorically hyped-up collage of his earlier publications, with some additional historical material to illustrate points he had been repeating in a series of critically essays since the 1950s.

With the publication of *Against Method* in 1975, Feyerabend became famous beyond the bounds of philosophy of science. He gained the reputation of being well acquainted with topics ranging from physics to the history of witchcraft, from the rituals of voodoo to the nuances of Dadaism. He challenged many age-old platitudes inherited by contemporary philosophy. Most notably, he attacked the widespread assumption that science is special because of its unique method consisting of universally binding rules. He claimed there is no such scientific method. He was not only historically well informed. His arguments were complex and challenging, and seemed inevitably to lead to unacceptable consequences. He proposed a radically different conception and evaluation of scientific knowledge and scientific advance. He challenged the legitimacy of science's status as an objective body of knowledge. For these reasons, many see Feyerabend as a founding father of the constructivist movement that now fights in today's 'Science Wars'. His slogan 'Anything goes!' marks a turning point toward post-modernism in the history of philosophy of science.[2]

Even if post-modernist interpretations of *Against Method* (1975a) may not have captured Feyerabend's true intentions, the book proved to be extremely controversial, and it received many severe condemnations. Feyerabend believed that these were based on misunderstandings due to incompetence, and he reacted by radicalizing his ideas even further, and by becoming even more polemic. He began to incorporate lessons that he learned while teaching in the late 1960s in Berkeley into his philosophical outlook. He contributed to the tumultuous atmosphere at the University

the manuscript again in 1976, but had trouble letting go of it. See (forthcoming [1971/1976]).

2 It is widely believed that Feyerabend introduced the slogan "Anything goes" in *Against Method* (1975). However, it appeared already in (1970d, p. 26) and in (1970b, p. 278). As (1970d) refers to criticisms of the slogan "anything goes", and as Feyerabend referred to his (1970b) in his (1970d, p. 100, fn. 25), it would appear that the less well-known (1970b) instance, cited only by Machamer (1973), is the first. This conclusion is also supported by the fact that in (1970b), Feyerabend had not yet radicalized and rhetoricized his position into "epistemological anarchism", as he does in (1970d), which is the genuine beginning of the *Against Method* series. See and compare (1970d), (1972), (1975a), (1976a), (1983a), (1988), (1993a), and (1993c). These are at least eight significantly different version of *Against Method* in English and German. In any case, (1970b) was the result of lectures series that Feyerabend gave at the London School of Economics in the fall of 1967. So the slogan dates back to well before 1975.

of Berkeley by supporting reforms in university policies concerning minorities (and correspondingly in Berlin during the famous student strikes at the Free University). He was not merely an active participant at the center of these discussions. As a teacher of Western philosophy, was he just propagating the myths of dead white men? Teaching minorities philosophy made him feel like he was contributing to cultural imperialism.

In 1976, Feyerabend first began explicitly to endorse some limited forms of relativism. In 1978, Feyerabend published *Science in a Free Society*, which embraces some aspects of relativism and further develops the political consequences of his 'epistemological anarchism'. Feyerabend called for a separation of the state and science, modeled on the separation of the state and religion. The book also contains several dismissive responses to reviews of *Against Method*. This period marks a major transition in his philosophical outlook. Up until the late 1970s, Feyerabend always assumed the superiority of the empirical sciences in producing knowledge over other forms of life. But now he began to ask: What's so good about science? How can we know that science improves our lives, if we have never had the chance to make an objective comparison between what it is offering to us and other forms of life? Also see (1980). His publications take on an increasing political character, leaving behind the often more technical and abstract considerations of his earlier career. *Farewell to Reason* (1987) collects many of Feyerabend's publications in political philosophy that appeared between 1981 and 1987.

There appears to be a general consensus that Feyerabend's contributions to political philosophy were of far less worth than his contributions to epistemology and the philosophy of science, and his 1980s publications have had far less impact than his earlier work. In any case, they are not the focus of this investigation.[3] Eventually, Feyerabend came to regret his stint with relativism. When asked why he gave up on relativism, he answered that when relativist slogans can be found on the walls of every university bathroom, it is time to move on. Apparently, he did not want *Science in a free Society* or *Farewell to Reason* to be reprinted after his death.

In 1984, Feyerabend set out the view that in science, like in art, there is no progress, only change. Feyerabend repeatedly reflected on the arts and their relation to science throughout his career. He even argued that the theatre, and not essays, is the better medium to treat philosophical issues. As his career progressed, Feyerabend was also increasingly critical of professional philosophy of science, especially its inclination toward abstraction. In 1970 he published a paper entitled "Philosophy of Science:

3 For an excellent discussion of Feyerabend's political philosophy, see Preston (1997), chapter 12. For an outstanding discussion of the values based form of rationality that Feyerabend developed in the late 1970s and 1980s, see Farrell (2003).

A Subject with a Great Past" pointedly implying that it had no future. At a conference in 1972, Feyerabend delivered a paper entitled "Die Wissenschaftstheorie - eine bisher unbekannte Form des Irrsinns?" or "Philosophy of Science – A Thus Far Unknown Form of Insanity?" (1973, my translation). Here he argued that professional philosophy of science had completely lost touch with actual scientific practice. By the end of the 1980s, Feyerabend contemptuously asked, "Who needs the philosophy of science?" and in the 1990s, in "Concerning an Appeal for Philosophy", he publicly rejected a call for money for education in philosophy, arguing it would be better spent elsewhere.

In the early 1990s, Feyerabend gave a five-day free-speaking lecture *Conflict and Harmony* in Italy, where he was very popular. It was subsequently published in Italian and German. He used the Rodney King riots to illustrate how disconnected contemporary philosophy has become to the meaningful events in people's lives, and he blamed this disconnectedness on developments that began in Ancient Greece – developments that are usually celebrated as the birth of Western philosophy.

But this did not prevent Feyerabend from his own philosophical work. He attempted a last book *The Conquest of Abundance* (1999b), which investigates how we create an 'objective' reality out of the richness of being, drawing on the invention of perspective in the fine arts. Although the book was never finished, several reworked copies of the first few chapters were compiled and published posthumously, together with a collection of short pieces he wrote towards the end of his career. In these latter works, Feyerabend retracted his views on relativism and cultural incommensurability arguing, "potentially every culture is all cultures". Instead of viewing cultures as closed units that can only be evaluated according to their own standards, Feyerabend argued that cultures benefit from interaction, and can indeed be evaluated from an outside, humanitarian perspective. In many respects, he returned to a more temperate redevelopment of his earlier ideas.

Paul Feyerabend died on the 11th of February 1994 of an inoperable brain tumor in a hospital in Genolier. He finished his autobiography, *Killing Time*, on his deathbed. This poignant tale of an extra-ordinary life has been an inspiration even outside the bounds of academia. His favorite photo of himself is captioned 'The philosopher at work'. He is doing the dishes.

This brief chronological overview reveals three closely related aspects of his philosophical development. First, in every stage of his development – from the positivism of the 1950s, the historical turn in the philosophy of science in the early 1960s, the radical student movements in Berkeley in the late 1960s, the relativist boom in the 1970s and 1980s, and his more compassionate, humanitarian outlook in the 1990s – Feyerabend reinvigorated his philosophy so that it reflected the concerns of the respective period. He was like a chameleon, or like Woody Allen's *Zelig*. He had a remarkable ability to adapt to changing interests and attitudes over half a century. Second, Feyerabend always managed to be directly connected to genuine developments in science and philosophy. He thought that philosophy should make a difference, and he tried to make a difference with his philosophy. Even his abstract theoretical considerations in the 1950s have a practical application. He was trying to stem the rising tide of dogmatism associated with the surging dominance of the Copenhagen Interpretation of Quantum Mechanics. He was always right where things were hottest; shaping what has become our philosophical heritage. Third, Feyerabend did not argue from a single coherent point of view. He was a philosophical pluralist. One main goal of this investigation is to examine and develop his pluralism. Feyerabend was fond of denying that he was a philosopher, and of claiming that he had no philosophy. How then can there be a book called *Feyerabend's Philosophy*, when obviously he did not have one?

Part I. Feyerabend's Philosophical Development

Chapter 1
Facing Feyerabend
Some preliminary problems

There are a number of potential obstacles to approaching and appreciating Feyerabend's philosophy. The aim of this chapter is to explain them, and to clarify how their potential for creating misunderstandings will de handled. This will lay the groundwork for the investigation of the three main influences on Feyerabend's philosophical development in Part I, the reconstruction of Feyerabend's critiques based on the notion of incommensurable concepts in Part II, and the reconstruction of Feyerabend's pluralistic philosophy in Part III.

1.1. The wide range of interpretations

Over the last forty years, there has been a notoriously wide range of interpretations of Feyerabend's views, including many diverse misunderstandings of them. Munévar even goes so far as to claim that, "few philosophers have been as misunderstood as Feyerabend" (Munévar 2000b, p. 58). Feyerabend's philosophical ideas were so scandalously controversial that they contributed to his pop-star status within the community of academic philosophy (Hoyningen-Huene 2000b).

His ideas are sometimes associated with intellectual movements that he explicitly renounced, and they have been taken to support political movements of which he explicitly disapproved. Examples of the former include epistemological relativism and social constructivism, and an example of the latter is political anarchism.[4] For example, Culp portrays Feyerabend as a part of the tradition that led to the social constructivist movement (Culp 1995, p. 439). Feyerabend is often perceived to have been a founding force of the social constructivist movement, and for vari-

4 See e.g. Gellner (1975). For a response from Feyerabend, see (1976c).

ous reasons: Feyerabend supposedly "argue[d] that there is no such thing as raw data since data only counts as data within the context of a theory" (Culp 1995, p. 439, fn. 1); Feyerabend was skeptical about the utility of some traditional epistemic values such as the notion of truth; Feyerabend's late 1970s and 1980s publications do indeed express some seemingly 'post-modernist' views (Hoyningen-Huene 2002, Preston 1998); Feyerabend has often been lumped together with social constructivist by influential commentators (e.g. Kitcher 1992, p. 80, fn. 75); and social constructivists explicitly adapt ideas from Feyerabend into their agenda. Some social constructivists reinterpret Feyerabend (and Kuhn's) notion of incommensurability and use it to explain sociological phenomena in the interaction between members of scientific communities. Shapin criticized Cantor's view that "Incommensurability, once identified, can then purport to be an explanation of prolonged disagreement among knowledge communities," because he believes that "incommensurables are not eccentricities, . . . they are normal features of discourse, [which] deserve explanation in terms of the social, and [once] explained, [make] better sense of the phenomena of distinguishable communities disagreeing about matters of concern to them" (Shapin 1975, p. 220).[5]

But even though Feyerabend's views may have contributed to the moment of the social constructivist movement, Feyerabend was not much of a social constructivist. This should come as no surprise. Over the years, Feyerabend collected many diverse labels, some of which are blatantly incompatible. These include positivist and post-modernist; realist and anti-realist as well as and rationalist and relativist.[6] He was also called any number of things in between or beyond; such as 'Anarchosurrealist', (Machan 1982) and 'methodological onanist' "whose aim is such pleasure as brings forth nothing else" (Hattiangadi 1978, p. 5).

Aspects of Feyerabend's philosophy have been understood and developed as Popperian (Preston 1997, pp. 9ff.), Wittgensteinian (Preston 1997, pp. 23ff.), neo-Kantian (Boyd 1992, Oberheim and Hoyningen-Huene 1997), Hegelian (1970d), and Machian (Hentschel 1985, Zahar 1981). Some see him as a modern proponent of Pyrrhonian scepticism, while others suggest that he was unfamiliar with Sextus Empiricus (Neto 1991, p. 144). As Kadvany has pointed out, Feyerabend had an attitude similar to a Pyrrhonian skeptic, "A book, a film, a theatrical performance, or a casual remark could move me in any direction", (Feyerabend cited in Kadvany 1996, p. 144).[7] The title 'Against Method' may even gave been

5 Compare Cantor (1975b) and Cantor (1975a). For a compilation of articles concerned with a sociological notion of incommensurability, see Harris (2005).

6 Cf. Preston (1997), Preston (1998), Hoyningen-Huene (2002) and Farrell (2003).

7 For similar remarks, also see (1995a), e.g. pp. 96-97.

modeled on Sextus Empiricus who called his treatise "Against the Logicians", Against the Physicists", "Against the Mathematicians", etc.

Feyerabend's remarks about himself are of little help. He called himself an epistemological anarchist (1970d) and a Dadaist (1975a), and this led to the charge of irrationalism (Giedymin 1971, Grunfeld 1984). He called himself so many different things, and drew from so many different sources, that it is no wonder that he has come to represent so many different things to so many different people.

In addition to the wide range of issues and positions Feyerabend is perceived to have championed, there have been an equally wide range of appraisals of his ideas.[8] To some, he was like a court jester to the philosophy of science. To others, he was one of the 'worst enemies of science'.[9] Agassi began a review of *Against Method* (1975a):

> How do you read a book which extols lies? Do you at least admire its author for his excessive honesty and take literally what he says? Or do you consider him a mere con-man? (Agassi 1976, p. 165)

Others celebrate Feyerabend as one of the most important philosophers of science of the second half of the twentieth century, and as a twentieth century champion of pluralism and tolerance. Munévar has gone so far as to claim that chapters three and six of Against Method "are among the most brilliant passages ever written in philosophy" (Munévar 2000a, p. vi).

The diverse interpretations and evaluations are due to a combination of interrelated factors: the intricacy of the texts, the obscurity of the subjects they handle, and ambiguity of expression, just to name a few. One contemporary critic emphasizes that Feyerabend and Kuhn's writings on incommensurability "are so notoriously various and vague as to leave plenty of room for disagreement over interpretation" (Devitt, p. 144). The tendency to conflate the views of Feyerabend and Kuhn on incommensurability leads to misunderstanding of both. Feyerabend explicitly warned of this danger and contrasted his own views on incommensurability from Kuhn's at length (1977b, pp. 363-367).[10] Feyerabend's (and Kuhn's) claims about incommensurability were at first widely rejected as incoherent or self-refuting. This assessment is due largely to influential criticisms made by Putnam (1981) and Davidson (1984). Putnam argues:

> The incommensurability thesis is the thesis that terms used in another culture, say the term 'temperature' as used by seventeenth-century scientists, cannot be

8 For examples, see Achinstein (1964), Putnam (1965), Gellner (1975), Hattiangadi (1978), Machan (1982), and Watkins (2000).

9 See e.g. Theocharis and Psimopoulos (1987). For a discussion, see Hoyningen-Huene (2000b), p. 9.

10 For discussions of the differences in their views, see Hoyningen-Huene (2000a) and Hoyningen-Huene (2001).

> equated in meaning or reference with any terms or expressions we possess . . . if the thesis were really true then we could not translate other languages – or even past stages of our own language – at all. And if we cannot interpret organisms' noises at all, then we have no grounds as regarding them as thinker, speakers, or even persons. In short if Feyerabend (and Kuhn at his most incommensurable) were right, then members of other cultures, including seventeenth century scientists, would be conceptualizable by us only as animals producing responses to stimuli. . .To tell us that Galileo had 'incommensurable' notions and then to go on to describe them at length is totally incoherent (Putnam 1981, pp. 114-115).

Davidson puts a similar point in terms of a paradox according to which the meanings expressed by the terms of an untranslatable language are supposed be expressible in the very language into which translation allegedly fails (Davidson 1984, pp. 183-184). However, Putnam and Davidson's incoherence arguments badly misrepresent Feyerabend (and Kuhn's) notions of incommensurability, and Putnam's argument also contains at least two obvious non-sequiturs. To conclude that we cannot translate a term of another language into our own because none of the terms of our language have exactly the same meaning or reference assumes that a one to one match of meaning and reference between terms in the two languages is necessary for a successful translation. Once this missing premise is delineated, it is immediately clear that it is untenable. There is no such criterion on a successful translation. For a start, the relevant unit of translation is a statement, not a term. More over, a translation of a term from an object language into a target language often requires explaining the meaning of a term that has no exact equivalent in the object language in the terms of the target language. What is worse, even if the two languages were untranslatable in the sense that there is no term in the target language that has an equivalent concerning meaning and reference in the object language, it does not follow that the object language cannot be understood or even recognized as a language. One can simply learn to speak the object language in its own terms, and understand it without translating it. For example, although there is no equivalent in meaning and reference in the language of contemporary chemistry, one can learn the principles of phlogiston chemistry in their own terms, and then understand the term phlogiston. One can even explain why there is no exact one-term to one-term translation of the term 'phlogiston' into the language of modern chemistry. This is exactly what historians of science do.

Both Kuhn and Feyerabend made such responses to these kind of incoherence objections against incommensurability, and other commentators, even those eager to distance themselves from the anti-realist implications of the incommensurability thesis, have accepted that these incoherence objects due not undermine the incommensurability thesis.[11] As we shall see, the entire discussion misrepresents Feyerabend's notion of incommensurability in that he did not think that a consequence of incommensurability was untranslatability, and he did not think that translation was necessary for theory comparison.

The incommensurability thesis gradually became more respected as a serious threat to some orthodox accounts of scientific advance. For example, in the preface to the first edition of Chalmers's introduction to philosophy of science, *What is This Thing Called Science?*, it says:

> since I have no time for obscurantist nonsense about the incommensurability of frameworks . . . (Chalmers 1978, pp. xii).

In the preface to the next edition of the same book, Chalmers was forced to retract this wholesale dismissal:

> I have also been obliged to separate the important sense from the 'obscurantist nonsense about the incommensurability of frameworks' (Chalmers 1982, pp. xiii).

Some of the misunderstandings of Feyerabend's ideas can be attributed to sources *outside* the philosophy of science from which Feyerabend drew. Some of these were completely unknown inside the philosophy of science, where they may even have been considered downright exotic. To quote John Watkins, "would any other philosopher's research have led to the discovery that the Devil has an ice-cold member?" (Watkins 2000, p. 48). As we shall see, Feyerabend's research took him down many of the roads less traveled to unwanted conclusions.

Obscure historical material was often interspersed with subjects considered unworthy by the philosophical establishment; such as the history of witchcraft[12] and voodoo rituals. He also drew from or taboo political sources, such as Lenin and Mao (e.g. 1970d, p. 94 and p. 109, fn. 48). But if the topic is the relationship between theory and observation, then testimony about seeing the Devil, or even feeling his ice-cold member, may indeed be relevant. Feyerabend thrived on sources widely considered by some to be irrelevant to the philosophy of science, such as theatre and the fine arts. Feyerabend even argued that theatrical performance is better suited to dealing with philosophical problems than verbal discussion or scholarly research.[13]

Often these sources, along with important but frequently overlooked qualifications and clarifications, appear in the numerous and lengthy notes

11 See example, Sankey (1994), chapter 4 for a rebuff of Putnam and Davidson's incoherence objections to incommensurability as implying untranslatability.

12 See e.g. (1970d), p. 101, fn. 27. In Feyerabend's home in Switzerland, there is a thoroughly worked, two volume encyclopedia on witchcraft.

13 See e.g. (1967b) and (1975b), p. 205. For interesting discussions, see and compare Couvalis (1986) and Bearn (1986), pp. 150ff.

attached to the main text. In the first version of *Against Method*, published as a long essay (1970d), there is a confusing tangle of 221 cross-referenced endnotes that comprise about 37 pages of small print.[14] All of these factors contribute to the wide range of interpretations of Feyerabend's texts.

To surmount the challenges presented by the wide range of interpretations, several strategies will be pursued. First, although many important misunderstandings of Feyerabend's ideas will be discussed, no attempt to provide a comprehensive list of every instance of these misunderstandings has been made. For example, Feyerabend's slogan 'Anything goes' has been misrepresented so many times, and this has been pointed out so many times, that it would be difficult even to catalogue all of them.[15] Second, in order to support claims about Feyerabend's ideas, I will draw directly from Feyerabend's *original* texts and compare them to the later reprints, as well as to similar passages in different publications. The symbol "; rp" is used to signify a later reprint of an earlier source. Whether or not passages cited have been revised will be noted with comments such as 'slightly revised', 'significantly revised', 'removed', etc. Comparing the reprints to the originals often exposes revealing aspects of Feyerabend's philosophical development. As we shall see, many of Feyerabend's publications are collages of earlier work that contain similar passages. Notorious examples of such 'cut, copy and paste' jobs include, but are not restricted to, (1963a); rp (1999a), pp. 50-77, (1965b) and the infamous 'collage' (1975a). Most importantly, untangling the various misunderstandings of Feyerabend's texts requires setting Feyerabend's claims into their proper context. As we shall see, this is especially important for understanding Feyerabend given the *ad hominem* nature of his criticisms. A combination of these strategies offers the only tenable approach to Feyerabend's philosophy given the wide range of interpretations of it.

1.2. Rhetoric, polemic and provocation

Another difficulty with interpreting Feyerabend's philosophy, which is also a contributing factor to the wide range of interpretations of his texts, is Feyerabend's literary style. His texts are filled with rhetoric, polemic and intentional provocations. His theatrical prose makes it difficult to realize when he was expressing a serious conviction of his own, and when he was

14 Also see e.g. (1965b), which has 180 footnotes, one of which spans six pages.

15 Corrections to misinterpretations of Feyerabend's 'Anything goes' began as early as 1977. See Lugg (1977). Also see e.g. Kadvany (1996), p. 142, Sokal and Bricmont (1998), pp. 78-80 and Weber (1993).

simply being sarcastic to drive a point home. While Feyerabend's rhetorical and provocative flair may have contributed to his fame,[16] unfortunately it seems also to have been a source of much of the confusion about his ideas. Misinterpretation is not unusual in philosophy, and Feyerabend often complained bitterly about being misread.[17] He even developed a theory about misinterpretation in philosophy. One ascribes absurd views to someone, criticizes those views, and then adopts that figures actual views as one's own. He called this the 'Lessing-effect' (1984).

Feyerabend was extremely disappointed with the reception of *Against Method*, which he believed had been badly misunderstood. In one response to a critic, he argued that it is quite usual for the views of philosophers to be 'mutilated' by the distortions of superficiality and incompetence (1984).[18] The bitterness in the allegations of incompetence that Feyerabend launched against his critics is somewhat surprising. He should have anticipated the confusions that he caused. After all, he explicitly claimed that one of his favorite pastimes was to confuse rationalists by inventing compelling reasons for unreasonable doctrines (1975a, p. 189), and defended the use of propaganda in the advancement of knowledge (e.g. 1993a, pp. 106ff.). Problems of interpretation are accentuated by a combination of his radical changing views, provocative style and theatrical aspirations. He liked to push unusual ideas to the extreme:

> You see, when I come across some unusual ideas I try them out. And my way of trying them out is to push them to the extreme (1988, p. 34, fn. 2, and 1993a, p. 34, fn. 2).

His rhetorical style has even led some critics to suggest that he sometimes used it *deliberately* in order to mislead his readers.[19]

Feyerabend's rhetorical remarks were sometimes rather demeaning. For example, he titled a response to John Watkins' Popperian criticism of his work: "The Gong Show—Popperian Style" (1978d). The joke plays on a television comedy talent show ('The Gong Show') popular at the time. Members of the jury could end one of a number of competing acts by simply banging a large bronze gong, in effect, gonging the act off the stage. The act that survives the longest wins. This is supposed to be a critical parody of Popper's falsificationism. According to Feyerabend, theories cannot simply be gonged off the stage by empirical evidence.

---

16 Koertge extravagantly praises Feyerabend's style; see Koertge (1972), p. 279.

17 For a discussion, see e.g. Hoyningen-Huene (2000b), p. 8.

18 Also see Part II of (1978a).

19 For example, see Watkins (2000), pp. 52-54. Somewhat ironically, Feyerabend made a similar accusation about Kuhn's style in *The Structure of Scientific Revolutions*. See (1970c); rp (1981b), pp. 131-161 and (1995b), as well as Hoyningen-Huene (2000a) for a discussion.

Other clear sources of provocation can be drawn from Feyerabend's attacks on the philosophy of science itself, or what in 1963, he called "that bastard child of epistemology" (1963e, p. 323). Early on, Feyerabend was quite critical of what he perceived of as *bad* philosophy of science. For example, "too much of contemporary philosophy of science is nothing but bad, because unnecessarily technical, popularization!" (1961h, p. 246, similarly 1959, p. 337). Or, "I really wish that writers in the beautiful and exciting field of the philosophy of science would stop appealing to the reason of ignorance even if appeal to such reason should happen to be very fashionable" (1961h, p. 248). But by 1970, Feyerabend began to imply that although the philosophy of science was a subject with a great past, it had no future (1970g; rp 1999a, pp. 127-137). At the tenth conference for philosophy of science at Kiel in 1972, Feyerabend delivered a paper with the shocking title, 'The Philosophy of Science – until now, a little known form of insanity' (1973, my translation). By the end of the 1980s, Feyerabend published a paper with the title, 'Who needs the philosophy of science?' (1989b, my translation). In 1994, Feyerabend publicly denounced a call for money for education in philosophy, arguing that the money could easily be spent in more worthwhile ways (1994b). In his autobiography, he chose to caption a picture of himself washing the dishes, 'the philosopher at work' (1995a, pp. 152-153). At the time he was struggling to finish a manuscript on the role of abstraction in conquering the richness of being – the process through which human beings produce a conception of what is real (see 1999b).

Feyerabend's use of polemic and rhetoric contributed to such a wide range of misunderstandings of his views that he began to criticize his critics in harsh and sometimes inappropriately personal terms. For example, part three of *Science in a Free Society* (1978) responds to reviews and criticisms. It is entitled "Conversations with Illiterates". He also repeatedly criticized Popper personally, sometimes even criticizing how Popper treated his wife.[20] He also often disapproved of intellectualism more generally, for example, in "From Incompetent Professionalism to Professionalized Incompetence - The Rise of a New Breed of Intellectuals" (1978c; rp 1978b, pp. 183-209 and 1987). Other titles include "More Clothes from the Emperor's Bargain Basement. A review of *Progress and its Problems*. By Larry Laudan" (1981c; rp 1981b, pp. 231-246), and "Academic Ratiofascism. Comments on Tibor Machan's Review" (1982).

The most famous mistaken interpretation of Feyerabend concerns his slogan "Anything goes!", which has been taken to suggest that science is

20 See (1988), p. 34, fn. 2, (1993a), p. 34, fn. 2, and (2000), p. 164.

irrational.[21] As many have already pointed out, Feyerabend was not suggesting that anything actually goes. He was making an immanent criticism. According to Feyerabend, on the view that there are methodological principles that are universally valid, the only one will be that 'Anything goes' (e.g. 1970d, p. 105, fn. 38). By this, he was not claiming that science is irrational. He was merely arguing that methodological rules have a limited validity; hence the title of the first German version of the Anything goes series (see 1972 and 1999c). They are not historically invariant. The rule 'anything goes' buys its universality at the cost of being completely empty.

Instead, according to Feyerabend, for any purportedly universal methodological rule, there are instances in which it is advisable, for the sake of progress, to suspend it temporarily. For example, although it of often argued that sound 'scientific method' requires testing theories against experience and abandoning ideas that have been empirically refuted, Copernicus' success required ignoring the fact that this theory had been considered empirically refuted for two thousand years. After all, Copernicus was not the first to consider the possibility that Earth is moving around the sun. He had seen the idea in Cicero, and perhaps others (Duhem 1969, p. 63ff.) Ignoring the obvious facts of immediate perception until they can be reinterpreted is sometimes a necessary step on the road to theoretical improvement. There are no universally valid methodological rules independent of historical context; or put positively, methodological rules are context dependent and vary historically.

Now my point is not to harp on this apparently never-ending misunderstanding, but to illustrate how a little attention to detail could have prevented it in the first place. In the original version of *Against Method* (1970d), Feyerabend attached a note directly to the phrase "*anything goes*" (p. 26, see note 38, p. 105). The note begins with a citation from Lenin's *Left Wing Communism*, which says more or less the same thing, i.e., that there are no universal methodological rules. For a paragraph, the note proceeds to qualify and severely restrict Feyerabend's endorsements to some very limited aspects of the ideas of Marx, Lenin, and Mao (insults Popper in passing); and then states:

> Some of my friends have chided me for elevating a statement such as "anything goes" into a fundamental principle of epistemology. They did not notice that I was joking. Theories of knowledge as I conceive them *develop*, like everything else. We find new principles, we abandon old ones. Now there are some people who will accept an epistemology only if it has some stability, or "rationality" as they are pleased to express themselves. Well, they can have such an epistemology, and "anything goes" will be its only principle (1970d, p. 105, fn. 38).

21 See e.g. Grunfeld (1984).

A little more attention to detail and a better appreciation of Feyerabend's rhetoric's could have prevented at least three decades of a perpetuating misunderstanding.

Now let us turn to a specific example of how even just a small stroke of sarcasm led to all sorts of confusions about Feyerabend's ideas and how they developed. The misinterpretation concerns Feyerabend stance on the notion of truth, and I will develop it at some length as it is instructive in a number of respects. It not only highlights how a touch of sarcasm caused a great deal of confusion, but it also illustrates one type of precaution for preventing such confusions: The need to work closely from Feyerabend's *original* texts (notes included), especially when claims about changes in Feyerabend's views are concerned. This is apparent as the original version of the text had a long footnote that was removed in the 1981 reprint, and as we shall see, greater attention to this footnote would have prevented the particular misreading in question. It also highlights three important features of Feyerabend's philosophy: his use of *ad hominem* argument, his skepticism about the utility of the notion of truth, and his support for theoretical pluralism over monism. Lastly, I have chosen this example because it challenges the standard view that initially Feyerabend was a scientific realist, who believed that science discover objective truths about the world, who then became dissuaded concerning the utility of notion like truth.

The particular misinterpretation of Feyerabend that I have chosen was made by John Preston, and has already been criticized by Gonzalo Munévar.[22] Munévar begins the conclusion of his review of Preston (1997) with: "A little less fixation on language will do Preston and analytic philosophy some good, even when thinking about language" (Munévar 1999, p. 225). He argues that Preston has simply mistaken Feyerabend's rhetoric, taking sarcasm for the expression of a radical, mistaken view on the nature of truth. The example concerned a statement made in which Feyerabend was comparing the consequences of the use of a single theory or point of view to that of using a set of different theories (theoretical monism versus theoretical pluralism). In 1962, Feyerabend wrote:

> it will seem that at last an absolute and irrevocable truth has been arrived at. Disagreement with facts may of course occur, but, being now convinced of the truth of the existing point of view, its proponents will try to save them with the help of *ad hoc* hypotheses. Experimental results that cannot be accommodated, even with the greatest ingenuity, will be put aside for later consideration. The result will be absolute truth, but at the same time, it will decrease in empirical content to such an extent that all that remains will be no more than a verbal machinery which enables us to accompany any kind of event with noises (or written symbols) which

22 See Munévar (1999), pp. 225ff.

are considered true statements by the theory (1962a, pp. 69-70; rp 1981a, pp. 74-75, significantly modified, cited in Preston (1997, p. 192).

As Munévar pointed out, from this passage Preston concluded that in his early work Feyerabend identified "absolute truth with total immunity from falsification" (Preston 1997, p. 192).[23] In other words, Preston ascribed the view to Feyerabend that a theory that cannot be shown to be wrong is absolutely true. Given the radical nature of this claim, Preston briefly considered and then rejected the thought that perhaps Feyerabend's wording had just been sloppy:

> It is interesting that here the result is said to be that the theory *is* absolutely true. Maybe we could write this off as a sloppy turn of phrase, but it occurs more than one in writings of the same period (Preston 1997, p. 192, italics in original).[24]

However, as Munévar points out: "Feyerabend is not offering any radical doctrines here. By the use of the words 'absolute truth' in this context, he is merely being sarcastic" (1999, p. 225). Feyerabend did not believe in absolute truth, and he did not believe that absolute truth is the same as immunity from falsification. The context surrounding the passage Preston cites makes this clear:

> Persistence of a single point of view will lead to the gradual establishment of well-circumscribed methods of observation and measurement; it will lead to codification of the ways in which these results are interpreted; it will lead to standardized terminology and to other developments of a similar conservative kind . . . In the end, all the key terms may be fixed in an unambiguous manner, and the idea (which may have led to such a procedure in the first place) that they are copies of unchanging entities and that change of meaning, if it should happen, is due to human mistake—this idea will now be very plausible. Such plausibility reinforces

23 To be more precise than Munévar, Preston also cited from a second source (1961i); rp (1999a), pp. 50-77, where Feyerabend wrote: "The power of a myth is not at all exhausted by the psychological factors we have described so far. A myth can very well stand on its own feet. It can give explanations, it can reply to criticism, it can give a satisfactory account even of events which *prima facia* seem to refute it. It can do this because it is absolutely true. It has, therefore, something to offer. It has truth to offer, absolute truth" (1961i, p. 38; rp 1999a, p. 64). However, here too Feyerabend is speaking in an *ad hominem* fashion: The truth is absolute from the perspective of the holder of the myth, not according to Feyerabend's own views. Thus, Munévar's criticism of Preston still holds.

24 In a note, Preston contrasts these two passages about 'absolute truth' with similar passages in (1965b) and (1975a). In these two later sources, Feyerabend makes similar arguments against theoretical monism and for pluralism, but Feyerabend only claims, "it is only the 'semblance of absolute truth' that results". See Preston (1997), p. 22, fn. 1, italics inserted, as well as (1965b), p. 179 and (1975a), p. 45. However, there is no real contrast. These are simply more precise, less rhetorical, statements of exactly the same idea. The term 'semblance' is supposed to make it absolutely clear that it merely appears to be absolute truth from the perspective of the well-entrenched theory. Yet even these later clarifications, of which Preston is aware, do not prevent Preston from an analysis of usage of the earlier uses of the term 'absolute truth' resulting in a misreading, and subsequent illegitimate criticism of Feyerabend's earlier ideas.

> all the maneuvers, which may be used for the preservation of the theory (elimination of opponents included). The conceptual apparatus of the theory having penetrated nearly all means of communication, such methods as transcendental deduction and analysis of usage, which are further means of solidifying the theory, will be very successful. Altogether . . . (1962a, pp. 69-70; rp 1981a, pp. 74-75, significantly modified).

This context to Feyerabend's remarks about 'absolute truth' has been conveniently ignored by Preston, who left out the slur on transcendental deductions, and the criticism of the kind of analysis of usage on which Preston basis this very misinterpretation of Feyerabend's stance on the notion of truth.

A more thorough examination of Feyerabend's publications is often sufficient to identify such sarcasm. For example, that Feyerabend does not seriously take absolute truth to be immunity from falsification is clearly apparent in the footnote attached to the original 1962 cited passage that was removed from the 1981 reprint from which Preston cited. In the attached footnote of original text, Feyerabend wrote:

> The fact that all the features of absolute knowledge can be manufactured by exorcizing an absolute conformism I regard as the most important objection to any claim to finality: it would seem to show that where as hypothetical knowledge, being the result of repeated corrections in the light of criticism is at least in partial contact with the world, absolute knowledge is entirely man made and can therefore not raise any claim to factual content. Even the stability of a testable hypothesis cannot be regarded as a sign of its truth because it may be due to the fact that, owing to some particular astigmatism on our part, we have has yet overlooked some very decisive tests. There is no sign by which factual truth may be recognized (1962a, p. 70, fn. 86, italics inserted).

This passage makes it explicitly clear that Feyerabend did not think that absolute truth is immunity from falsification, as claimed by Preston.[25]

On Feyerabend's account, theoretical monism will lead to a false belief in absolute truth that is characteristic of mythical systems and religions, whereas the theoretical pluralism of the natural sciences helps prevent such unwanted stagnation in the growth of ideas. In this respect Feyerabend's notion of absolute truth is part of an *ad hominem* argument against theoretical monism, not a serious conviction of Feyerabend's own. An *ad hominem* argument adopts the assumptions of a position in order to criticize it in its own terms.[26] (The term is sometimes used to mean an argu-

25 A more comprehensive knowledge of Feyerabend's texts may also have helped, as Feyerabend repeatedly used the idea of 'absolute truth' in sarcastic and *ad hominem* fashions that appear to have been overlooked by Preston. See e.g. (1961i), p. 41; rp (1999a), p. 65 and (1969e), p. 53 and p. 55; rp (1981a), p. 154 and p. 155.

26 To see an instance of Feyerabend complaining bitterly about how his reviewers failed to appreciate the *ad hominem* style of his arguments, see e.g. (1977a). For a response to Feyerabend that misses the point, see Suchting (1978) especially p. 340.

ment directed at an individual or person as opposed to his ideas. This is not how the term is used in this investigation). An *ad hominem* argument is an 'immanent criticism'. As we shall see, the use of *ad hominem* arguments plays is an integral part of Feyerabend's philosophical method.

Feyerabend's stance on the status of claims to scientific knowledge is that they are only hypothetical. Even the fact that they may have withstood the most severe tests that anyone has been able to think up until now cannot be taken as an indication of their truth. In accordance with Feyerabend's rejection of inductivism: There simply are no such indicators. Knowledge is not something certain. It is something speculative, hypothetical and conventional.

Preston misunderstands Feyerabend's views about truth because of his failure to appreciate Feyerabend's sarcasm and the *ad hominem* nature of Feyerabend's argument. Such a simple misreading can have multiple consequences. In this case, it led to an inappropriate criticism of Feyerabend, and then to a subsequent misconception of how Feyerabend's ideas developed. For example, after misleadingly ascribing an identification of absolute truth with total immunity from falsification to Feyerabend, Preston immediately proceeded to criticize Feyerabend for doing so: "Such an identification, coming form a close student of Popper's work, is both astounding and misconceived" (Preston 1997, p. 192). He also contrasts his misrepresentation of Feyerabend's earlier views with Feyerabend's later views, concluding that a dramatic development must have taken place. According to Preston, Feyerabend's disenchantment with traditional epistemic ideals such as truth began to surface in earnest in the early 1970s when Feyerabend mockingly began to capitalize Truth or puts it in scare quotes (Preston 1997, p. 192). Apparently Preston is more attune to these more noticeable rhetorical devices. This led Preston to the conclusion that Feyerabend changed his views about truth. According to Preston, in the 1950s, he was allegedly a Popperian rationalist who "honored and invoked classical rationalist ideals like truth" (Preston 1997, p. 191); in the early 1960s, the first signs of his disenchantment with Popper and truth began to appear, and in the early 1970s, they eventually flourished. Preston concludes his presentation of Feyerabend's developing views on the epistemic ideal of truth by pointing out that, although Feyerabend did use the concept of truth in his later publications, he claimed to have used it only in an *ad hominem* capacity (Preston 1997, p. 193).

Instead of recognizing Feyerabend's sarcasm about 'absolute truth' in his early 1960s publications, and thereby realizing that Feyerabend's use of 'absolute truth' in this context is also *ad hominem*, Preston misread Feyerabend, and mistakenly attributed radical views about truth to him. Then, he incorrectly criticized Feyerabend for holding such 'astounding' and 'misconceived' views about truth. And finally, he mistakenly contrasted Feyerabend's early, middle, and later views on the subject, *which are actually more or less the same.* Feyerabend was just as skeptical about the utility of the concept of truth in the passages Preston cites from the early 1960, as he was in his later publications. There is no substantive difference in Feyerabend's rejection of the utility of the notion of truth from the 1950s, when Feyerabend had been heavily influence by logical positivism, through to his later more relativist views of the 1980s. There is only a difference in how obvious Feyerabend was about it.

Preston was unable fully to develop Feyerabend's ideas *because of his particular approach to them.* Preston approached Feyerabend's ideas from within a philosophical school called 'analytical' or linguistic philosophy, or what Feyerabend called 'Oxford philosophy'. The reason that approaching Feyerabend's ideas from this point of view is *not* conducive to the development of Feyerabend's ideas is that this kind of philosophy begins with different presuppositions about the legitimate goals and methods of philosophy itself. To compound the problem, Feyerabend repeatedly rejected this kind of philosophy, often sharply criticizing it.[27] In fact, rather ironically, Feyerabend was criticizing exactly the kind of linguistic philosophy that Preston practiced on the very passage that Preston poorly analyzed.

My strategy for attempting to avoid the misunderstandings that Feyerabend's rhetoric can cause is to inspect the original texts very closely. This may sound simple, but it includes providing ample amounts of Feyerabend's own words to support interpretations of them, working form original sources and comparing them to the reprints, cross-checking and comparing cited passages with other similar passages in different publications, and checking the qualifications and clarifications in the notes, which often provide important clues to the real nature of his ideas. As we shall see, behind the outrageousness and the intentional provocations, there was often and intriguing insight, or a devastating argument. It may seem that I have belabored a minor point, but keep in mind that in the first edition (but not the subsequent editions) of *Against Method*, the index suggests seeing pp. 1-303 for 'rhetoric'.

27 Such criticisms are dispersed throughout Feyerabend's publications. See e.g. (1956a), pp. 95-96, (1995b), p. 385, written 1960/61, (1961g), p. 442, (1961hi), pp. 247ff., (1961i), p. 66; rp (1999a), p. 75, (1965b), p. 186ff. and p. 226, fn. 9, (1967a), p. 37, fn. 7, and p. 39, "Oxford-Academic-Gossip-English (OAGE)", and (1961i); rp (1999a), pp. 50-77.

1.3. Referencing and reliability

There are two further complications with approaching Feyerabend's ideas that are very relevant to this investigation. First, Feyerabend was not always forthcoming about the sources of his ideas. As we shall see, it would appear that Feyerabend borrowed the basis of his notion of incommensurability, perhaps even term "incommensurable" itself, from separate sources, neither of which he ever properly acknowledged. To take another example, Feyerabend has become well know as an ardent champion of pluralism based on his often repeated argument that the use of radically different sets of concepts is methodologically preferable to the exclusive utilization of just one set of concepts. It is not nearly as well known that Feyerabend developed this idea directly from Bohm (1957). In his review of Bohm's *Causality and Chance in Modern Physics*, Feyerabend stated:

> *it is Bohm's conviction* that in all fields the alternative use of opposite sets of concepts is to be preferred to the exclusive utilization of only one of them (1960e, p. 331; rp 1981a, p. 228, italics inserted).

This quickly became one of Feyerabend's main convictions, and it is one of the most fundamental assumptions of Feyerabend's philosophy.

Feyerabend did not usually identify Bohm as the source of the insight behind his main methodological argument for pluralism, and Feyerabend's enormous debt to Bohm has gone largely unnoticed in the hefty secondary literature on Feyerabend's philosophy.[28] As we shall see, it appears that Feyerabend was in the habit of adopting and then developing insights made by renowned scientists such as Bohm, Duhem, and Einstein, without identifying them as his sources. This would seem to suggest that Feyerabend was a plagiarist. But that is not quite right either, as Feyerabend repeatedly emphasized that none of the ideas presented in texts were actually his own. For example: "I for one am not aware of having produced a single idea that is not already contained in the realistic tradition" (1965c, p. 251). Or in preface to the German version of *Against Method*:

> I want to emphasize yet again that the views in this book are not new – for physicist like Mach, Boltzmann, Einstein and Bohr they were a triviality. But the ideas of these great thinkers have been distorted beyond recognition by the rodents of the Vienna Circle and the gnawing critical rationalists (1983a, p. 12, my translations).

28 In the original article version of Against Method (1970d), Bohm is only mentioned three times (p. 114, fn. 71, and p. 115, fns. 77 and 89) all of which are brief criticisms of some other views of Bohm, and *not* acknowledgements of Feyerabend's debt to Bohm concerning his pluralism. Bohm is not acknowledged as the source of Feyerabend's methodological views concerning pluralism in (1975), (1988), and (1993) either.

The second major obstacle to appreciating Feyerabend's ideas concerning referencing and reliability is the sometimes rather dubious status of much of Feyerabend's own testimony about himself and his own work. As Feyerabend became increasingly critical of the philosophy of science, he became increasingly dismissive about his own ideas in the field. But, as we shall see, even early on (before he began to reject the philosophy of science), sometimes his own claims about his own views are more than just a little misleading, and his statements about his own earlier views sometimes contain inconsistencies that do not cohere with the published record. For this reasons, statements he made about himself are met with caution. For example, having written many papers on meaning, Feyerabend brashly declared:

> As far as I am concerned even the most detailed considerations about meaning belong in the gossip columns and have no place in the theory of knowledge (1965c, p. 230; rp 1981a, p. 113).

This was a typical over-reaction to unwarranted criticism of his ideas.

The most extensive case of difficulties with the reliability of Feyerabend's testimony concerns his relationship to Popper. In the 1950s and early 1960s, Feyerabend can be seen as a proponent of many aspects of Popper's philosophy. Feyerabend celebrated Popper's "non-authoritarian epistemology" claiming that:

> The power of Popper's nonauthoritarian point of view . . . becomes very evident from a perusal of these essays [*Conjectures and Refutations. The Growth of Scientific Knowledge*], which are a major contribution to philosophy and whose publication constitutes a major event in the history of this subject (1965d , p. 88).

There is no doubt that Feyerabend was heavily influenced by Popper, and that Feyerabend adopted and developed many of Popper's ideas. After they first met in Alpbach, Popper recommended Duhem to Feyerabend. From 1952 to 1953, Feyerabend worked as an assistant to Popper, and later he translated the German versions of Popper's *The Open Society and its Enemies*, which appeared in 1957 and 1958. But sometime toward the end of the 1960s, Feyerabend seems to have dramatically changed his attitude. In 1974, he published a long, scathing review of Popper's *Objective Knowledge. An Evolutionary Approach.* The review contains a series of harsh criticisms of Popper's ideas, and even belittles Popper's contribution to philosophy. The review concludes with a critical parody based on one of Popper's student's work (Lakatos), "with *Objective Knowledge* Popper's research program has entered its degenerating phase" (1974b, p. 500; rp 1981b, p. 201). Popper, once the beneficiary of Feyerabend's highest praise, eventually became the target of his most scathing criticisms.

The original versions of many of Feyerabend's texts contain copious praise of Popper and often acknowledge the influence that Popper's ideas had on them. However in 1980, Feyerabend edited his earlier publications before they appeared in his two volumes of collected papers. Sometimes, but not always, he explicitly identified new passages that had been added.[29] He was less forthcoming about what he deleted or changed. A comparison of the original texts to the reprinted versions reveals that Feyerabend systematically purged the papers of his praise of Popper.

In the course of this investigation, we will confront many examples of changes to the reprints, especially concerning Feyerabend's relation to Popper. Here, I will provide just two representative examples. In the 1958 version of "An attempt at a realistic interpretation of experience", in the third footnote, after explaining that his use of the term 'instrumentalism' had been taken from Popper, Feyerabend wrote,

> Personal Statement.-I should like to use this occasion for pointing out that my own discussion of instrumentalism in ["Physik und Ontologie"] employs ideas which Prof. Popper has kindly explained to me in conversation before his own views had appeared in print (1958a, p. 143, fn. 3).

In the version that appeared in his collected papers in 1981, this statement is conspicuously absent. In 1965, Feyerabend wrote:

> We need the historical background . . . for raising the level of discussion to the heights already achieved by Democritus, Galileo, Descartes, Faraday, Kant, Boltzmann, Maxwell, Einstein, *Popper*. . . (1965c, p. 251, fn. 1, italics inserted).

But in the version that appeared in his 1981 collected papers, 'Popper' had vanished. 'Duhem' had taken its place (1981a, p. 104, fn. 1). In terms of a general trend, Feyerabend often acknowledged Duhem, and then increasingly Mill, as the source of views that he had once credited to Popper. Ironically, in this particular case, Feyerabend did in fact owe more to Duhem than to Popper.

Another example of a misleading claim that Feyerabend made about his own work concerns the relationship between his doctoral thesis (1951) and his "Attempt at a realistic interpretation of experience" (1958a). In the 1981 reprint (but not in the original 1958 version), Feyerabend claimed that the it is "a very much abbreviated version" of his dissertation (cf. 1981a, p. 17, fn. * and 1958a, p. 143). This would suggest a certain structural similarity, and topical overlap between the two. However, a comparison of the two reveals that while they both may concern some of the same topics and make some of the same points, they have completely different structures. They also state different aims and cite

29 For a few examples, see the first footnotes of chapters 9 and 17, p. 146 and p. 298 respectively, as well as fn. 6 of chapter 4, p. 47, and fn. 14 of chapter 8, p. 143. For a brief discussion, also see Preston (1997).

different sources. The latter even makes arguments that are not found in the former, and vice versa.

Although Feyerabend may not always have been completely up front about his intellectual past, and although his difficult style has led to any number of misinterpretations of his ideas, it is possible to reconstruct his philosophy. As we shall see, there is a central theme and a coherent method, but developing it will require a very close cross-examination of his texts.

Chapter 2
Ludwig Wittgenstein
Meaning and Ontology

The nature and extent of Wittgenstein's influence on the development of Feyerabend's ideas is as complex and contentious as Wittgenstein's own philosophical writings. For example, how did Wittgenstein shape Feyerabend's philosophical development? Did Feyerabend develop his semantic views from a poor reading of Wittgenstein, as has been suggested in the secondary literature? How did Wittgenstein's discussions of Gestalt Psychology impact Feyerabend's developing notion of incommensurability? What lasting impact, if any, did Wittgenstein have on Feyerabend? This chapter aims to investigate Feyerabend's relationship to different aspects of Wittgenstein's philosophy. Of course, there is ample evidence that Feyerabend's ideas were indeed heavily influenced by Wittgenstein's writings. Feyerabend spilled a considerable amount of ink over Wittgenstein's ideas, including an extensive attempt to rewrite what he understood to be the main argument of the *Philosophical Investigations* (Wittgenstein 1958).[30] As is well known, Feyerabend had even planned to go to Cambridge to study under Wittgenstein.[31] Moreover, Feyerabend *sometimes* credited an insight central to his development of the notion of incommensurable concepts to ideas he found in Wittgenstein's *Investigations.* He paid tribute to Wittgenstein as a forceful twentieth century defender of the contextual theory of meaning that he used to flesh out his claims about incommensurability in his landmark "Explanation, Reduction and Empiricism".[32] Feyerabend also paid tribute to Wittgenstein's ideas for having helped him redevelop his notion of incommensurability in the 1970s.

30 See (1955a); rp (1981b), pp. 99-130, also reprinted (1966d) and (1967f). Also see (1954a), (1954b), (1954d), (1960g), and (1966d). Compare Feyerabend's rewrite of the *Investigations* (1955a); rp (1981b), pp. 99-130 to Waismann's rewrite of it. See Waismann (1965).

31 See e.g. (1995a), pp. 86, and (1987), p. 312, as well as Preston (1997), pp. 3. Also see Hoyningen-Huene (2000b). For more of Feyerabend's own comments on his relation to Wittgenstein, see (1966a), pp. 4ff., and (1978a), p. 66.

32 See (1962a), p. 68, fn. 83; rp (1981a), p. 74, fn. 68. Also see (1965b), p. 252, fn. 137. But note that Feyerabend also denied the Wittgensteinian basis of exactly these ideas in (1965b), p. 227, fn. 20.

Feyerabend applied this later notion of incommensurability to cultures, world-views, and historical traditions. For example:

> It was Wittgenstein's great merit to have spotted and criticized this procedure [assuming that explicitly formulated structural elements are the only operative ingredients in a language, and then restricting discussions of scientific theories to such structural elements] and the mistake that underlies it and to have emphasized that science contains not only formulae and rules for their application but entire traditions (1978a, p. 66. See similarly 1993a, p. 224, fn. 7 and accompanying text, as well as 1978a, p. 26, fn. 11 and accompanying text).

One could even argue that the basic idea of mutually exclusive conceptual perspectives of the same basic material, which is central to the notion of incommensurability, was developed from insights gleaned from Feyerabend's reading of Wittgenstein on Gestalt figures. As Feyerabend much later put it:

> why should it not be possible to say conflicting things about the same situation and not yet be right? A picture that can be seen in two different ways (Wittgenstein's duck-rabbit is an example) can be described in two different ways – and both parties will be right. It is a matter of research and not philosophical fiat to decide whether the world we inhabit resembles a duck-rabbit picture (1987, p. 81).

One could also make a strong case for the claim that in some publications, Feyerabend analyzed philosophical problems in a strikingly Wittgensteinian manner. At least two of Feyerabend's publications are best understood as the kind of conceptual analysis of philosophical paradoxes and problems that Wittgenstein's later philosophy suggested. For example, Feyerabend's "Note on the paradox of analysis" is a perfect example of a Wittgensteinian analysis that dissolves the apparent paradox by analyzing the meaning of the terms that give rise to it.[33] Moreover, the opening discussion of "The problem of the existence of theoretical entities" is another excellent example of a Wittgensteinian linguistic analysis (See 1960f, translated into English as 1999d). In this paper, Feyerabend attempts to dissolve the merely apparent philosophical problem of the theoretical entities through an analysis of the concepts that give rise to it.[34] These publications are perfect examples of Wittgensteinian analysis of philoso-

33 See (1956a) and (1957e). Feyerabend's use of conceptual analysis in his dissolution of the paradox of analysis is also an excellent example of Feyerabend's *ad hominem* argumentative strategy. For a detailed reconstruction of Feyerabend's argument, see Oberheim and Hoyningen-Huene (2000).

34 In order to see exactly what I mean by a linguistic analysis in the spirit of Wittgenstein's later philosophy, see Waismann (1965), especially chapters 1 and 2, pp. 3-43, which contains clear illustrations of philosophical problems and their dissolution through analysis. The only difference is that Feyerabend was also willing to try to change and improve the concepts that he was analyzing, whereas this is apparently forbidden to a Wittgensteinian (in Feyerabend's view). See (1956a) and (1957e).

phical problems because the method employed in both pieces is aimed at clarifying the meanings of the terms that give rise to the apparent paradox, in order to dissolve it (see Waismann 1965, p. 5). There are also other examples of Feyerabend explicitly embarking on linguistic analysis (e.g. 1961b, pp. 82-83).

However, even though Feyerabend did occasionally adopt the latter Wittgenstein's philosophical method to handle philosophical pseudo-problems and even though Wittgenstein did significantly contribute to Feyerabend's philosophical development, I will argue that Feyerabend was no Wittgensteinian. As we shall see, not only did Feyerabend reject what is widely considered to be one of Wittgenstein's main contributions to philosophy (his treatment of meaning as use), he was also increasingly cautious about crediting or attributing ideas to Wittgenstein. Moreover, from the 1950s to the 1970s, he was extremely critical of the philosophical legacy that Wittgenstein left behind.

In the secondary literature, there are at least two opposing views of Feyerabend's relation to Wittgenstein:

> At least two different views can be taken of the overall significance of Feyerabend's developing ideas on the nature of science. The first [Preston's] is that he wrong-headedly took various important changes in science to show that there are no *a priori* truths and misleadingly blurred the boundary between science and philosophy. *Seduced by Popperians at an early stage, he failed to gain enough important insights from Kantians and Wittgensteinians. Much that is important in his work is there because he continued to be influenced by Kant or Wittgenstein despite his inclinations.* A second view [Couvalis's] is that Feyerabend was a prescient philosopher who gradually realized how deeply we have been misled by various kinds of *a priorism*. As his views developed, he pushed Popper's fallibilism to its logical conclusion (except perhaps in the case of induction). *Popper brilliantly showed how developments in physics and geometry render Kantianism and its relatives deeply implausible. But, unlike Feyerabend, he mistakenly continued to assume that there are analytic truths* . . .(Couvalis 1999, p. 252 italics inserted).[35]

According to Couvalis, Feyerabend gave up the idea of analytic truth, and with it, he rejected Wittgensteinian philosophy. According to Couvalis, Feyerabend also rejected some central aspects of Kantianism: Both the *a priori* and the *analytic* are suspect. In *this* respect, Couvalis and Preston take almost opposing views about the nature and extend to which Wittgenstein influenced Feyerabend's ideas. According to Preston, who apparently approaches Feyerabend from within a Wittgensteinian perspective, Feyerabend's important insights can be traced to what little he managed to understand correctly about Wittgenstein's later philosophy; whereas ac-

35 Preston approvingly signs on to this summary of "where he is coming from", and the contrast to Couvalis. See Preston (1999), p. 235.

cording to Couvalis, Feyerabend flatly rejected Wittgensteinian views and their basis.[36]

The following considerations will resolve some of the tension between these opposing views on the nature of Wittgenstein's influence on Feyerabend's developing ideas. The next section begins with a detailed development of Feyerabend's semantic views, and then contrasts them to those of Wittgenstein. Then there is a more general discussion of Feyerabend's relation to Wittgenstein's philosophy. Thereafter, I will develop what I take to be the most significant influence that Wittgenstein had on Feyerabend' philosophy, and conclude with a summary of their relations.

2.1. Feyerabend's distinction between meaning and use

A central aspect of Feyerabend's relationship to Wittgenstein's philosophical insights concerns the nature of the relationship between meaning and experience. In order to compare Feyerabend's views on meaning and experience to Wittgenstein's, and then to assess Wittgenstein's influence on Feyerabend in the realm of semantics, we will first need to develop Feyerabend's semantic views. Although he was reluctant to call his considerations a 'theory of meaning', Feyerabend discussed, developed and defended a specific notion of meaning in many of his early publications. The most extensive development can be found in "An Attempt at a Realistic Interpretation of Experience" (1958a), which was based on the ideas developed in his 1951 Ph.D. thesis. The 1958 paper, unlike his Ph.D. thesis, is an explicit attack on three prominent views: Bohr's view that the task of science is both to extend the range of experience and to reduce it to order (Bohr 1934, p. 1), Hempel's view that science is ultimately intended to systematize the data of our experience (Hempel 1952, especially p. 21, fn. 7), and Carnap's view that the meaning of scientific theories is purely the result of their connection to experience (Carnap 1956, especially pp. 40-47). The best way to develop Feyerabend's semantic views is to reconstruct the arguments he developed against these views in that paper. Only then will we be in position to compare them to Wittgenstein's pronouncements about meaning.

Logical positivists called the atomic sentences used to test theories 'basic sentences' or 'basic statements'; hence the title of Feyerabend's dissertation "Zur Theorie der Basissätze", or "On the Theory of Basic

36 This controversy between Couvalis and Preston as to the influence of Wittgenstein's views on meaning on Feyerabend's ideas began with Preston's review of Couvalis (1989). Also see Preston (1992).

Statements", (1951, my translation). They were also sometimes called 'protocol sentences' – a term apparently adapted from the practice of keeping experimental protocols: Systematic, step by step, records of how the test was conducted, including any and all relevant observations of the results; for example, 'After being submerged in the solution, the litmus paper turned red'. The epistemic significance of such basic sentences derives from the assumption that they are immediately inter-subjectively verifiable, and thus they were believed to provide an empirical *foundation* for scientific knowledge claims.

In his (1958a), Feyerabend attacked the foundational role that protocol statements play in logical positivist models of testing theories against experience. To launch his attack, Feyerabend distinguished between an observation language's *characteristic*, which consists of four pragmatic (psychological, sociological) conditions that determine how that language is used, and its *interpretation*, which is an additional semantic act that confers meaning.[37] The four pragmatic conditions (the characteristic) stipulate the *causal* relationship between the behavior of human beings (acceptance or rejection of particular *atomic sentences*) and a set of physical situations. To continue with our example, presented with a piece of litmus paper that has just been submerged in solution, a biologist might accept the sentence 'The litmus paper turned red'. In Feyerabend's view, the physical situation (the litmus paper turning red) *causes* the biologist to accept the sentence 'The litmus paper turned red'. In this sense, the use of the sentence is a pragmatic act. However, the *interpretation* of the sentence 'The litmus paper turned red' as demonstrating that the solution is acidic is an additional act.

Although Feyerabend never mentioned the background or context of this distinction between a language's *characteristic* and its *interpretation*, perhaps it may be helpful to attempt to develop its historical context. The distinction was part of a popular movement in psychology at the time Feyerabend wrote his dissertation: Radical Behaviorism. According to Radical Behaviorism, scientific psychology should only be concerned with observable physical phenomena such as stimuli and response, and not with unobservable mental processes and mechanisms.[38] Some logical positivists, following Skinner, had endorsed such a view in psychology. Using Feyerabend's terms, one can say that the observable pragmatic functions make up a language's characteristic; while the unobservable processes and mechanisms (such as attention, intention, memory and motivation) are part of its interpretation. Now although there is a need for extreme caution when reading into Feyerabend's texts, it seems to me this is a fair

37 For the following, see (1958a), pp. 144-147; rp (1981a), pp. 18-19.

38 For some basic philosophical background on radical behaviorism, see Gallistel (1999).

match, and a probable source of Feyerabend's distinction between a languages characteristic and its interpretation. Moreover, Feyerabend repeatedly published several defenses of a form of eliminative materialism in the philosophy of mind, which can be understood as a form of Radical Behaviorism as it also distinguishes between pragmatic and other mental properties.[39]

The four pragmatic conditions that determine the use of such basic *sentences* (not 'statements') are:[40]

(i) *the condition of decidability*, according to which, for every atomic sentence of the language considered, there is an appropriate situation such that when an observer is presented with the atomic sentence in the appropriate situation, the observer runs through a series of states and operations that terminates either with the acceptance or rejection of the atomic sentence. For example, the sentence "After being submerged in the solution, the litmus paper turned red", uttered in front of a particular piece of litmus paper that had just been submerged in the solution, will result in either the acceptance or rejection of the sentence by an observer. In other words, either one will agree or one will disagree with the claim that the litmus paper turned red.

(ii) t*he condition of quick decidability*, according to which the series of states and operations terminating in the acceptance or rejection of the atomic sentence is passed through fairly quickly. To continue with our example, an observer will agree that the litmus paper turned red fairly quickly.

(iii) *the condition of unanimous decidability*, according to which if an atomic sentence is accepted (or rejected) by some observer, then it will be accepted (or rejected) by nearly every observer in the same situation. For example, if I would assent to the sentence 'The litmus paper turned red' when presented with the litmus paper that was submerged in the solution, *then* nearly everyone in the that situation would agree that the litmus paper turned red, and nearly nobody would dissent by offering a different verdict, such as that the litmus paper turned blue, or did not change its original color.

And (iv) *the condition of relevance*, according to which the decision to accept or reject the atomic sentence is *causally* determined by the situation, and not by the utterance of the atomic sentence alone, or by the internal state of the observer alone. For example, the reason that I accept the sen-

39 See e.g. (1963b), (1963d), (1967a), (1969c); rp (1981a), pp. 161-175, and (1970f).

40 Feyerabend emphasized that he considers them to be *sentences* and not *statements* because through his afore mentioned distinction, he has separated out those semantic properties that underlie their assertive function, i.e. that make them statements with prepositional content, instead of mere physical events (1958a), p. 145; rp (1981a), p. 18.

tence 'the litmus paper turned red' is that I see the litmus paper and it appears as if the litmus paper turned red to me, and not, for example, that I have been conditioned to say 'Yes' whenever someone says 'did this litmus paper turn red?' — regardless of the color of the litmus paper, or because all I ever do is agree with everything anyone happens to say about the color of litmus papers. Together these four pragmatic conditions stipulate an observation languages 'characteristic'. They determine the use of each atomic sentence (not statement) of an observation language, without making any stipulations as to what the sentences are supposed to assert.

According to Feyerabend, in order to have a full-fledged language (i.e. statements, not just sentences), further conditions are required (). Feyerabend called any complete class of such further conditions an 'interpretation'. An observation language is completely specified by its characteristic and its interpretation. At this point, Feyerabend offered no specific, concrete suggestions about what conditions determine an interpretation (in other words, he offered no specific theory of meaning beyond distinguishing a languages interpretation from its characteristic). He was, however, clear that an interpretation differs from the pragmatic properties of a language in that the latter are determined by *causes*, whereas the former are *conventions* (1958a, pp. 146-147; rp 1981a, pp. 18-19). He merely made what he considered to be a "clear and unambiguous" distinction between the pragmatic, causal properties of a language (i.e. its characteristic) and its semantic, conventional properties (i.e. its interpretation) (1958a, p. 146; rp 1981a, p. 19). Meanings are conventions, according to Feyerabend, because they depend on the theories we decide to use to confer interpretations. When we decide to adopt new theories, we also decide to adopt new meanings.[41] An existing sentence may take on a new meaning if new theories are used to reinterpret it.

Feyerabend then proceeded to illustrate his claims with examples, and his use of concrete examples, instead of merely abstract considerations, is a hallmark of his philosophical work, especially concerning meaning and the notion of incommensurability (Oberheim and Hoyningen-Huene 1997, pp. 447ff., and especially p. 450, and fn. 8 on p. 462). Even so, to bolster his distinction between the pragmatic and semantic characteristics of a language, Feyerabend invoked elements of what he called the 'pragmatic theory of observation':[42]

41 Feyerabend explained his reasons for treating meanings as conventions very clearly in his (1954c).

42 Feyerabend attributed the pragmatic theory of observation to Popper, Carnap, and Neurath, see (1962a), pp. 34ff.; rp (1981a), p. 49ff. Also see (1954c), and (1961b), p. 83,

> What the observation situation determines (causally) is the acceptance or rejection of a sentence, *i.e.* a physical event. In so far as this causal chain involves our own organism we are on a par with physical instruments. But we also interpret the indications of these instruments (i.e. either the sensations which occur during observation, or the observational sentence uttered) and this interpretation is an additional act, whether the instrument used is some apparatus or our own sensory organization (our own body) (1958a, pp. 146-147; rp 1981a, p. 19).

Thus the pragmatic theory of observation makes the same type of distinction, between pragmatic and semantic properties. As we shall see, Feyerabend used these ideas to attack the stability thesis.

Based on the distinction between the pragmatic and the semantic properties of a language, Feyerabend argued:

> Any philosopher who holds that scientific theories and other general assumptions are nothing but convenient means for the systemization of the data of our experience is thereby committed to the view (which I shall call the stability thesis) that the interpretations (in the sense explained above) do not depend upon the status of our theoretical knowledge (1958a, p. 147; rp 1981a, p. 20).

In a footnote, Feyerabend gave two examples that illustrate why such positivistic interpretations of science are committed to the stability thesis. First, according to instrumentalism, theories are tools for the prediction of *certain kinds* of observable events. It follows that instrumentalism requires an observation language whose sentences describes observable events and have an *interpretation* (stipulating the kind of events they are). However according to instrumentalism, theories are un-interpreted, i.e. they have no descriptive meaning (as they are merely tools for the prediction of observable events). Therefore, the interpretation of the observable events cannot be provided by theories. Consequently, according to instrumentalism, "whatever interpretation an observation-language may possess, it will *not* depend on theoretical 'superstructure' (1958a, p. 147, fn. 8; rp 1981a, p. 20, fn. 7, italics inserted).[43]

Second, according to Carnap's method of reconstructing the language of science with a duel scheme consisting of an interpreted observation language and a theoretical language, descriptive terms receive their interpretation through their connection with experience. There is no independent interpretation for theoretical terms.[44] Feyerabend concluded:

> This implies that the interpretation of a theory depends upon the interpretation of the observation-language used, but not the other way round. And as it is stipulated that the observation language be completely interpreted it follows also that

(1965b), p. 152 and pp. 212ff., (1965c) p. 243 and pp. 246-247; rp (1981a), p. 125. For a discussion of the pragmatic theory of observation, also see Preston (1997), chapter 3.

43 The reason that it is fn. 8 in the original and fn. 7 in the 1981 reprint is that fn. 3 of the original 1955 publication was removed. It referred to Popper.

44 Feyerabend based these attributions on Carnap (1956), p. 40 and p. 47.

> Carnap's more sophisticated account [than instrumentalism] is based upon an observation-language whose interpretation has been introduced independently of the state of theoretical 'superstructure' (1958a, p. 147, fn. 8; rp 1981a, p. 20, fn. 7).

Feyerabend proceeded to attack the stability thesis on two grounds: One descriptive and one normative. Such a *two-pronged argumentative strategy* is common to many of Feyerabend's major challenges to the ideas proposed by his contemporaries. Such two-pronged attacks argue historically (or descriptively) that some philosophical thesis does not match actual scientific practice, and methodologically (or normatively) that the thesis would have undesirable consequences. As we shall see, such a two-pronged argumentative strategy is representative of many of Feyerabend's major arguments. This two-pronged attack on Carnap's method of reconstructing scientific language is just one example of Feyerabend challenging a view on the grounds that it supports the stability thesis. Feyerabend also challenged traditional 'school' philosophies such as the realism of Platonism and the idealism of Berkeley, everyday language philosophy, some particular views of Bohr and Hempel, as well as Nagel's account reduction, and the Hempel-Oppenheim account of explanation, all on the basis that they maintain the stability thesis.[45] In this particular case, Feyerabend argued that the stability thesis misrepresents the relationship between theory and experience in actual scientific practice; and that methodologically (or normatively) it would inhibit scientific progress. Following Feyerabend's lead, we will begin with the normative prong of his argumentative strategy, and then move on to the descriptive prong.

Feyerabend's methodological argument against the stability thesis is concise, (the entire section is about a page long), but because of its technical style, and because of Feyerabend's unhappy choice of a difficult and rather opaque example, it initially appears anything but straightforward. Like so many of Feyerabend's arguments, it is an *ad hominem* argument or immanent criticism. The basic idea is that the stability thesis has consequences that are undesirable to those who uphold it. The undesirable consequence, in this case, is that "*every positivistic observation-language is based upon a metaphysical ontology*" (1958a, p. 149; rp 1981a, p. 21). In other words, every positivistic observation language is based on assumptions about the kinds of things that there are (i.e. ontology) which are not testable (i.e. metaphysical).[46] This is undesirable to positivists, because they wish to allow only empirically testable elements into scientific knowledge claims. It is important to emphasize the *ad hominem* nature of this argument, as

45 For some further criticisms of the stability thesis, see e.g. (1954c), (1960d), p. 247, (1962a); rp (1981a), pp. 44-96, and (1965b), pp. 164ff.

46 See e.g. (1960a), pp. 158-159; rp (1981a), p. 42. For a discussion, see Harré (1959).

Feyerabend argues, here and elsewhere that metaphysical (i.e. non-testable) elements, *pace* positivism, do actually play a positive role in the advancement of knowledge claims. According to Feyerabend, "*the interpretation of any physical theory contains metaphysical elements*" (1960a, pp. 158-159; rp 1981a, p. 42, italics in original).[47] Why does the stability thesis lead to the consequence that every observation language makes untestable assumptions about the kind of things there are?

Feyerabend's argument runs as follows:

> Assume (a) that the observational language has ontological consequences; (b) that it satisfies the stability-thesis ...; and (c) that it is applicable, was applicable and will always be applicable. Then it follows, (1) that those ontological consequences cannot have emerged as the result of empirical research (for if this were the case, the stability-thesis would have been violated at some time in the past); (2) nor will it ever be possible to show by empirical research that they are incorrect (for if this were the case, the stability-thesis would be violated at some time in the future). Hence, if the ontological consequences of a given language are not all logically true statements (in which latter case the language would be applicable for purely logical reasons which seems implausible) we arrive at the result that *every positivistic observation language is based upon a metaphysical ontology* (1958a, pp. 148-149; rp 1981a, p. 21).

The last clause, set out as (c), is too strong, and consequently somewhat uncharitable. None of the targets of Feyerabend's attack mentioned above would subscribe to such a strong condition, which denies that there ever has been, or ever will be, changes in the meaning of the observational language of the natural sciences. For example, from Carnap's view that the meaning of on observational language is determined by experience, and is independent from theoretical superstructure, it does not necessarily follow that there can never be changes to it. This follows on the assumption that only changes in theoretical superstructure can change the meaning of the observation language. Perhaps, for example, changes in behavior, such as either how we use observation language or how we interact with experience, or even the possibility of new kinds of experience (through new technologies or psycho-tropes) could affect changes in the meaning of observation language, independently of changes in theoretical superstructure. As (c) appears to be doing the argumentative work supporting the conclusion that every positivistic observation language is based on a metaphysical ontology, Feyerabend has not made a convincing case against Carnap with this argument. He has not shown that Carnap actually needs to, or even would, accept (c). In order to strengthen Feyerabend's argument, it needs to be shown either that only changes in theoretical

47 See (1958a), p. 153 and section 7; rp (1981a), p. 24 and section 7. Similarly, see (1954c), p. 465. Feyerabend also discussed the positive roles metaphysics play in science in (1961h), pp. 245-246.

superstructure can affect changes in the meaning of the observation language, or that changes in theoretical superstructure do in fact affect changes in the meaning of observation language.

In any case, as Feyerabend's reasoning may not be immediately clear, let me rephrase his point, allowing for some repetition for emphasis, some explications of Feyerabend's use of terms like 'metaphysical', and 'ontological consequence', an illustration in the form of a reconstruction of Feyerabend's own example, and some filling in of the gaps. Thereafter, I will try to tease out the source of the tension in logical positivist position.

The reason that the stability thesis leads to the consequence that every observation language makes untestable assumptions about the kind of things that there are follows from the positivists' acknowledgement that observation languages have interpretations. By this, they admit that observation sentences have *ontological consequences*. In other words, observation languages contain *assumptions about the kinds of things there are*. This is because they are supposed to be applicable to reality. Feyerabend illustrated ontological consequence with the use of natural numbers for counting actual objects. Using natural numbers to count actual objects implicitly assumes that the objects counted are discrete entities that can always be arranged in a series. This assumption is not true *a priori*. For example, given the discovery that this assumption is sometimes false, that there are objects that are not discrete entities, one must conclude that the natural numbers can no longer be applied to reality for the purpose of counting all kinds of objects (1958a, p. 148; rp 1981a, p. 20).

Now concerning the status of such ontological consequences, according to Feyerabend, there are just two possibilities. Either these assumptions about the kinds of things that there are follow from the observation sentences logically (possibility one), in which case a particular sentence applies to a particular situation for logical, i.e. non-empirical, reasons, or these assumptions about the kinds of things that there are follow from observation sentences empirically (possibility two).

Feyerabend claimed that the first possibility is implausible, but he did not specify why case one is implausible. This has contributed to some confusion about his argument, and some unwarranted criticism of it (see e.g. Preston 1997, pp. 38-39). For example, in reconstructing Feyerabend's argument, Preston cited the same passage that I do, but he omitted the clause '(in which latter case the language would be applicable for purely logical reasons which seems implausible) we arrive at the result that'. He then assed Feyerabend's argument as follows:

> Such a consequence would be unacceptable to positivists, who are allergic to metaphysics. But the only reason Feyerabend gives why the ontological consequences of an observation-language should not be logically true is that if they

> were, that language would be applicable for purely logical reasons, and this, he says, is implausible. If we discover that such ontological consequence are false, this amounts to a discovery that such a language is 'inapplicable to reality'. This is a *very strange* objection. Languages are, *by definition*, 'applicable to reality', and *logically true statements cannot be discovered to be false.* Feyerabend errs here in assimilating languages to theories, thinking that some language 'fit' reality better than others, and in thinking that *a change in whether we accept a sentence which expresses a logical truth amounts to its falsification* (Preston 1997, pp. 38-39, italics inserted).

The reason that it is implausible that a particular sentence applies to a particular situation for logical reasons, however, is so simple and straightforward that Feyerabend simply neglected to state it explicitly. The reason is that the question as to whether or not a sentence applies to an actual situation is an empirical, not a logical, issue. The assumptions about the kind of things that there are that are built into an interpretation of an observation language are empirical, not logical issues. Preston has misunderstood Feyerabend's argument. Having omitted the relevant segment from the passage he cited, he claimed that ontological consequences follow from observation statements *logically* (although Feyerabend explicitly says that they do not), and then he criticized Feyerabend's objection as 'very strange' (a sign that he simply does not understand the point), based on his own preconceived *definition* of a 'language' (which apparently he is not willing or able to set aside for the sake of understanding Feyerabend's point), seemingly regardless of Feyerabend's views on languages or those of the positivists under consideration. In any case, Preston proceeds to ascribe the idea, "that a change in whether we accept a sentence which expresses a logical truth amounts to its falsification" to Feyerabend, although it is nowhere to be found in the text. As I have stated, Feyerabend's claim was that the ontological consequences of observation sentences are *not* logical truths, but empirical commitments, and this is *explicitly* stated exactly in the passage that Preston replace by " . . ." in the citation that he used to support his claim that ontological consequences are *logical* truths. Feyerabend's point is that using an observation language involves conjecturing that its terms are applicable to reality, and this conjectured applicability has in fact sometimes been falsified in the course of scientific advance. This is supposed to undermine the stability thesis, and hence the foundational role of observations in the logical empiricists bottom-up relationship between experience and theory.

This is just one of many examples of how Preston misunderstood Feyerabend, because he approached Feyerabend's ideas from outside, not from within, Feyerabend's perspective. This led him to mistaken interpretations, the feeling that something 'strange' or 'bizarre' is going on (an often repeated accusation in his book), and then consequently to illegitimate criticisms of Feyerabend's ideas and arguments.[48]

Possibility one is ruled out because whether or not a sentence applies to a particular situation is supposed to be an empirical, not a logical, issue. There is no *a priori* reason, to use Feyerabend's example, that one of the ontological consequences of using natural numbers for counting, i.e. that objects are discrete entities that can always be arranged in a series, should be true. It is an empirical issue. The latter case (that the ontological consequences follow for empirical reasons) is the source of the real difficulty for the positivist: "The existence of ontological consequences which are not logically true leads to the first difficulty for the stability-thesis" (1958a, p. 148; rp 1981a, pp. 20-21). The difficulty is that according to the positivist, those assumptions (i.e. the ontological consequences) *are not testable.* They cannot have been made in the past on the basis of some theoretical knowledge, because testable theories do not play a role in the interpretation of observation sentences according to the positivists. And moreover, they cannot be reinterpreted in the future on the basis of theoretical advancements, again because testable theories do not play a role in the interpretation of observation sentences. Thus, if the stability thesis is true, then the ontological consequences are simply untestable.

The result is that according to a combination of the positivist idea that observation languages are interpreted (i.e. applicable to reality) together with the stability thesis; we are just plain stuck with our primitive non-theoretical interpretations of the sentences of our observation language *once and for all.* Observation sentences have not, and cannot, be subject to empirical tests, as such tests could have, or would potentially, change their interpretation, breaching the stability thesis. Therefore, such observation languages are based on a 'metaphysical ontology'. By 'metaphysical', Feyerabend basically means non-empirical or un-testable. He does not mean ontology or on the nature of things. Feyerabend drew this notion of 'metaphysical' as non-empirical from the logical positivists and from the adherents of the orthodox interpretation of Quantum Mechanics, who meant "neither mathematical, nor empirical" (1960e, p. 326, fn. 1; rp 1981a, pp. 223-224, fn. 11).[49] Feyerabend followed Heisenberg in criticiz-

48 See, for example, Oberheim and Hoyningen-Huene (2000) and Oberheim and Hoyningen-Huene (1999), as well as Couvalis (1999) and Munévar (1999), all of which make basically the same point, i.e. that Preston would have done better, if he had launched immanent criticisms of Feyerabend, instead of Wittgensteinian ones.

49 This notion of metaphysical as non-empirical should be distinguished from other meanings of metaphysical that do not have the implication of un-testability. Devitt even argues for a naturalized or "empirical metaphysics" (Devitt 1997, chapter 5).

ing the Copenhagen interpretation as metaphysical in this sense. By 'ontology', he means assumptions about the kinds of things there are.

The source of the tension in the logical positivist position is now readily clear without Feyerabend's rather convoluted and badly illustrated argument. On the one hand, logical positivists wanted knowledge claims to be verifiable, and this means that they should be subject, or at least subjectable, to empirical test. But on the other hand, empirical tests are supposedly made on the stable basis of inter-subjectively verifiable observations, which are crucial to positivism as they provide the empirical foundation for such tests. In other words, knowledge claims have a special epistemic status, because they are supposed to be tested against experience. But in order for the observation sentences to play this foundational role for testing knowledge claims, they must be stable and inter-subjectively verifiable. Consequently, they cannot themselves be open to tests. Otherwise their foundational role would be undermined. Hence, a non-empirical (or metaphysical) element has seeped in through the floor, as it were, by way of the observations sentences themselves. The real tension in the positivist's position is between their maxim of testability and their foundational use of observation claims.

Feyerabend, on the other hand, tried to extend and enforce the maxim of testability across the whole system – not just to the knowledge claims, but to the observations supporting them as well as. Not only should the theoretical knowledge claims be testable, but so should the observations sentences that serve as the basis for testing them. However, the stability thesis, according to Feyerabend, prevents observations from being tested, and consequently, for the positivist *every positivistic observation language is based upon a metaphysical ontology*.

Before he moved on to the second (descriptive, historical) prong of his argumentative strategy, which merely consists of two examples of how the stability thesis has indeed been breached in the course of scientific advance, Feyerabend considered and criticized two ways in which positivists justify their interpretation of observation languages. They are the *principle of phenomenological meaning* and the *principle of pragmatic meaning*. According to the principle of phenomenological meaning, the interpretation of an observation term is uniquely and completely determined by its use (or its characteristic). According to the principle of pragmatic meaning, the interpretation of an observation term is uniquely and completely determined by what is immediately given directly before the acceptance or rejection of the observation sentence, i.e. the phenomena experienced. While Feyerabend offered a number of interesting and illuminative arguments against both of these possibilities, criticizing some particular views of Bohr and

Russell, I will pass over these arguments, as they are not directly relevant to developing Feyerabend's semantic views.

As we have seen, Feyerabend's two main contentions, which are central to his views generally are with the assumptions that theories derive their meaning *solely* through their connection with experience,[50] and that experience itself is a stable foundation on which theoretical meaning can be based.[51] In other words, according to the empiricists, the meaning of a term is grounded on experience, either directly in the case of observation terms, or indirectly in the case of theoretical terms. For example, the meaning of 'red' is determined by observations of the color red, while the meaning of a theoretical term such as 'mass' is determined by its connection to other terms such as 'weight', whose meaning is determined directly by observations (which need not be perceptual). Feyerabend repeatedly challenged the epistemological and semantic significance of the distinction between observational and theoretical terms. Instead of such a bottom-up version of the relation of experience and theoretical knowledge, according to which experience determines meaning of our theories, Feyerabend argued for a top-down version, according to which our theories determine the meaning we attach to our experiences – and *not* the other way around.[52] According to Feyerabend, experience cannot be taken for granted. It takes on its particular character in light of the theories we bring to it.

Feyerabend did not think that this claim was particularly new, and decades later he ridiculed Popper for taking credit as the first one to question the priority of experience over theory:

> In his *Objective Knowledge*, Oxford 1973, p. 8, Popper wrote with characteristic modesty: 'Neither Hume nor any writer on the subject before me has to my knowledge moved from here (impossibility of justifying reasoning from experienced to unexperienced instances) to the further questions: can we take the "experienced instances" for granted? And are they really prior to theories?' (1978a, p. 59, fn. 98).

Feyerabend also listed a number of individuals who had challenged the priority of experience over theory prior to Popper:

> It is surprising and a sign both of the historical illiteracy of most contemporary philosophers and of their low standards of hero worship that statements such as these are taken as historical evidence and as indications of philosophical profun-

50 See e.g. (1954c), p. 470, (1958a), especially pp. 164-165; rp (1981a), pp. 32-33, (1958b), (1960a); rp (1981a), pp. 37-43, especially (1960f), (1961b), p. 82, (1961g), p. 443, (1962a), pp. 160-163; rp (1981a), pp. 29-31, (1965b), p. 198, and especially note 8, pp. 219-224, and (1969b); rp (1981a), pp. 132-135.

51 See Couvalis (1989) for an excellent analysis of Feyerabend's attack on foundationalism.

52 See (1993a), chapter 15, especially pp. 149ff. for a later characterization of his views on the relation of experience and theory and its relation to incommensurability.

> dity. But Newton corrected phenomena 'from above' [. . .] Mill requested that there be a discussion of experience in order to determine both its content and its force [. . .] . . . in Goethe's *Maximen and Reflexionen* we find the statement 'Das Höchste wäre zu begreifen, dass alles Faktische schon Theorie ist' [The greatest insight would be to realize that everything factual is already theory – my translation] [. . .], expressing the very same idea which Popper here claims for himself. Boltzmann has often quoted Goethe's dictum that experience is only half experience [. . .]. and then there is, of course, Mach's observation that already the 'name "sensation" entails a one-sided theory'. . . (1978a, p. 59, fn. 98).

Feyerabend's immanent criticisms are often followed by positive suggestions. In this case, Feyerabend proposed an alternative to the stability thesis, which he stated as *thesis I*:

> [t]he interpretation of an observation language is determined by the theories, which we use to explain what we observe, and it changes as soon as those theories change (1958a, p. 163; rp 1981a, p. 31, also see 1961g, pp. 442).

This thesis is a version of what is sometimes called a 'contextual theory of meaning', according to which "the meaning of a term is not an intrinsic property of it but is dependent on the way in which the term has been incorporated into a theory..." (1962a, pp. 68-69; rp (981a, p. 74). It is also sometimes presented in a slightly different form and as a consequence of what Feyerabend calls a 'pragmatic theory of observation' (1965b, p. 180 and p. 184). The ideas expressed in *thesis I* are basis of much of Feyerabend's philosophical work. By "an interpretation of an observation language", Feyerabend meant the meaning of statements about observable phenomena used to test a theory. While these statements purport to describe observations, according to Feyerabend, their meanings are *not* determined by how they are used (the pragmatic theory of meaning, nor by the immediate experiences of the describer (the phenomenological theory of meaning).[53] *Their meanings are determined by the theories which we use to explain what we observe* (a contextual theory of meaning). As these theories often include *terms* used in the observation language for testing a theory, it follows[54] that theories determine the meanings of the statements used to describe observable phenomena used to test the theories. By emphasizing that observation statements are used to test theories, Feyerabend was not restricted his claim to the meanings of observation statements only when they are being used to test theories. Rather, according to Feyerabend, the

53 Feyerabend repeatedly specifically rejects these two other possibilities. See e.g. (1958a), pp. 146-147; rp (1981a), pp. 18-19.

54 To be more precise, it follows given the assumption that interpretation of the terms used to make a sentence determine the interpretation of the sentences as a statement, i.e. that the meanings of the terms determine the statements meaning. The additional assumption is required because *thesis I* concerns statements and the contextual theory of meaning concerns terms.

meanings of observation statements are always determined by the theories to which they belong. The fact that these same statements are actually used to test theories is what causes difficulties for some accounts of theory testing and reduction. Feyerabend's point is that interpretations of observation sentences, i.e. statements of observable facts, *are fully theoretical.*

Consider just how much theory is needed to defend and explain a factual statement about an observable, physical state of affairs. For a start, the terms in the observation language have to pick out things and properties that correspond to the facts that they state.[55] *Thesis I* is developed from the view that *using certain terms to explain or to describe some feature of the observable world involves presupposing that the particular theoretical assumptions that determine their meaning can be used to correctly describe some features of the world.*[56] For example, using the term 'phlogiston' in a description of a burning candle, such as "The phlogiston in that candle is being taken away' presupposes that phlogiston theory of combustion. If it turns out that the phlogiston theory of combustion is incorrect, then such statements must be abandoned.[57]

As we have seen, according to the view expressed in Feyerabend's *thesis I*, some scientific statements receive their meaning from the theories whose truth is presupposed when the statement is used to describe or explain some feature of the world. The 'some' in the above statement is somewhat misleading. *Thesis I* makes an explicit claim about what determines the meanings of observation statements, which are distinguished from the theories that determine their meaning; i.e., it specifically applies to observation language and not to theory. What about theoretical statements? According to the orthodox accounts of the logical positivists and logical empiricists, theoretical terms receive their meaning through their connection to observation terms, which receive their meaning from experience. Feyerabend explicitly presented his views as an inversion of this orthodoxy,[58] and he argued that the inversion in the standard accounts of experience determining meaning follows form his pragmatic theory of observation:

> The most important consequence of the transition to the pragmatic theory of observation is the reversal that takes place in the relation between theory and observation. The philosophies that we have been discussing so far assumed that observation sentences are meaningful per se, that the theories which

55 In addition, some logic and mathematics will be needed.

56 For a discussion, see Hattiangadi (1977), pp. 296ff.

57 See Couvalis (1989), p. 62 for an illuminating discussion.

58 Feyerabend took this idea to its extreme, arguing for the possibility of science without experience altogether. See (1969b); rp (1981a), pp. 132-135.

> have been separated from observation are not meaningful, and that such theories obtain their interpretation by being connected by some observation language that possess a stable interpretation. According to the point of view that I am advocating, the meaning of observation sentences is determined by the theories with which they are connected. Theories are meaningful independent of observations; observation statements are not meaningful independent of theories . . . It is therefore the observation sentence that is in need of interpretation and not the theory (1965d, p. 213).

The standard bottom-up view of experience as determining the meaning of our observational terms (and then also our theoretical terms) is replaced by a top-down model according to which our theories, stated with theoretical terms, determine the meaning we attach to our experiences – and not the other way around. According to Feyerabend, *theories are not interpreted on the bases of independently fixed or 'given' experiences.*[59] *Experiences are interpreted on the basis of our improving theories.*[60]

By specifying that *thesis I* holds for observation statements, Feyerabend was not making an exception with regards to observational language as opposed to theoretical terms as they are used in scientific language. Rather, he was emphasizing that even the meaning of our observation terms are not determined by experience, but by our theories – let alone the meanings of our theoretical terms. On Feyerabend's view:

> the existential problems of observable entities are identical with the existential problems of the so-called 'theoretical terms'. An example is the problem of the existence of the devil which was decided *on the plane of theories* and not on the plane of observational practices . . . Logically speaking all terms are 'theoretical' (1958a, p. 164, fn. 23; rp 1981a, p. 32, fn. 22).

Consequently, *thesis I* should be understood as an *extended* version of what is sometimes called the 'contextual theory of meaning', according to which:

> the meaning of a term is not an intrinsic property of it but is dependent upon the way in which the term has been incorporated into a theory (1962a, p. 68; rp 1981a, p. 74).

This basic thought runs throughout Feyerabend's philosophy.[61] Combined with Feyerabend's view of an interpretation as distinct form its characteristic, the idea is that *scientific terms receive their meaning from the theories whose truth is presupposed when the term is used to describe or explain some feature of the world.* It is exactly this conception of the language of scientific theories that

59 Feyerabend also argued this point at length in e.g. (1963b), pp. 50ff.

60 Compare to Kuhn (1970a), e.g. p. 102.

61 See e.g. (1951), pp. 90, (1954c), (1958a), p. 163; also see (1981a), p. 31, (1962a), p. 68; rp (1981a), p. 74, (1965c), p. 259, fn. 27; rp (1981a), p. 114, fn. 27, and (1965b), p. 180 and p. 184. It is also sometimes discussed in the secondary literature as the "doctrine of implicit definition" and "the cluster theory of meaning". See e.g. Musgrave (1979), p. 345, fn. 7.

Feyerabend used to develop and support the normative views he expressed as 'realism' (not 'scientific realism') in the 1950s and 1960s.[62] In this regard, Feyerabend's views are quite comparable to Fleck's:

> This is how a fact arises. At first there is a signal of resistance in the chaotic initial thinking, then a definite thought constraint, and finally a form to be directly perceived. A fact always occurs in the context of the history of thought and is always the result of a definite thought style (Fleck 1935, cited in Golinski 1998, p. 33).

Simmons has also argued that Feyerabend's notion of incommensurability has much in common with Fleck's views (Simmons 1994). As one commentator put the point quite succinctly: "Theories are not functions of facts; facts are functions of theories" (Kenshur 1984, p. 376). *Thesis I* applies a contextual theory of meaning to the terms of our observation language, and concludes that when theories change,[63] the meanings of observation terms change with it.[64]

The difference between *thesis I* and the contextual theory of meaning is not as straightforward as it may appear, and the relationship between *thesis I*, and the contextual theory of meaning is easily misunderstood.[65] It appears that the contextual theory of meaning is more general than *thesis I*, because it applies to the meanings of scientific terms generally, and not just to observation language. This has led Preston to present *thesis I* as a logical consequence of the contextual theory of meaning: "The contextual theory of meaning, being more general than Thesis I in applying to any kind of language, logically implies this thesis" (Preston 1997, p. 30). This is both *incorrect*, and also rather *misleading*. The contextual theory of meaning, in Feyerabend sense, did not apply to *any* language, but to the meanings of scientific terms. Feyerabend distinguished scientific terms, for example, from everyday terms, and he distinguished a scientific observation language from 'everyday language', which in Feyerabend's view, is not really a proper language at all. For example, according to Feyerabend:

> It has been frequently asserted that that the language in which we describe our surroundings, chairs, tables and also the ultimate results of experiment (pointer-reading) is fairly insensitive towards changes in theoretical 'superstructure'. It seems somewhat doubtful whether even this modest thesis can be defended; first, because a uniform 'everyday language' does not exist. The language used by the 'everyday man' (whoever that may be) is a mixture of languages, as it were, i.e. it

62 For a discussion, see Preston (1997), p. 30, and see chapter 6 of this volume.

63 Note, again, that *thesis I* concerns observation languages and the contextual theory of meaning concerns terms. The missing assumption is that the meanings of terms (as determined by the theories of which they are part) determine the meaning of the observation statements in which they occur.

64 For a discussion of the implications of this idea to the possibility of historical research, see e.g. Kuppermann (1975).

65 See e.g. Preston (1997), pp. 25ff.

> is a means of communication which has received its interpretation from various and often incompatible and obsolete theories. Secondly, it is not correct that the mixture does not undergo important changes: terms which at sometime were regarded as observational elements of 'everyday language' (such as the term 'devil') are no longer regarded as such. Other terms such as 'potential', 'velocity', etc. have been included in the observational part of everyday language, and many terms have assumed a new use. . . (1958a, pp. 162-163; rp 1981a, pp. 30-31. See similarly e.g. 1962a; rp 1981a, pp. 44-96 and 1965a; rp 1981a, pp. 97-103).

The point is not merely pedantic. The point is that *thesis I* does not follow logically from the contextual theory of meaning. For example, it has not been shown that the terms of an observation language are scientific terms (and not for example 'everyday terms'). On the logical positivist conception that Feyerabend was attacking, the basic idea was that the terms of the observation language could be readily distinguished from the theoretical terms used to state the theories. The terms of the observation language were to receive their meaning from immediately given experience, whereas the theoretical terms were seen as piggy-backing on the more commonly understood, inter-subjectively verifiable, observation terms used in the protocol statements. They receive their meaning by being defined on the basis of observational terms. Consequently, on the logical positivist conception, *thesis I* does not follow logically from the contextual theory of meaning, because the two claims apply to terms that have their meanings determined in different ways, and that belong to different language (observation language and theoretical language).

What about on Feyerabend's conception? Feyerabend rejected the logical positivist distinction between the observation language and the theoretical language,[66] so does *thesis I* follow from the contextual theory of meaning logically from Feyerabend's perspective?

While Feyerabend distinguished between scientific and everyday terms, he did not distinguish between observation terms and theoretical terms. In Feyerabend's view, the observational terms of observation statements are just as theoretical as the theoretical terms of scientific theories. However, even on Feyerabend's conception, where there is no distinction between the two languages, *thesis I* still does follow from the contextual theory of meaning. To see this, note that *thesis I* claims that the interpretation of an observation language is determined by *the theories we use to explain what we observe.* According to the contextual theory of meaning, on the other hand, the meanings of scientific terms are determined by *how they have been incorporated into a theory.* The set of theories used to explain a

66 See e.g. (1954c), p. 470, (1958a), especially pp. 164-165; (1981a), pp. 32-33, (1958b), (1960a); rp (1981a), pp. 37-43, *especially* (1960f), (1961b), p. 82, (1961g), p. 443, (1962a), pp. 29-31, rp (1981a), pp. 45-47, (1965b), p. 198, and pp. 219-224, fn. 8, and (1969b); rp (1981a), pp. 132-135.

particular observation need not be identical to the set of theories in which that term has been incorporated. In fact, the same term may be incorporated into different theories in different ways (e.g. the term 'gene'). It may be that a term is incorporated into the theories which we use to explain what we observe in a different way than it is incorporated into a theory which explains something else, and which may be based on entirely different observations. This is not the place to develop arguments or an account of the different ways that a term can be incorporated into different theories, or the different kinds of relations that these theories might have to each other. However, the term 'gene' may serve as an obvious example. Here, it suffices to point out that the simple fact that there are such possibilities undermines Preston's claim that *thesis I* follows logically from the contextual theory of meaning. Moreover, as I have mentioned in two footnotes, *thesis I* applies to observation *statements* and the contextual theory of meaning applies to *terms*. From this alone, it follows that one does not follow from the other *logically*.

While logical implication is much too narrow to capture the relation between the contextual theory of meaning and *thesis I*, the basic ideas behind *thesis I* and the contextual theory of meaning are very similar. As I understand it, and as I imagine Feyerabend conceived of it, *thesis I*, while less general than the contextual theory of meaning, actually *extends* the contextual theory of meaning to include the interpretation of the terms in the observation statements. By stressing that the meanings of terms in the observation language are determined by the theories used to explain what we observe, Feyerabend was actually *extending* the contextual theory of meaning from applying to just the theoretical terms of scientific theories, which are determined by their connection to other terms on the positivists own view, to cover the terms used in the observation language as well. Thus, *thesis I*, while more specific than the contextual theory of meaning, should be understood as an *extension*, not a *restriction* of the contextual theory of meaning. In summary, not only is Preston's characterization of the type of relationship between *thesis I* and the contextual theory of meaning (logical implication) incorrect, it also characterizes the relation the wrong way around (the contextual theory of meaning as a generalization of *thesis I* instead of *thesis I* as an extension of the contextual theory of meaning).

This brings us to the second (descriptive, historical) prong of Feyerabend's argument against the stability thesis. As is often the case, rather than just present theoretical reasons supporting his thesis, Feyerabend illustrates his point with examples of changes that have taken place in the interpretation of an observation language in the actual course of scientific advance.[67] These examples are supposed to undermine the stability thesis and to support Feyerabend's alternative: *thesis I*. In his (1958a; rp 1981a, pp. 17-36), Feyerabend illustrated *thesis I* with just two examples. The first example of a change in the interpretation of an observation term is the color predicates. For the sake of clarity, I will simplify Feyerabend's highly technical presentation of it. Feyerabend invited us to consider an observation language that ascribes colors to self-luminescent objects (like stars), and to assume, first that the *characteristic* of these color predicates has been defined, and second that the methods of observations implied by this characteristic involve only velocities and masses that are encountered in normal everyday circumstances, which can be produced and handled with relative ease. In this situation, human beings will *interpret* the color predicates according to their general ideas about things and their properties. For example, a person may think that a color predicate designates a property of an object, and that the object has this property regardless of whether or not it is being observed. Some stars appear to be blue, and thus they are believed to be blue, and this regardless of whether or not they are being observed by us.

However, as it turns out, given improved theories about the nature of light, color predicates, and self-luminescent objects, *according to Feyerabend* this everyday interpretation is incorrect.[68] According to an improved theory of light and color predicates, given the discovery of the Doppler-effect, the wavelength of light, as measured by an observer *in accordance with the usual characteristic* of color predicates, depends upon the relative velocity of the observer and the light source (among other things). According to this improved theory, the very same use of a color predicate to describe an object is now an instance of a *relation* rather than an instance of a *property*. Thus, the adoption of a new theory can lead to a new interpretation of the observation language. For example, as it turns out, some stars that may appear blue to us, are not actually themselves blue. The light that they emit only appears blue to us given their relative motion to us. Moreover, the new interpretation does not necessarily change the *use* of the color predicate in normal, everyday circumstances.[69] We still talk about blue stars. However:

> an expression does not cease to designate a relation even if for all situations inside the domain of its applicability this relation depends upon one term only and

67 Also see (1961b), p. 83, and (1961h), pp. 246ff.

68 Some contemporary scientific realists would disagree with this claim. They attempt to reinterpret colors as complex dispositional properties of objects.

69 From this point alone, it follows that Feyerabend's semantic views are not the same as Wittgenstein's in the *Investigations*.

> remains completely insensitive towards all changes of the remaining terms (1958a, pp. 161-162; rp 1981a, p. 30).[70]

The second example is more technical, but supposedly more straightforward. It appears in a footnote from which I will simply cite:

> the magnitudes (properties) of classical physics can be determined at any time with any required precision. On the other hand, quantum-mechanical entities are complementary in the sense that at a given time they are able to possess only some of their possible properties. Now classical mechanics is a special case of quantum mechanics, which means that all the objects of the macroscopic level obey the laws of quantum mechanics. Hence, we must re-interpret the signs of classical physics as designating properties, which apply to their objects (the objects of the macroscopic level) in almost all circumstances (whereas according to classical physics proper they apply strictly in all circumstances). This means that *having adopted quantum mechanics we must drop the classical interpretation of classical physics* (1958a, p. 161, fn. 20; rp 1981a, pp. 30, fn. 19).

To summarize the point, according to Feyerabend, the meaning of an observation terms, such as a color predicate like 'blue' is not determined either by our everyday use of color terms, or by our phenomenal experience of colors. Rather, the meaning of the term 'blue' is determined by our best scientific theories, according to which blue is not a *property* of objects, but a *relation* between objects and observers. *The interpretation of an observation sentence depends upon the theories we use to interpret it.* This is an inversion of the logical positivist (and logical empiricist) idea that the interpretation of theories depends solely on observations used to support it. From this inversion of the relation of theory to experience, it follows that *theoretical advance has, and can, alter how we interpret our experiences; and consequently, experience is not a stable foundation for testing theoretical knowledge.*[71]

Having developed Feyerabend's views on the meaning and its relationship to experience concerning theory testing at some length, we are finally in position to examine how these semantic views compare to Wittgenstein's later philosophy. In his 1955 review of Wittgenstein's *Investigations* (1951),[72] Feyerabend found a "new (instrumentalist, nominalist, or whatever you like to call it theory of meaning" (1955a, p. 479; rp 1981b, p. 125) which he characterizes as part of an "intuitionist (pragmatist, con-

---

70 This example, according to Feyerabend, is itself enough to refute the principle of pragmatic meaning. See (1958a), p. 163; rp (1981a), p. 31.

71 The same ideas also feature prominently in (1954c) and in (1962a); rp (1981a), pp. 44-96.

72 It was not originally intended as a review. It was written on the basis of German manuscripts of the *Investigations*, provided by Elizabeth Anscombe, in London in 1952. Anscombe, who also translated the *Investigation* from German into English, also partially translated (1955a) into English, see (1978a), pp. 114-115, (1993a), p. 261, as well as (1995a), pp. 92-93.

structivist) theory of language" (1955a, p. 462, fn. 12; rp 1981b, p. 111, fn. 12). By 'constructivist', Feyerabend meant constructivism in mathematics:

> Wittgenstein's theory of language can be understood as a constructivist theory of meaning, i.e. as constructivism applied not only to the meanings of mathematical expressions but to meanings in general (1955a, p. 462, fn. 12; rp 1981b, p. 111, fn. 12. Also see 1966a, p. 8).

As with many things, he credited Mach as the forerunner of this, and other, ideas central to his developing ideas: "In this [rejecting Einstein's characterization of scientific research as an extension of common-sense] he [Mach] became the one of the forerunners of Gestalt psychology, constructivism in mathematics, Piaget, Lorenz, Polanyi and Wittgenstein" (1987, p. 206).

Feyerabend also thought he found some version of the contextual theory of meaning in Wittgenstein's *Investigations*, according to which the interpretations of observations sentences are determined by theories, and not that the other way around.[73] Preston claims this is how Feyerabend developed his contextual theory of meaning, according to which the meaning of a term is determined by its theoretical context; hence, the name 'the contextual theory' of meaning. On this view, theoretical context consist of a concepts conceptual relations to the other concepts in the theory. For example, the theoretical context of the concept of mass in Newton's theory involves statements such as 'Force equals mass times acceleration', which relates the concept of mass to other concepts such as force and acceleration. Supporting Preston's claim, sometimes, Feyerabend partially attributed the idea of the contextual theory of meaning to his reading of Wittgenstein. For example, he wrote:

> the idea (which may be found in Wittgenstein, ***but which has been held by many nominalists before him***) that the meaning of a word is a function of the (theoretical) context in which they occur) . . . (1960d, p. 248, italics inserted).

Thus, one *could* join Preston and conclude that a key idea behind Feyerabend's attack on logical positivism, and the idea that he used to support the notion of incommensurable concepts (see chapter 5) was at least partially a result of his reading (or misreading) of Wittgenstein's *Philosophical Investigations*. Preston argues that Feyerabend's semantic views are the result of Feyerabend's *mis*reading of Wittgenstein (Preston 1997 pp. 23ff), and Preston appears to be right in so far as the contextual theory of meaning does not represent Wittgenstein's views. Whatever the exact nature of Wittgenstein's views on meaning, which is a controversial issue, it seems clear that they are some version of the principle of pragmatic meaning,

73 See and compare (1955a), pp. 462-463 and pp. 473-474 and p. 483; rp (1981b), p. 111 and pp. 120-121 and p. 130.

according to which meaning is determined by use, rather than a version of a contextual theory of meaning. As Feyerabend clearly and repeatedly rejected the principle of pragmatic meaning in favor of a contextual theory of meaning, we can safely conclude that Feyerabend's views on meaning are not the same as Wittgenstein's.

Although Feyerabend almost consistently defended such contextualism about meaning, in one place Feyerabend does criticize Hanson's claim that the discovery of the positron was actually the discovery of three different particles as a consequence of an extreme contextualism of meaning. (See 1964a, p. 266 and Preston 1997, p. 213, fns. 8-9). Feyerabend rejected such an extreme contextualism of meaning because it leads to the conclusion that independent tests of a physical theory are logically impossible. It follows that Feyerabend's contextual theory of meaning is not supposed to rule out the possibility of using evidence to test a theory. Preston, cited *thesis I* from (1981a, p. 31) and immediately claimed that Feyerabend "explicitly ascribe[d] the idea to Wittgenstein, Galileo and other scientists" Preston (1997, p. 30). However, neither Wittgenstein, Galileo or any other scientists appear on p. 31 of (1981a), nor did Feyerabend attribute *thesis I* to Wittgenstein, Galileo or any other scientists anywhere in his (1958a); rp (1981a), pp. 17-36, nor in the (1981a) reprint cited by Preston. As we have seen, in other publications, Feyerabend sometimes referred to Wittgenstein as a forceful defender of the contextual theory of meaning *in the 20th century*, and it would not be surprising in the least that he ascribed the idea to Galileo or to other scientists, but as Preston has failed to provide any citations to back up his claims, it has not been able to follow up on these suggestions.

However before joining with Preston and concluding that Feyerabend developed his semantic views as a result of a misreading of Wittgenstein, we should also consider some other things that Feyerabend said about his relationship to Wittgenstein concerning the contextual theory of meaning. And moreover, we should be aware that Feyerabend later went to great lengths to deny the claim that incommensurability, which is based on the contextual theory of meaning (see chapter 5) should be seen *as a consequence* of the contextual theory of meaning.

In (1958a; rp 1981a, pp. 17-36), Feyerabend did follow Wittgenstein in arguing against phenomenological accounts of meaning and essentialism, although he offered his own *ad hominem* argument different from Wittgenstein's.[74] But he certainly did not follow Wittgenstein to the famous conclusion that "the meaning of a word is its use in the language"

74 See (1958a), pp. 153ff.; rp and (1981a), pp. 24ff. for his *reductio*, and Preston (1997), pp. 34ff. for a discussion of it. The same argument also appears in (1965b), pp. 204-205.

(Wittgenstein 1958, p. 43). As we have seen, Feyerabend maintained that there is "a clear and unambiguous" distinction between the pragmatic properties of a language and its semantic properties, rejecting the idea that meaning is use.[75] In making the case that the central concepts of Newton's theory are incommensurable with those of impetus theory, Feyerabend clearly and unambiguously distinguished between meaning and use: "meaning and application are not the same thing" (1962a, p. 57, rp 1981a, p. 66). Feyerabend also explicitly warns of the "pitfalls in the principle that difference of use indicates difference of meaning" (1960d, p. 250). In fact, Feyerabend spent considerable time arguing that meaning change does not necessarily require changes in use, and concluded that:

> The occurrence of such changes [conceptual revisions for which no direct observable reason is available] quite obviously refutes the contention of some philosophers that the invariance of *usage* in the trivial and uninteresting contexts of the private lives of not-too-intelligent and inquisitive people indicates invariance of *meaning* . . . (1965b, p. 180).

Moreover, Feyerabend acknowledged that both Newton's theory and impetus theory sometimes apply a term such as 'force' under exactly the same testing conditions. He claimed that the *operational* characteristic is sometimes the same even for two incommensurable theories. In other words, a single sentence can be used to describe the result of an experiment, but can be interpreted according to two different theories. Take for example the sentence "When dropped from a height of seven meters, the lead ball penetrated 2.2 centimeters into the wax." This one sentence can be used to test two incommensurable theories (Newton's theory and impetus theory). The single experimental result will be interpreted differently, because of the inconsistency between some very basic theoretical principles of the two theories used to interpret it.

In the original review, Feyerabend traced the origins of Wittgenstein's nominalist approach to meaning to Gomperz, Mach, D'Alembert, and Popper, (1955a, p. 482, fn. 23; rp 1981b, p. 128, fn. 23). In (1962a), he qualified the credit he gave to Wittgenstein as a forceful defender of the contextual theory of meaning to "*In the twenty century*" (1962a, p. 68, fn. 83; rp 1981a, p. 74 fn. 68, italics inserted). Later, Feyerabend also emphasized that these idea were not particularly Wittgensteinian:

> I feel I must object to the tendency . . . to forego historical accuracy in favor of hero worship. The arguments demanding inter-subjectivity of observation statements (and corresponding high theoretical content) *are not at all original with Wittgenstein*; they were formulated before him, and in a much clearer and simpler way. History, and not fashion, should decide who has achieved what and who is therefore worth listening to (1965b, p. 227, fn. 20, italics inserted).

75 See (1958a), p. 146, rp (1981a), p. 19, and (1954c), pp. 471-473.

In the same spirit, in the early to mid 1960s, Feyerabend often chastised Wittgenstein and Wittgensteinians for their lack of historical perspective:

> Wittgenstein only articulates (if that is the right word) in a very esoteric fashion an attitude which existed long before his time in many different fields . . . and of whose independent existence he and his pupils do not seem to be aware (1964b, p. 250).

Similarly, Feyerabend rebuked Wittgenstein:

> Many contemporary philosophers are so convinced of the revolutionary character of their enterprise that they believe that they can safely neglect everything that happened before, say, 1900 (or perhaps 1889, the birth-date of Wittgenstein). Wittgenstein himself has brought a completely unhistorical atmosphere into philosophy. Although dealing with problems which have a firm foundation in the history of philosophy he has presented these problems in a fashion that suggested that they had just been born and were completely unrelated to whatever else had been thought before (except perhaps the *Tractatus* and Russell's *Principia*) (1963e, p. 323).

Moreover, sometimes it is not even a contextual theory of meaning itself that Feyerabend actually credited to Wittgenstein. For example:

> just as a concept can be made clear only from the function it plays in a statement—a fact that has been seen, among others by Wittgenstein who therefore in his *Tractatus* took sentences as the fundamental units of discourse—in the very same matter a sentence, its structure, the way it will be replaced, in the course of scientific progress, by a different sentence can only be made clear by pointing to the role it plays in a wider context of a theory, or of an explanation, that last an final unit of scientific knowledge (1961h, p. 247).

Here, Feyerabend made it clear that his views on the meaning of a statement in relation to a theory can be explained through an *analogy* to Wittgenstein's claims about how a statement provides the context that affects the meaning of a term. The real link between Feyerabend's contextual theory of meaning and Wittgenstein's views on meaning is an *analogy* about the affects that context has on meaning. According to Feyerabend, the meaning of an observation *statement* depends on the context provided by the *theory* to which it belongs. According to Wittgenstein, the meaning of a *term* partially depends on the context provided by the *statement* in which it is used. In my view, that is all there is to the positive relation between the contextual theory of meaning and Wittgenstein's views on meaning. Feyerabend simply was not a Wittgensteinian concerning semantics.

In conclusion, Feyerabend's views on meaning, expressed as a form of the contextual theory of meaning, are definitely not those of Wittgenstein in the *Investigations*. While Feyerabend may *sometimes* have misread the contextual theory of meaning into Wittgenstein's philosophy, he did not always do so, and he certainly did not think that the basic idea was particularly Wittgensteinian. As Preston claims, Wittgenstein did influence Feyerabend's key insight about the role of theories in determining the meaning of observational terms. However, Feyerabend sharply distinguished between meaning and use, thereby rejecting Wittgenstein's semantic views. Moreover, as we shall see in the next section, Preston's positive portrayal of Feyerabend's relation to Wittgenstein simply ignores the often repeated and sometimes scathing criticisms of Wittgensteinianism.

2.2. Feyerabend's critique of Wittgensteinianism

From the 1950s through the 1960s, Feyerabend was often extremely critical of the movement known as analytical or linguistic philosophy,[76] or what Feyerabend scathingly mocked as 'Oxford philosophy' that Wittgenstein's texts spawned.[77] For example, in a 1961 letter to Kuhn on a draft of *The Structure of Scientific Revolutions*, Feyerabend wrote:

> People ask me to talk about Wittgenstein, because he is an influential contemporary philosopher, and they also ask me to talk about Ayer, etc. etc. I refuse to do this. And if they ask "shall we read the PHILOSOPHICAL INVESTIGATIONS?" I reply "If you want to waste your time, yes" (1995b, written 1961, p. 384).

Munévar is absolutely correct in his claim that Feyerabend rejected the basic distinction between analytic and synthetic on which Wittgenstein philosophy is based, as Feyerabend repeatedly applauded efforts to undermine the distinction between the two.[78] Most significantly, Feyerabend criticized specific dogmatic elements of linguistic philosophy[79] and he was clearly keen to distance himself from what he considered to be its dogmatic tendencies:

> Considering this feature of the argument [that experience does not contain a factual core independent of the theories need to interpret it] I cannot understand why Feigl should find "neo-Wittgensteinian traces in my reasoning . . . "Wittgen-

76 For some background on ordinary language philosophy, see e.g. Warnock (1999).

77 The following list is far from comprehensive, but includes some of the more amusing name calling, as well as the major arguments against linguistic philosophy that Feyerabend launched. See e.g. (1955a), pp. 480-483; rp (1981a), pp. 127-130, (1956a), p. 96 fn. 4), (1957e), p. 238, fn. 3, and p. 244, (1960d), p. 247, (1960e), p. 322; rp (1981a), pp. 219-220, (1995b), written 1960/61, p. 385, (1961i), especially pp. 65-67; rp (1999a), p. 75, (1962a), e.g. p. 61 fn. 61; rp (1981a), pp. 68-69 where it has been removed, (1963e), p. 320, (1964b), pp. 250ff., (1965b), p. 226, fn. 10, p. 246, fn. 114 and p. 247, fn. 116, p. 249 fn. 118, p. 251, fn. 132, p. 252, fn. 137, pp. 256-257, fn. 151, and (1966a), p. 9.

78 See e.g. (1962a), p. 68, fn. 82; rp (1981a), p. 74, fn. 67, (1966b), p. 259, fn. 27, (1991b), p. 535. For a discussion, also see Preston (1999), p. 236.

79 See specifically (1969e); rp (1981a), pp. 146-160.

steinian" analyses are characterized by the fact that they are restricted to a single "language game"; they are *monistic*. Alternatives are brought in, not in order to arrive at a better theory through a criticism of the existing one, but rather *in order to get better insight into the existing theory*. And getting better insight into the existing theory (the existing "language game") means revealing its hidden strength, i.e., its capability to deal with the problems that have arisen and to remove the impression that a revision might be needed. A Wittgensteinian uses alternatives with a *dogmatic*, or a *conservative*, purpose. Knowing where the argument is going to lead, he does not really take them seriously, although he may admit that they possess an important therapeutic function. Our own point of view is radically different. Alternatives *are* taken seriously. They are all candidates for a better theory in the future. Their function is *critical* and *progressive*, that is they are used with the purpose of finding the *weakness* of the customary ideas. The discussion of such weakness does not have an exclusively dramatic purpose, viz., to let the accepted point of view shine forth the brighter once the philosophical fog has been removed. A greater difference can hardly be imagined (1965b, p. 227, fn. 20).

Feyerabend repeatedly criticized Wittgenstein's conception of philosophy as being limited to the inside of language games, claiming that it implies that progress from one theory to another is impossible. For example, according to Feyerabend: "Wittgenstein is inclined to restrict his theory to the inside of his language games: Platonism of concepts is replaced by Platonism of (theories or) games" (1962a, p. 68, fn. 83; rp 1981a, p. 74, fn. 68). According to Feyerabend as philosophers, we should not just leave everything as it is.[80] Feyerabend often based his criticism of the dogmatic and conservative element of Wittgenstein's philosophy on § 124 of the *Investigations*:

Philosophy may in no way interfere with the actual use of language; it can in the end only describe it. For it cannot give it any foundation either. It leaves everything as it is. . . (1958, p. 49e).

While early on, Feyerabend often took the opportunity to launch this criticism of Wittgenstein, later he thought it somewhat uncharitable. For example, the pointed criticism of linguistic philosophy in his (1962a, pp. 61-62 and fn. 71) was removed from the reprint in (1981a, pp. 68-69).

In addition to rejecting Wittgenstein's conception about the methods and goals of philosophy, Feyerabend consistently rejected Wittgenstein's conception of the relationship between philosophy and science, according to which philosophy must stand above or below science, but not on a par with it.[81]

80 See e.g. (1962a), pp. 61-62; rp (1981a), p. 68, removed entirely.

81 See e.g. (1955a), pp. 481-482; rp (1981a), pp. 128-130 and e.g. (1969e), pp. 54-55, fn. 25 as well as pp. 61-62, fn. 35; rp (1981a), pp. 154-155, fn. 25, removed, and p. 159, fn. 29.

Feyerabend did not sharply distinguish between science and philosophy,[82] often emphasizing that "empirical arguments and speculation go hand in hand" (p. 465).[83] In "Physik und Ontologie" (1954c), he discussed the significance of revolutionary developments in physics for philosophy at length, arguing that revolutionary developments in physics marked the advent of universal theories capable of providing worldviews. This effectively *ended the separation of philosophical knowledge from scientific knowledge.* Such a separation had been enforced since Plato (but not by the Ionian philosophers before) by the "ontological prejudice".[84] According to this prejudice, philosophical knowledge must be certain, while scientific, or empirical, knowledge is not. Feyerabend's view is that "The institution of ontology is no business that philosophers could do independently of scientists" (1954c, p. 467). By ontology, Feyerabend explained that he meant "that what is" (p. 467). For example, the ontology of the positivists includes 'sense data' while for common sense realists it will include everyday objects. An ontology provides a *Bedeutungslehre*, according to which a *sentence* is a *statement.* It tells us what is real, as for example ideas according to Platonists, experiences according to phenomenalists, and material objects according to scientific realists. As a result of the rejection of the ontological prejudice (according to which knowledge is certain), philosophical claims have a speculative and fallible status just like claims in the empirical sciences. Moreover, according to Feyerabend's conception of the relationship between philosophy and science, philosophy must take account of advancements in the sciences, and in this respect it is subsidiary to it. For example, Kant's speculative philosophy was tailored to Newtonian Mechanics. As the latter has been shown to be inadequate empirically, the former has also been refuted. In other words, philosophical interpretations of science can be overtaken by events on the ground. Even so, philosophy can make positive contributions. On Feyerabend's view it can be a "motor" for scientific change (e.g. 1963e, p. 319).

To support his reading of Wittgenstein on the relationship between philosophy and the sciences, Feyerabend cited from the *Tractatus* 4.III "the word 'philosophy' must mean something that stands above or below, not beside the natural sciences" (Wittgenstein 1922). Feyerabend often

82 See for example (1955a), pp. 481-482; rp (1981b), pp. 127-128, (1956a), (1961i), pp. 6ff.; rp (1999a), pp. 52ff., (1964c), pp. 349ff.; rp (1981a), p. 203ff., (1965b), p. 150, (1966a), p. 9, (1969e), pp. 54-55, fn. 25 as well as pp. 61-62, fn. 35; rp (1981a), pp. 154-155, fn. 25, removed, and p. 159, fn. 29, (1975a), p. 48, fn. 2, (1988), p. 34, fn. 2, as well as (1993a), p. 34, fn. 2.

83 He also explicitly addressed the issue in his doctoral thesis (1951). Also see (1958a), section 7, pp. 166ff.; rp (1981a), pp. 33ff.

84 Also see especially Feyerabend's Conflict and Harmony (forthcoming [1993]).

criticized such separations of philosophy from the sciences,[85] although he himself does sometimes distinguish between the two (see e.g. 1958c, pp. 346ff.).

However, as is so often the case with Feyerabend, there are two sides to the story. Feyerabend also praised and recommend some passages of Wittgenstein's work. For example on the very same page of the letter to Kuhn in which Feyerabend claimed that he flatly refused to discuss Wittgenstein, and had claimed that reading Wittgenstein, in his view, is a waste of, (see the quote above), Feyerabend also wrote:

> Hanson has added nothing at all to Wittgenstein whom he imitates (all you need is to read, in this connexion, and really should read, are Wittgenstein's investigations on seeing (1995b, written 1961, p. 384, underlined in original).

Feyerabend (as usual) was not always consistent in his expressions of his views, not even within the same page of a letter he wrote to Kuhn in the early 1960s. Moreover, somewhat surprisingly, by 1980 Feyerabend had relaxed his extremely negative views toward analytical philosophy. This can be made extremely clear through a comparison of the negative attitude toward analytical philosophy expressed in (1969e, pp. 54-55, fn. 25), to the reprint (1981a, pp. 154-155, fn. 25 in which the footnote reads "Added 1980: I no longer agree" and refers the reader to fn. 1, which was lengthened to include a more tolerant attitude toward the kind of analytical philosophy that he had repeatedly criticized in his earlier work.

To summarize, prior to 1980, Feyerabend was extremely critical of many core aspects of Wittgensteinian philosophy, especially what he perceived as an unjust conservativism with regard to the task of philosophy itself. According to Feyerabend, philosophers are not restricted to conceptual analysis that leaves meanings unchanged. They should change the meanings of the terms through their analyses.[86] Such analytical philosophy (that does not try to change or improve the meanings of terms, but accepts the utility of concepts already given in everyday language and merely tries to clarify them) epitomizes the kind of conceptual conservativism that Feyerabend absolutely despised, and which, as we shall see, he repeatedly attacked on basis of the notion of incommensurable concepts. Feyerabend thought the task of philosophy was to help to improve existing concepts, and to develop *new* concepts, not merely clarify what is already given from the slide lines. After 1980, Feyerabend somewhat moderated his anti-Wittgensteinianism.

---

85 See e.g. (1954c), pp. 464ff., and (1961g), p. 441.

86 See e.g. (1955a), pp. 480-483; rp (1981b), pp. pp. 127-130, including the passage labeled "Added 1980" p. 129, (1956a), (1957e), and his other criticisms of Wittgenstein cited above.

2.3. Feyerabend's adoption of nominalism

While Feyerabend rejected Wittgenstein's theory of meaning as use, and his conception of the task of philosophy, Wittgenstein's philosophical ideas did have a deep and lasting impact on Feyerabend's philosophical development. The most lasting and significant influence that Wittgenstein's later philosophy had on Feyerabend was nominalism. As Feyerabend recalled in his 1995 autobiography, "In a sense, I have become a Wittgensteinian (or a nominalist, to use a more traditional term)" (1995a, p. 94). As we have seen, from very early on, Feyerabend understands Wittgenstein's theory of meaning, as some "new (instrumentalist, nominalist, or whatever you like to call it theory of meaning" (1955a, p. 479; rp 1981b, p. 125).

What kind of nominalism did Feyerabend developed? Wittgenstein's famous example of the use of the general term 'game' attacked the idea that there are 'universals', i.e. properties or some set of properties which are shared by all of those things to which we apply a term. In so doing, he was suggesting a nominalist approach to concepts. According to some forms of nominalism, the way in which we organize our experiences by classifying objects into kinds does not reflect real divisions in nature. Such a nominalism is a form of non-realism according to which we classify objects into kinds primarily in order to make sense of our experiences and to establish a means of reliable prediction and control. Our classifications are conceptual tools which aid our understanding, but that do not reflect real divisions in nature.[87] *They do not reflect natural kinds.* Feyerabend's nominalism is a form of anti-realism about natural kinds.

The contemporary debate about nominalism derives from Kant's idea that the everyday world is empirically real but only transcendentally ideal.[88] Our concepts do not reflect how things are in themselves. According to nominalism, how we experience the world is partly the result of the concepts we use to have our experiences. It is the result of our imposition of concepts on experience. In other words, *pace* scientific realism, we do not succeed in discovering the true nature of things, or some underlying essence through our conceptual classifications. Moreover, *pace* Kant, the categories we use to organize our experience are not necessary. They can change through revolutionary developments in the sciences. Feyerabend, like the latter Wittgenstein, was a nominalist about concepts and their relation to nature. As we shall see, it is from such a non-realist, neo-Kantian perspective that Feyerabend launched his critic of logical empiri-

87 For a development of Kuhn's views into a form of nominalism, see e.g. Hacking (1993).

88 See Kant (1781), pp. 65-91, as well as McDowell (1994), pp. 3ff.

cism when he introduced the idea of incommensurable concepts (1962a; rp 1981a, pp. 44-96).

Feyerabend's rejection of the notion of natural kinds and adoption of nominalism is not only a fundamental aspect of his early philosophy of science. It also underwrites one of his last major works, *Conquest of Abundance. A Tale of Abstraction versus the Richness of Being* (1999b). The leading idea in that work is that 'Being' responds differently to different conceptual approaches. To take Feyerabend's extreme example, we can conceive of the world as consisting of Gods or of atoms. The sets of concepts we bring will determine the kinds of experiences we have. This represents an *extension* of Feyerabend's early nominalism from applying to concepts within the natural sciences to applying to worldviews more generally (a pattern that fits Feyerabend's philosophical developments more generally, e.g. his extension of the notion of incommensurability from scientific theories to worldviews and cultures). For example, in the 1950s, Feyerabend's investigations of Quantum Mechanics and the dualities he found there led him to the belief that how we experience the world depends on the concepts we bring to it. For example, we can learn about some aspects of light by conceptualizing it as particles, whereas we can learn about other aspects of light by conceptualizing it as waves. While both are helpful for investigating some of the properties of light, the distinction between waves and particles is an artifact of our conceptual approach that does not mark a real distinction in nature. Feyerabend's later notion of our continuing conquest of abundance extends the ontological lessons that ground his early philosophy of science to a more general view of experience in his later works.

2.4. Conclusions

While Feyerabend's relation to Wittgenstein ideas are complicated and contentious, and while Feyerabend's pronouncements on Wittgenstein are not always consistent, the following conclusions are warranted. Feyerabend developed his notion of incommensurability around the same time that he attempted to interpret Wittgenstein. Wittgenstein's ideas influenced the development of Feyerabend's semantic views. Feyerabend thought he found some version of a contextual theory of meaning in Wittgenstein's *Investigations*, according to which the interpretation of an observation sentence is determined by theories, and not the other way around. In this respect, Feyerabend's interpretation of Wittgenstein's insights about the contextual nature of meaning did significantly influence the development of Feyerabend's ideas on meaning. However, Feyerabend repeatedly emphasized that these ideas were not original to Wittgenstein, who was merely a forceful defender of them in the *twentieth century*, and Feyerabend was extremely critical of the tendency to credit Wittgenstein as the original source of these ideas. Moreover, Feyerabend did not share Wittgenstein's views on meaning. Wittgenstein developed a view according to which meaning is use. Feyerabend sharply distinguished meaning from use. In fact, as we shall see in chapter 5, Feyerabend's theory of meaning (which distinguishes between meaning and use), is the basis of the development of the notion of incommensurable concepts at the very heart of Feyerabend's philosophy.

Feyerabend did sometimes praise and recommend Wittgenstein's philosophy, and he was not averse to using conceptual analysis for the purpose of improving our concepts.[89] However, he rejected the idea that philosophy is limited to conceptual analysis. In fact, Feyerabend was extremely critical of an inherent conservatism he found in Wittgenstein's conception of the task of philosophy. Feyerabend repeatedly argued against Wittgenstein that the task of philosophy is not merely to clarify existing concepts, but also to *improve* existing concepts, and to help invent *new kinds of concepts*. For the same reason, Feyerabend repeatedly and scathingly criticized ordinary language philosophy.

The deepest and most lasting impact that Wittgenstein's ideas had on Feyerabend's philosophical development concerns nominalism. The real lesson Feyerabend drew from Wittgenstein is that out concepts are not uniquely correct descriptions of the real structures of the world. Instead, the concepts we choose to use to organize experience determine how we experience the world. This lesson plays a major role throughout Feyerabend's philosophical publications.

89 Moreover, as we shall see, he even sometimes practiced a form of conceptual analysis to dissolve apparent philosophical paradoxes and problems.

Chapter 3
Karl Popper
Using and abusing critical rationalism

The most significant influence on Feyerabend's philosophical development was undoubtedly Karl Popper and his critical rationalist philosophy. In his early work, Feyerabend often praised Popper in glowing terms, and he used and developed many of Popper's ideas and arguments. This has lead to the standard view that Feyerabend was a Popperian. However in his post-1960s work, Popper was the target of much of Feyerabend's pointed criticism. As with the case of Wittgenstein, there are both positive and negative aspects of Feyerabend's relation to Popper and to Popper's influence on Feyerabend. However, whereas the negative and positive aspects of Wittgenstein's influence on Feyerabend are synchronic, both appearing in his pre-1970s work, in Popper's case the situation can be more easily temporally differentiated. Even so, Feyerabend's evolving relationship to Popper was multi-faceted and complex, and so it should come as no surprise that it is also a somewhat contentious issue. Some philosophers interpret Feyerabend in the 1950s and 1960s as just one more proponent of Popper's critical rationalism. For example, John Preston has argued that Feyerabend was a Popperian until the late 1960s (Preston 1997). This also holds for some Popperians who seem eager to claim Feyerabend as one of their own, as for example John Watkins, who has also argued that Feyerabend was a Popperian (Watkins 2000).

However, beginning in the second-half of the 1960s, Feyerabend became increasingly vocal about his eagerness to distance himself from Popper and Popperianism. As we shall see, at some point, he even began to deny his Popperian roots. The situation is further complicated by the fact that Feyerabend removed or altered many, but not all, of the footnotes praising or even just pertaining to Popper from the reprints of his papers in the two volumes of his collected works published in 1981. As per usual, Feyerabend was not even consistent in this respect, as in at least one case; Feyerabend added a footnote to a 1981 reprint acknowledging Popper as the source of an argument that he developed. For the most part, the many instances of praise and acknowledgment of Popper as an inspiring source, and as an authority on any number of topics, either simply disappear or are replaced by acknowledgements to and praise for Duhem or Mill. This fact, together with the later Feyerabend's increasingly venomous denials of the Popperian origins of his ideas, has led some to make the accusation that Feyerabend was somewhat dishonest about his relationship to Popper (Preston 1997, p. 11).

Preston supports the widespread view that, "Feyerabend was, *until the late 1960s* a Popperian" (Preston 1997, p. 11, italics inserted). He basis this interpretation on Feyerabend's adherence "to three crucial views: normative epistemology, falsificationism, and inductive skepticism", as well as on his interpretation of Feyerabend as a contributor to Popper's project of developing a rational model of science (see Preston 1997, p. 12). Preston's account of Feyerabend's relationship to Popper does not accurately reflect the true nature of Popper's influence on the development of Feyerabend's philosophical ideas. Feyerabend did very often adapt and use many important elements of Popper's critical rationalism, but Feyerabend was never a member of the *Popperian school.* Moreover, Feyerabend rejected critical rationalism at the beginning of the 1960s, not at the end them. The main goal of this chapter is to reassess Popper's influence on the development of Feyerabend's ideas, with special attention to those aspects directly relevant to the development of the notion of incommensurable concepts that I take to be central to Feyerabend's philosophy.

The account of their deteriorating relationship presented here differs significantly from the accepted view of Feyerabend as a Popperian. Feyerabend did praise, support and temporarily adopt many elements of Popper's critical rationalist philosophy in the 1950s and early 1960s. However, Feyerabend was also very uncomfortable with his relation to Popper *as early as 1952.* Feyerabend's "Explanation, Reduction and Empiricism" (1962a) marks a turning point after which Feyerabend sometimes continued to develop and extend Popper's critical rationalist ideas, but had also begun to revise and even reject many of their main tenets. The original version of that article explicitly took Popper's ideas as a starting point, but clearly developed them in ways that no longer cohered with Popper's critical rationalist philosophy. Sometime in the mid- to late 1960s, an outright hostile phase began. Feyerabend dramatically publicly rejected Popper's philosophical ideas and their significance by the late 1960s. Feyerabend's malevolence toward Popper culminated in his aggressively unsympathetic 1986 review of several of Popper's books.[90]

Section 3.1. begins with a brief account of how the two first met, as well as an account of Feyerabend's initial reaction to Popper's philosophy. Unfortunately, I will often have to rely on Feyerabend's statement in later

90 For details, see (1986) and (1987), pp. 162-191.

publications, especially Feyerabend's 1995 autobiography, to support some aspects of my account of their developing relationship. However, my account also draws heavily from letters written in the 1960s, and I cross-check Feyerabend's claims from different versions of *Against Method* (1975, 1988, 1993) as well *as Science in a Free Society* (1978a) and *Farewell to Reason* (1987), and compare these later claims both with his earlier publications and their reprints, as well as his published correspondence. Section 3.2. provides a very brief synapses of the most relevant aspects of Popper's philosophy to Feyerabend's philosophical development. The sketch of these critical rationalist elements is neither complete nor particularly detailed.[91] It is only intended to help locate Popper's influence on Feyerabend and his development of the notion of incommensurable concepts at the heart of Feyerabend's philosophy. This includes Popper's conceptions of falsificationism, criticism, and conventionalism about scientific methodology. As we shall see, Feyerabend often, but not exclusively, adopted and developed these ideas in his publications, and this sometimes even after he had rejected some of the main tenets of Popper's critical rationalist philosophy. Section 3.3. assesses some of the evidence supporting the claim that Feyerabend's notion of incommensurable concepts should be considered to be a Popperian thesis. While the notion of incommensurable concepts may have been motivated by Popperian considerations, it does not belong within the critical rationalist tradition. The section concludes by addressing the issue as to whether Feyerabend was ever really a member of the *Popperian school*. This last question has special importance to my reconstruction of Feyerabend as a philosophical pluralist who avoided philosophical positions and isms, including Popperianism. The last section documents and explains Feyerabend's increasing eagerness to distance himself from Popper, critical rationalism, and the Popperian school.

But before beginning, it is perhaps appropriate to note that Feyerabend would have been scathingly dismissive about any such attempt to assess the nature and extent of Popper's influence on him (as perhaps he would have been scathingly dismissive of this attempt to reconstruct his philosophy as a form of pluralism). He implied as much himself on a number of occasions: "Little did I know that one fine day such friendly gestures [mentioning Popper in footnotes] would give rise to serious dissertations about 'influences'" (1979, p. 62, also see 1976b and 1981d), or:

> Yesterday someone else sent me a clipping from a new lexicon where it says that first I followed Popper and then fell out with him. That was my entire life – he followed an ass, and then stopped following him. What a surprise that I do not want to give interviews and do not want to appear on television as they always want to investigate me in terms of such a life, and nobody is interested in my fe-

91 For an excellent detailed reconstruction of Popper's philosophy, see Keuth (2005).

ces, for example, which is much more important to me (1995c, Letter from 14 June 1985, p. 255, my translation).

3.1. Falling for falsificationism

Feyerabend and Popper first met in 1948 in Alpbach at the Austrian College Society international summer seminar. Feyerabend recalled this first encounter:

> I admired [Popper's] freedom of manners, his cheek, his disrespectful attitude toward the German philosophers who gave the proceedings weight in more sense than one, his sense of humor . . . [and] his ability to restate ponderous problems in simple and journalistic language. Here was a free mind, joyfully putting forth his ideas, unconcerned about the reaction of the 'professionals' (1978a, p. 115; rp 1993a, p. 260, and similarly, 1995a, pp. 71ff.).

Ironically, Popper's talent for putting points simply would become a target of Feyerabend's later criticisms, as witnessed by the title of Feyerabend's rather condescending review of Popper's books: "Trivializing Knowledge" (1986; rp 1987, pp. 162-191).

Feyerabend had planned to go to Cambridge to study under Wittgenstein, but in April of 1951 Wittgenstein died before Feyerabend had reached England. Feyerabend's second choice was the London School of Economics where he planned to study the philosophy of Quantum Mechanics under Popper's supervision (1995a, p. 86). Once there, he admitted to have uncritically swallowed Popper's falsificationism as a triviality:

> The falsificationist philosophy was taken for granted in the physics seminar . . . and so we did not understand what all the fuss was about. 'Philosophy must be in a desperate state' we said 'if trivialities like these count as major discoveries' (1978a, p. 115).

Feyerabend joined Popper in rejecting inductivism based on an insight about the relationship between general theories and the empirical laws that are supposedly derivable from them:

> Then came attacks on "inductivism", the idea that theories can be derived from, or established on the basis of, facts. Those who emphasize empirical backing, said Popper, advise us to stay close to the facts, which means they try to be as *ad hoc* as possible. *Ad hocness* is out; hence the requirement of a close empirical fit must be replaced by the requirement to go as far beyond the evidence as possible. Metaphysics does go beyond the known facts and often contradicts them. Does this mean that science is metaphysics? It does not; scientific hypotheses can be refuted, metaphysical systems cannot. *The argument that finally convinced me that inductivism is a sham* and which Popper presented at a talk to the British Society for the Philosophy of Science (the argument is Duhem's—but Popper didn't say so) *was that higher-level laws (such as Newton's law of gravitation) often conflict with lower-level laws*

(such as Kepler's laws) and therefore cannot be derived from them, no matter how many assumptions are added to the premises (1995a, p. 89, italics inserted).[92]

Although Popper had claimed the idea as his own (Popper 1972, pp. 204ff.), Feyerabend correctly traced its origins back to Duhem, as can be seen from a comparison (Duhem 1954, p. 193) to (Popper 1972, pp. 198-200) and (Popper 1975, pp. 82-83).

In *Against Method* (1975a), Feyerabend responded to Popper's contention that it was his idea, not Feyerabend's, that theories may correct an observational law that they are supposed to explain. According to Feyerabend, Popper's contention is incorrect in two respects. First, Popper had not realized that Feyerabend's references to Popper were "just friendly gestures", and not "historical evidence for his priority". And second, the idea was not even original to Popper (which Feyerabend was aware of from the early 1950s). Feyerabend pointed out that the idea in question had appeared in Duhem, Einstein and Boltzmann, "who anticipated every philosophical observation of Popper's 'The Aim of Science'" (where Popper made the point).[93]

In order to appreciate the significance of this argument against inductivism, one should recall that according to inductivist theories of science, one of the major achievements of Newton's theory of gravitation was considered to be *its capability to explain* both Kepler's three laws of planetary motion and what Nagel calls 'Galilean science' (Nagel 1961 [1966], p. 58 and pp. 336ff). Kepler's three laws are (1) The planets travel in paths that are ellipses with the sun in one focus; (2) The areas swept out in any orbit by the straight line joining the centers of the sun and a planet are proportional to the times the planets take to describe their orbits; and (3) The squares of the periodic times, which the different planets take to describe their orbits, are proportional to the cubes of their mean distances from the sun.[94] To flesh out the argument, I will briefly attempt to illustrate the basic idea of how such laws are supposedly explained by Newton's theory of gravitation, and then explain why Duhem, Popper and Feyerabend rejected this view. As the basic idea is the same for all three of Kepler's laws and Galilean science, focus will be on the simplest case: Kepler's first law.

92 Also see (1961f), (1962b), pp. 138-139, (1965b) and every version of *Against Method.* For short discussions, see Preston (1997), p. 3 and Hoyningen-Huene (2000b) p. 6. For a detailed discussion, see Dilworth (1981), pp. 135ff.

93 For more on Boltzmann, Feyerabend refers to his (1967d). See similarly footnote 6 added 1980 of (1981a), p. 47. Feyerabend cites Duhem on the point already in 1963 (1963a; rp 1999a, p. 85, fn. 18), but had encountered the point in Duhem much earlier.

94 See Dampier (1929), p. 140, Nagel (1961 [1966]), pp. 57ff., and Dilworth (1981), pp. 135ff.

According to Newton's theory, given any two mass-points that differ significantly in size and are taken as non-accelerating, the smaller will describe a conic with the larger mass at its focal point. The shape of this conic will depend upon the velocity of the smaller mass. In order to explain Kepler's first law, Newton's abstract model must be applied (with the aid of correspondence rules) to the concrete situation in the solar system. For example, one associates the larger immobile mass with the sun, and the smaller mass with an individual planet such as Mercury. Having set these conditions, Mercury's velocity can then be independently established empirically. According to Newton's theory, other things being equal, given its velocity, Mercury should describe a perfect ellipse with the sun at one focal point, which is exactly what is expected given Kepler's first law. In this way, Newton's theory of gravitation supposedly explains Kepler's first law. In other words, the concrete situation described by Kepler's laws is exactly the result obtained when the sun and the planets are individually plugged into Newton's abstract theory, which should hold for any such masses.

Now Popper, as Duhem before him, rejected the view that Newton's theory explains Kepler's laws. Instead, Popper, as Duhem before him, suggested that Newton's theory, far from explaining Kepler's laws, actually formally contradicts Kepler's laws. The basic idea is that if the concrete situation in the solar system is plugged into Newton's theory, then given that more than just a single planet must be taken into account, as well as the gravitational forces of the planet on the sun, the path of a given planet will *not* be a perfect ellipse as suggested by Kepler's first law. Instead, some other (ellipse-like) curve will result. Consequently, Newton's theory does not explain (by derivation) Kepler's laws. To cite from Popper:

> [F]rom a logical point of view, Newton's theory, strictly speaking, contradicts both Galileo's and Kepler's . . . For this reason it is impossible to derive Newton's theory from either Galileo's or Kepler's or both, whether by deduction or induction. For neither a deductive nor an inductive inference can ever proceed from consistent premises to a conclusion that formally contradicts the premises from which we started (Popper 1972, p. 198, first published as Popper 1957b; rp Popper 1975, pp. 82-83).

This is clearly the same point that Duhem had made:

> The principle of universal gravity, very far from being derivable by generalization and induction from the observable laws of Kepler, formally contradicts these laws. If Newton's theory is correct, Kepler's laws are necessarily false (Duhem 1954, p. 193, italics in original).

Thus, it does appear that Popper appropriated Duhem's argument.

Moreover, Popper used it to convince Feyerabend of falsificationism. Confronted exactly with this Duhemian/Popperian point, Feyerabend recalled having fallen for falsificationism in Popper's seminar (directly continuing the citation from (1995a) above:

> *Falsificationism now seemed like a real option, and I fell for it.* I occasionally felt a little uncomfortable, especially when talking to Walter Hollitscher; there seemed to be a worm somewhere in the woodwork. Still, I applied the procedure to a variety of topics and made it the centerpiece of my lectures when I started teaching (1995a, p. 89, italics inserted).

Immediately thereafter, Feyerabend continued:

> Today I regard this episode as an excellent example of the dangers of abstract reasoning. There are lots of dangerous philosophies around. Why are they dangerous? Because they contain elements that paralyze our judgment (1995a, p. 89).

A better argument for the merits of falsificationism over verificationism, and one that fits quite comfortably within Feyerabend's pluralistic philosophy (which emphasizes the roles that alternatives play in conceptual advance) emphasizes the epistemic consequences of the fact that potential there are always many other alternative accounts of any verification. Popper's basic point in favor of falsificationism builds on an asymmetry between falsifications and verification. Popper emphasized that while a single black raven falsifies the generalization that all ravens are black, no matter how many white ravens have been observed, the generalization that *all* ravens are black cannot be verified. The argument I am suggesting emphasizes a different asymmetry. It runs as follows. If a theory predicts that an object will move with a certain velocity, and then an experimental results shows that actually the object moves with a different velocity (a falsifying instance), then that provides at least some grounds on which to abandon the theory as false (one could alternatively challenge the veracity of the experimental result itself). However, if the velocity measured is what the theory predicted, it does not provide any good reason for believing the theory to be true. This is because there are always potentially many competing theoretical accounts that predict that very same result. The verifying instance can only be taken as good grounds to believe the theory if all of the other alternatives that make the same prediction have been considered and ruled out. As this is impossible, verification is always pure speculation. Feyerabend never made such a pluralistic argument in favor of falsificationism over verificationism, but perhaps he should have.

In any case, in his autobiography, Feyerabend claimed to have had some suppressed reservations about falsificationism from the very beginning. One might argue that this assumes that Feyerabend's 1995 autobiographical recollections are an accurate source of his 1950s self, but given that he turned down Popper's offer to continue on with him as his assistant in 1952, we have some independent grounds to suspect that there were some early sources of tension between the two. By the end of 1953, Feyerabend was well versed in Popper's critical rationalism. That is the year Feyerabend began his translations of Popper's two-volume *Die offene Gesellschaft und ihre Feinde* (Popper 1957a and Popper 1958, *The Open Society and its Enemies*).[95] As we shall see, the well-known joke that the books should have been called *The Open Society BY ONE OF its Enemies* captures the spirit of some of Feyerabend's later pointed criticisms of Popper, his work, and his followers. However, in order to assess the nature and extent to which Feyerabend can be said to have been a Popperian, we will need a sketch of some of the central ideas that Feyerabend learned from Popper. Some other less central Popperian ideas that Feyerabend used and discussed will be introduced along the way.

3.2. Critical rationalist elements in Feyerabend's philosophy

This section presents some of the main evidence in favor of the view that Feyerabend should be considered to have been a member of the Popperian school. The leading question is whether or not Feyerabend's philosophical ideas are best understood as part of Popper's critical rationalist philosophy. In order to address this question, Popper's critical rationalism needs to be briefly introduced, and then its three main elements need to be sketched. As we shall see, the evidence in favor of the view that Feyerabend was a Popperian is two-fold: First of all, Feyerabend often extravagantly praised and acknowledged Popper as a main source of his ideas. Second, Feyerabend often adopted the central tenets of Popper's critical rationalist philosophy. The main reasons for rejecting the view that Feyerabend was a member of the Popperian school will be presented in the following section. As always, the focus will be on aspects of Feyerabend's relationship to Popper that bare directly on Feyerabend's development of the notion of incommensurability.

The most obvious indication that Feyerabend's work is properly understood as Popperian is that it contains generous and abundant praise of Popper and his contribution to the philosophy of science, as well as many acknowledgements to Popper as an influential source of his ideas. To begin with, on the very first page of (1958a), Feyerabend credited Popper with introducing the term 'instrumentalism', and added:

95 Popper was not particularly pleased with the quality of the translations; see (1995a), p. 98.

> Personal Statement.—I should like to use this occasion for pointing out that my own discussion of instrumentalism [in (1954c)] employs ideas which Prof. Popper had kindly explained to me in conversations before his own views had appeared in print (1958a, p. 143, fn. 3; rp 1981a, p. 17, removed).

Feyerabend also reported that Feigl believed Popper's *Logik der Forschung* could have changed the course of the Vienna Circle:

> It was Feigl's belief that an earlier publication of the *Logik der Forschung* or of some other abridged version of the "Grundprobleme" would have decisively changed the history of the Vienna Circle. "We made a mistake," said he, "to choose Wittgenstein as our bible" (Private communication to Karl R. Popper) (1966a, p. 7, fn. 3).

In his early publications, there are countless examples of Feyerabend praising Popper, many of which were removed from the reprints.[96] Feyerabend's praise of Popper continues through the mid-1960s. As late as 1965, Feyerabend wrote a glowing review of Popper's *Conjectures and Refutations* (1965d). And in the same year, when replying to criticisms, Feyerabend wrote, "I for one am not aware of having produced a single idea that is not already contained in the realistic tradition *and especially in Professor Popper's account of it*" (1965c, p. 251, fn. 1; rp 1981a, p. 104, fn. 1, where the italicized portion of the citation has been removed). Such blanket acknowledgements, and their subsequent removal in the reprinted versions, clearly lend strong support to the view that Feyerabend, once a Popperian, was eager to hide his real roots. As we shall see, Feyerabend did not only praise Popper and acknowledge him as an influential source of his ideas, he also repeatedly adopted the three central elements of critical rationalism.

Popper's critical rationalism is shaped by two epistemological problems (see Popper 1968): The problem of induction and the problem of demarcation. The first concerns the relation between theory and experience. The second concerns the relation between empirical science (such as physics and chemistry) and non-science, such as metaphysics, logic and mathematics. Critical rationalism reverses the standard inductivist view of the relation between theoretical knowledge and facts, and it rejects the standard demarcation of empirical science from non-science as based on this inductivist method.

The first central component of Popper's critical rationalism is anti-inductivism or falsificationism. Critical rationalism reverses the standard inductivist view according to which experience leads to scientific hypotheses (or theories). Instead, hypotheses, or 'conjectures', precede experience against which they are tested. As mentioned at the end of the last section,

96 Additional examples include (1960e), p. 336; rp (1981a), p. 234, (1961i); rp (1999a), pp. 50-77, (1966a), p. 7, (1961h), p. 244 and p. 248, and see especially (1965d).

Popper argues that while such conjectures cannot be inductively confirmed, they can be refuted. The reason is due to a fundamental disparity between confirmation (or verification) and falsification. It is that while no amount of true observations can ever definitively establish a general claim, a single observation can falsify a general claim. For example, no matter how many badly-dressed philosophers one meets, the claim that all philosophers dress badly might still be false, given that the next philosopher observed might not be badly dressed. Yet a single stylishly dressed philosophers would suffice to refute the general claim that all philosophers dress badly (see e.g. 1974b, p. 499; rp 1981b, p. 199). According to Popper's falsificationism, scientists suggest general hypothesis and test them against observations. In short, we propose, and nature disposes. Hence, the title of Popper's book *Conjectures and Refutations* (Popper 1963). Only singular hypotheses deduced from general thesis that have survived empirical tests constitute current scientific knowledge. According to Popper, a deductive, not inductive, relation holds between scientific knowledge and experience. Falsificationism also purportedly solves the demarcation problem. Only hypotheses that are empirically falsifiable count as scientific. Although Popper clearly distinguished between metaphysics and empirical sciences, he also envisioned a positive role for metaphysics in science.[97] In this way, Popper developed his epistemology in contrast to logical positivist ideas about meaning and verifiability. He rejected their denial of a positive role for metaphysical speculation in the science. Instead, he conceived of contemporary natural science as an extension of the early Greek speculations about the nature of the world.

A second central component of Popper's critical rationalism is conventionalism about methodology. Popper understood the demarcation between science and non-science as the result of a choice or decision, not as a difference due to the nature of things. According to Popper, methodological rules are norms governing behavior. Such methodological rules are based on decisions, not facts. Such decisions are made according to our epistemic aims and goals. It follows, for Popper, that a methodology should be judged according to whether it is fruitful in reaching those goals. According to Popper, the 'rules of scientific procedure must be designed in such a way that they do not protect any statement in science against falsification' (Popper 1935, cited from Popper 1968, p. 54).

The third, last, and most significant (for Feyerabend's philosophy) component of Popper's critical rationalism is its emphasis on criticism as the motor for progress. According to Popper, epistemology should use

97 As we have seen, Feyerabend also accepts a positive role for metaphysics in science; see e.g. (1966a), p. 10.

the Socratic method of posing questions and critically discussing the proposed answers (Popper 1934). Everything should be open to scrutiny. For the critical rationalist, no dogmatic element is allowable in science. As we have seen, according to Popper, scientists are in the business of proposing highly falsifiable hypothesis, and then testing them as rigorously as possible. For Popper, to stop testing hypotheses is to stop doing science. In this respect, science is at heart critical. Moreover, for Popper, scientific progress is the result of such a critical discussion of alternative views (1970c, pp. 216ff.; rp 1981b, pp. 149ff).

In summary, there are three main components of Popper's philosophy: falsificationism, conventionalism, and criticism. Now let us return to the issue of Popper's influence on Feyerabend, as it is evidenced in his publications before and including the mid-1960s when Feyerabend was purportedly a proponent of Popperianism. As we shall see, in these publications, Feyerabend clearly and repeatedly endorsed and employed all three of these central ideas of Popper's critical rationalism.

Feyerabend explicitly used the tools of falsificationism and the basic tenets of critical rationalism in any number of his pre-1970 publications.[98] First of all, two of Feyerabend's main early publications "Physik und Ontologie" (1954c) and "Knowledge without Foundations" (1961i; rp 1999a, pp. 50-77) repeat Popper's views about the pre-Socratics and their contribution to the development of the natural science.[99] According to both Popper and Feyerabend, science, as a critical enterprise, begins with the pre-Socratics. In these pieces, as well as in others, Feyerabend followed Popper in emphasizing that scientific progress is the result of the Pre-Socratic's rejection of dogmatism and adoption a of critical method. This is the element of Popper's philosophy that had the deepest influence on Feyerabend's developing ideas. As we shall see in detail Part III, his emphasis on the essential role of criticism is the central component of his pluralistic conception of knowledge and philosophy.

The clearest example of Feyerabend explicitly endorsing the falsificationist element of Popper's philosophy comes in his review of Bohm:

> The reason why I cannot accept Bohm's methodology of caution and why I prefer to it the methodology of falsification as it has been developed by Popper is therefore . . . (1960e, pp. 337-338; rp 1981a, p. 235).

Another particularly explicit example is (1963e, especially pp. 320-322).

98 There are many examples of Feyerabend adopting and defending any number of tenets of Popper's falsificationist philosophy. They continue at least through 1970. See e.g. (1968b), p. 253 and (1970c). For a brief criticism of Feyerabend's version of Popper's falsificationism, see Settle (1969).

99 Also see (1961i), pp. 3ff.; rp (1999a), pp. 51ff.

Feyerabend also cited and praised Popper on the importance of the distinction between nature and convention as applied to methodological questions, which he claimed is another form of his distinction between the pragmatic constraints on an observation language and its semantic properties, or interpretation.[100] As we have seen in chapter 2, this is the distinction that shaped Feyerabend's entire semantic outlook. From this, one can conclude that the distinction at the heart of Feyerabend's pragmatic theory of observation and his so-called contextual theory of meaning is traceable to Popper's influence. Feyerabend unequivocally traced his criticism of the conception of methodological rules as based on facts to "an excellent discussion of this doctrine, its history, and its shortcomings" back to Popper (1960e, p. 336, fn. 1; rp 1981a, p. 234, fn. 18). And he explicitly subscribed to Popper's notion of conventionalism about methodology many times over (see Preston 1997). However, concerning conventionalism about methodology, one should also take note that Feyerabend was following his Ph.D. supervisor Viktor Kraft just as much as he was Popper (1963e). Popper was not the only conventionalist about methodology who strongly influenced Feyerabend.

Feyerabend also often, but not always, followed Popper in adopting a normative standpoint about methodology.[101] From this perspective, the job of philosophy of science is not merely to describe what constitutes knowledge, or how knowledge is attained, or the conditions under which knowledge is attainable. Instead, it is to tell us what good science *should* be. For example, as early as 1958, in section seven "*The Logical Basis of the Argument in Sec. 6*" Feyerabend claimed that:

> the issue between positivism and realism is not a factual issue which can be decided by pointing to certain actually existing things, procedures, forms of language, etc., it is an issue between different ideals of knowledge (1958a, p. 166; rp 1981a, pp. 33-34).

Feyerabend explicitly acknowledged that this thesis is an extension of Popper's views on scientific method, and also credited Kraft for stressing the normative character of epistemology.[102] He stressed that the basis of his argument against the stability thesis in section 6 of (1958a; rp 1981a, pp. 17-36) should be interpreted in the context of the Popperian conception of normative epistemology. This is of special relevance because Feyerabend traced his notion of incommensurability back to exactly that section of his dissertation. In other words, Feyerabend claimed that he

100 See (1958a), p. 147, fn. 7; rp (1981a), p. 19, fn. 6. Also see (1961i), p. 50; rp (1999a), p. 69.

101 For a discussion, see Preston (1997), e.g. p. 14.

102 See (1958a), p. 166, fn. 25. Feyerabend deleted this acknowledgement to Popper, like so many others, in the reprint. See (1981a), p. 34, fn. 24.

originally developed his notion of incommensurability from the Popperian perspective of normative epistemology.

This can also clearly be seen in (1962a), where Feyerabend first explicitly introduced the term 'incommensurable concepts':

> scientific method, as well as the rules for reduction and explanation connected with it, is not supposed to describe what scientists are actually doing. Rather, it is supposed to provide us with normative rules which should be followed, and to which actual scientific practice will correspond only more or less closely. *Adopting this point of view*, one cannot regard the arguments of the last section as ultimately decisive (1962a, p. 60; rp 1981a, pp. 68-69, italics inserted).

Feyerabend originally cited Popper for having made this point most forcefully.[103] Moreover, in a passage deleted from the reprint, Feyerabend continued:

> It is very important nowadays to defend such a normative interpretation of scientific method and to uphold reasonable demands even if actual scientific practice should proceed along completely different lines. It is important because many contemporary philosophers of science seem to see their task in a very different light. For them actual scientific practice is the material from which they start, and a methodology is considered reasonable only to the extent to which it mirrors such practice (1962a, p. 60; rp 1981a, pp. 68-69, removed).

Even so, some caution is needed. These two passages have been partially cited by Preston in support of his claim that Feyerabend conceived of epistemology "as a *wholly* normative discipline" (Preston 1997, p. 14, italics inserted), and in support of his portrayal of Feyerabend as a Popperian. Preston's claim is correct in so far as Feyerabend, following Popper, often adopted a normative perspective in epistemology. However, Preston's characterization of Feyerabend's conception of epistemology as '*wholly* normative' is inaccurate. Feyerabend did sometimes adopt a *normative* perspective, but at other time his epistemology adopted a *descriptive* perspectives.

To see this, consider the text that Feyerabend removed from the reprint more closely, and consider its context. First of all, the passage is a *defense* of the need for normative epistemology, "even if actual scientific practice should proceed along completely different lines" (1962a, p. 60; rp 1981a, p. 68). It is not a *rejection* of descriptive or historical epistemology. Feyerabend used the example of the ideal that every explanation should also function as a prediction, and argued that this remains a worthy ideal, even though it has been shown to be false, as for example in the case of medicine (1962a, pp. 60-62; rp 1981a, pp. 68-69, removed).[104] That Fey-

103 Feyerabend refers the reader to section 10 of 1959 edition of Popper (1934).

104 Also see (1961d) where he makes the same argument about explanation and prediction in the course of defending the logical empiricist model of explanation.

erabend later deleted this *defense* of the need for normative epistemology, even in the face of discrepancies with actual scientific practice, does not support Preston's claim that Feyerabend was not one of the founders of the historical school in the philosophy of science. Rather, it supports the view that Feyerabend did not *exclusively*, or *even primarily*, belong to the historical school.

Against Preston's claim that Feyerabend attempted to practice a *wholly* normative epistemology, one should keep several further points in mind. Most relevantly, the normative element of Feyerabend's attack on the logical empiricists models of explanation and reduction is only half of his two-pronged argumentative strategy, which aims to establish that the logical empiricist model neither fits the facts (descriptively), nor promotes progress (normatively). Setting the preceding citations supporting the view that Feyerabend conceived of epistemology as a normative discipline into their original context (from which Preston has conveniently extracted them), reveals that Feyerabend made them directly after finishing a long criticism of the logical empiricist model of explanation on the grounds that it "does not give a correct account of actual scientific practice" (1962a, p. 60; rp 1981a, p. 68). Exactly this descriptive half of Feyerabend's argument has been misleadingly ignored by Preston in his attempt to squeeze Feyerabend's ideas into a *wholly* normative box. The descriptive half of Feyerabend's criticism can be properly placed within the historical school. It does not fit into a wholly normative conception of epistemology. As we have seen, Feyerabend explicitly emphasized that he was *switching* perspectives in the middle of his attack on the logical empiricists models of explanation and reduction. He even emphasized the point with the phrase: "Adopting this point of view". Feyerabend's investigations into the nature of theoretical advance were not tied to any single perspective – normative or descriptive. He used both.

Moreover, the passage removed is actually part of a limited defense of the logical empiricist model of explanation from Feyerabend's own descriptive attack on it. It is not part of Feyerabend's criticism of the logical empiricists models of explanation and reduction as Preston misleading suggests (Preston 1997, p. 60). Feyerabend emphasized that

> *the increased attention paid to the historical aspects of a subject*, and the attempts to breakdown the distinction between the synthetic and the analytic *must be welcomed as steps in the rights direction* (1962a, p. 68, fn. 82; rp 1981a, p. 74, fn. 67),

This does not sound like something someone who thought epistemology is, or should be, a wholly normative discipline would say. Feyerabend did insist on the need for a "methodological backbone". Yet Feyerabend *usually* relied on detailed examples from actual scientific practice, not just abstract considerations, to make his points. This further highlights the prominent descriptive element in Feyerabend's philosophy. While Feyerabend did often adopt a normative perspective, and did argue the need for and merits of a normative perspective *over* a *purely* historical approach, he also sometimes temporarily adopted a descriptive or historical perspective, defended the need to do so, and repeatedly emphasized the need for philosophy to be in closer contact to actual scientific practice. The point is not that Feyerabend practiced a 'wholly normative epistemology', but rather that Feyerabend warned against practicing a wholly descriptive epistemology. Feyerabend's attitude was that epistemology should make some difference to the subject it investigates.

The examples set out above are not isolated instances of Feyerabend adopting a normative standpoint or of criticizing a purely descriptive one, nor of him acknowledging that he was following Popper and Kraft in so doing.[105] For example, Feyerabend wrote:

> In about 1925 philosophers of science were bold enough to stick to their theses even in those cases where they were inconsistent with actual science. They meant to be *reformers* of science, and not *imitators*. . . In the meantime they have become rather tame (or beat) and are much more prepared to change their ideas in accordance with the latest discoveries of the historians, or the latest fashion of the contemporary scientific enterprise. This is very regrettable, indeed, for it considerably decreases the number of rational critics of the scientific enterprise. And it also seems to give unwarranted support to the Hegelian thesis (which is now implicitly held by many historians and philosophers of science) that what exists has a 'logic' of its own and is for that very reason reasonable (1963a; rp 1999a, p. 89, fn. 23).

However, this criticism of the deterioration of the normative approach comes directly on the heels of his own criticism of the consistency condition on the grounds that it does not conform to actual scientific practice. Moreover, Feyerabend also emphasized that

> True: there are empirically inclined philosophers who have derived some satisfaction from the assumption that they only make explicit what is implicitly contained in scientific practice. *It is therefore quite important to show that scientific practice is not what it is supposed to be by them* (1963a; rp 1999a, p. 89, italics inserted).

This means that just because the descriptive approach was becoming increasingly popular, it is important to use the tools of descriptive approach to attack its results. Here Feyerabend clearly sanctioned a descriptive or historical approach to the philosophy of science. Feyerabend thought that it is important to do so, because only by doing so could he engage those

105 See e.g. (1995b), written 1960/61, p. 366, p. 368, (1961i), pp. 2ff., and especially pp. 56-57, (1961h), p. 247, (1963e), p. 319ff., (1963a); rp (1999a), pp. 50-77, and (1964d); rp (1981a), pp. 161-175.

practicing a descriptive or historical approach to the philosophy of science in order to fruitfully criticize their ideas.

Much later Feyerabend explicitly claimed to have rejected normative epistemology in favor of the emerging descriptive school in the philosophy of science:

> Nor would I ever presume to legislate for scientists or, for that matter, to anyone else as Gellner implies . . . I did this in earlier papers when I was younger, more ignorant, more pushy and conceited. At the time my arguments for proliferation were indeed designed to show that a monistic life is not worth living and they urged everyone to think, feel, live through a competition of alternatives. Today the same arguments are offered with a very different *purpose* in mind, and they lead to a very different *result* (1978a, p. 144).

Feyerabend reported that the change was due to a conversation he had with von Weizsäcker in 1965:

> The change is due to a conversation that occurred at Professor von Weizsäcker's seminar in Hamburg 1965: Professor von Weizsäcker's gave a detailed account of the Copenhagen interpretation and showed how it could be applied to specific problems. I pressed the point that hidden variable theories are needed to increase the empirical content of the orthodox view when suddenly I realized how barren such an attitude is in the face of concrete research (1978a, p. 144, fn. 7, and p. 117 for the same point).[106]

As a result of this conversation, Feyerabend claims to have rejected Popper's normative conception of epistemology and joined with Kuhn and his historical approach to the philosophy of science. In 1981 he added the following remarks about his earlier work in a footnote to a reprint of them:

> Added 1980. My criticism in sections 10 and 11 is a paradigm case of abstract reasoning, and shows all its weaknesses. The argument assumes (a) that methodological stipulations can be introduced and enforced independently of what goes on in the world and (b) that developing them is just a matter of imagination, not of real relations that might hinder or further the work of the theoretician. In a word – it is assumed that the domain of knowledge construction and knowledge-modification is unencumbered by the laws of nature, it is a special domain in which the laws of nature are suspended, but whose products can transform nature. One only needs to formulate this assumption to see its absurdity. Yet philosophers of science who praise methodological rules because they circumvent some recondite logical problems (such as 'Hume's problem,') and regard them as adequate on those grounds alone make this absurd assumption (1981, p. 235, fn. 20).

106 Feyerabend also traced his change of attitude toward anarchism to this conversation with von Weizsäcker, see (1978a), p. 117, and also (1993a), p. 262.

Feyerabend, of course, is referring to Popper. Feyerabend also emphasized the magnitude of this change by comparing his experience to Popper's:

> Popper never faced a situation that forced him to revise his whole philosophy and is perhaps incapable of recognizing such a situation (1978a, p. 148).

However, although Feyerabend claimed that he gave up the normative standpoint and revised his whole philosophy, he returned to making methodological prescriptions in his later work, even arguing that ethics ought to be the measure of truth (1992; rp 1999b, chapter 9, pp. 242ff.), and he still considered methodology to be a matter of convention in the 1970s. For example:

> The attempt to judge cosmologies by their content may have to be given up. Such a development, far from being undesirable, changes science from a stern and demanding mistress into an attractive and yielding courtesan who tries to anticipate every wish of her lover. Of course, it is up to us to choose either a dragon or a pussy cat for our company. I do not think I need to explain my own preferences (1970c, p. 229; rp 1981b, p. 161).

Feyerabend also continued to use Popperian ideas in his criticisms as late as 1970.[107] This is significant because by then Feyerabend had already began to criticize Popper severely, and had explicitly broken with critical rationalist philosophy. The point worth emphasizing is that Feyerabend continued to use Popper's ideas in his attack on others (e.g. 1970c; rp 1981b, pp. 131-161), and to praise some of Popper's contributions to philosophy (e.g. 1968b, p. 253), even after he had already begun dramatically to criticize and reject Popperianism and its historical significance. This directly supports the pluralistic, opportunistic nature of Feyerabend's philosophy that will be developed in Part III.

In summary, Feyerabend did use and develop many ideas attributed to Popper. However, there are already some reasons for exercising caution in placing Feyerabend within the Popperian school. Although he often adopted a conventionalist stance toward methodology, and the perspective of normative epistemology more generally, at other times, even in the early 1960s, he also sometimes practiced and defended the type of descriptive philosophy associated with the historical school in the philosophy of science. Moreover, he continued to use critical rationalist ideas to criticize Kuhn even after he began to reject these very ideas and their philosophical significance. This suggests that his commitment to critical

---

107 See (1970c); rp (1981b), pp. 131-161. For a discussion, see Hoyningen-Huene who has also noted that Feyerabend used the tools of critical rationalism in his critique of Kuhn's notion of normal science as developed in ***The Structure of Scientific Revolutions***. See Hoyningen-Huene (2000a).

rationalism was pragmatic, not part of some deep epistemological conviction.

3.3. From critical rationalism to incommensurability

This section examines some of the evidence in favor of the view that Feyerabend's development of the notion of incommensurable concepts should be considered to be part of the critical rationalist tradition. Popper, Lakatos and John Watkins considered Feyerabend's 'Explanation, reduction and Empiricism' (1962a), where Feyerabend first used the term 'incommensurable', to be a mere repetition of Popper's ideas. Watkins has even gone so far as to suggest that the reason that Feyerabend turned away from critical rationalism and toward epistemological anarchism was because that he could not face up to his indebtedness to Popper (Watkins 2000, p. 49). This is incorrect. Feyerabend's move away from critical rationalism toward epistemological anarchism was the result of Feyerabend's explicit rejection of some of Popper's ideas. Feyerabend's 1962 discussion of incommensurable concepts was indeed motivated by Popper's discussion of Duhem's view of successive scientific theories, and Feyerabend did repeat the point that some successive scientific theories are empirically inconsistent. However, this should not be taken to support the idea that in his 1962 discussion of incommensurable concepts, Feyerabend was a Popperian proponent of critical rationalist philosophy, as has been suggested by Preston, Watson, Lakatos, and even Popper himself.

Hoyningen-Huene has also 'grouped' the Feyerabend of 1960-1961 with the critical rationalists on the grounds that an essential part of the arsenal of arguments that Feyerabend used in criticizing a draft of Kuhn [(1962a] 1970) was taken from the Popperian camp (Hoyningen-Huene 2000a, pp. 108-109). Hoyningen-Huene even calls Feyerabend a Popperian:

> Let me briefly add a word about Kuhn's own defense of the existence of normal science against Feyerabend and the other Popperians (Hoyningen-Huene 2000a, p. 113).

However as we shall see, just because Feyerabend may have used some particular Popperian ideas and arguments in his criticism of some aspects of Kuhn's, which Feyerabend clearly did, does not sanction grouping him with the critical rationalists or calling him a Popperian.

We have seen that Feyerabend rejected inductivism and fell for falsificationism based on a Duhemian/Popperian argument to the effect that *some* pairs of successive scientific theories, far from having a relationship of deductive subsumption or explanation, actually formally contradict one another.[108] Feyerabend developed this idea into an attack on what he later called the consistency condition:

> Only such theories are then admissible in a given domain which either contain the theories already used in this domain, or which are at least consistent with them inside the domain (1962a, p. 44; rp 1981a, p. 55).[109]

According to Feyerabend, the particular instance of the Duhemian / Popperian argument (Kepler's first law and Newtonian Mechanics) can be applied to Kepler's other laws. It can also be generalized to some other cases of purportedly successful reductions, such as the relationship between Galilean science and Newton's theory, especially some consequences of Newton's theory in the domain of validity of Galileo's laws. As we shall see in chapter 5 in more detail, this was Feyerabend's favorite example in 'Explanation, Reduction and Empiricism' (1962a, pp. 46ff.; rp 1981a, pp. 57-58). This is part of the descriptive half of his two-pronged argumentative strategy. He chose to attack this particular example because it was one of Nagel's most prominent examples of a successful reduction: The reduction of the laws of 'Galilean physics' to the laws of 'Newton's physics', or the reduction of 'Galilean science' to 'Newton's science'.

Watkins (a Popperian) recently chastised Feyerabend for not citing Popper as the source of this point about the logical inconsistency of successive scientific theories (Watkins 2000, pp. 48-49). However, Watkins also acknowledged that Popper had the argument from Duhem (although Popper did not initially cite Duhem). Feyerabend had chastised Popper for not citing Duhem as the source of this point (1995a, p. 89). Feyerabend was long familiar with the Duhem's work, see e.g. (1961c), but Feyerabend did not cite Duhem on this point in his original (1962a), nor in the 1981 reprint. Feyerabend merely opened the argument against logical empiricism based on this point with "As is well known" (1962a, p. 47; rp 1981a, p. 58), mentioning neither Popper nor Duhem as its source. Later, such as in *Against Method*, Feyerabend did cite Duhem, and not Popper as

---

108 This formulation is imprecise in that it ignores the logical empiricist distinction between laws and theories, treating Kepler and Galileo's laws as theories. However according to Nagel, Kepler and Galileo's laws are instantial, while Newton's theory is not. In other words, whereas Kepler's laws refer directly to actual objects such as the planets in our solar system, Newton's theory does not deal directly with any actually existing entities. Instead, it deals with abstract quantities such as mass-points, which can be assigned to actual particular objects, such the individual planets in our solar system. Feyerabend explicitly adopted this distinction between laws and theories at the very beginning of his (1962a), p. 28, fn. 1; rp (1981a), p. 44, fn. 1. This is relevant, as Popper did not make this distinction. He treated Kepler and Galileo's laws as theories.

109 Also see (1961a), p. 37, (1963a), pp. 13ff.; rp (1999a), p. 83, (1965b), pp. 164ff, (1975a), pp. 35ff., (1988), pp. 24ff., (1993a), pp. 24ff. Feyerabend also attacked the stability thesis in (1958a); rp (1981a), pp. 17-36, which is supposed to be a version of the consistency condition.

the source of this argument, and more generally as a main inspiration driving his ideas.[110] Feyerabend made this point most explicitly in the same essay the he first used the term 'incommensurable'. He used it to attack logical empiricist accounts of reduction and explanation. However, the fact that Feyerabend made this quasi-Popperian point in the same essay that he introduced the term 'incommensurable' does make incommensurability a Popperian thesis.

According to Feyerabend, by the Galilean science, or Galilean Physics, Nagel meant the body of theory dealing with the motion of material objects (such as falling stones, pendulums, and balls on an inclined plane), near the surface of the earth (1962a, p. 47; rp 1981a, p. 57). Feyerabend argued that it is impossible, *for quantitative reasons*, to establish a deductive relationship between Galilean Physics and Newton's Celestial Mechanics.[111] The reason is that while Galilean Physics assumed that vertical accelerations of falling objects in the close vicinity of the earth are constant over any finite interval of vertical distance, according to Newton's theory, this is not the case (1962a, p. 47; rp 1981a, pp. 57-58). An application of Newton's theory to the surface of the earth gives an acceleration that is slightly decreasing, although imperceptibly so (1965b, p. 168). This indeed sounds very much like a repetition of Popper's claim:

> Newton's theory unifies Galileo's and Kepler's. But far from being a conjunction of these two theories – which play the part of *explicanda* for Newton's – *it corrects them while explaining them* . . . Far from repeating its *explanadum*, the new theory contradicts it and corrects it (Popper 1957b, p. 202, italics in original).[112]

And Feyerabend eventually came to the conclusion that:

> from the point of view of scientific method [i.e. normatively speaking], T will be most satisfactory if it is *inconsistent* with T′ [T and T′ are successive scientific theories] in the domain where they both overlap;* and if it is *incommensurable* with T′ (1962a, pp. 91-92; rp 1981a, p. 91, slightly altered, italics in original).

Where I have inserted '*', there is a footnote:

> This condition has been discussed with great clarity by Popper in [(Popper 1957b)]. It was this discussion (*as well as dissatisfaction with* [(Nagel 1949)]) *that was the starting point of the present analysis of the problem of explanation* (1962a, p. 92; rp 1981a, p. 91, italics inserted).

110 See e.g. (1975a), p. 36, (1988), p. 24, main text and fn. 2, and (1993a), p. 24, main text as well as fn. 2. This begins as early as 1963; see (1963a); rp (1999a), p. 85, fn. 18.

111 Feyerabend discussed the same Duhemian point in (1961a), p. 37, (1963a), p. 13; rp (1989a), p. 108; rp (1999a) pp. 83ff., (1965b), pp. 164ff., and especially p. 168, and (1975a), pp. 36ff., (1988), pp. 24ff., (1993a), pp. 24ff.

112 Compare "The theories of Kepler and Galileo were unified and superseded by Newton's logically stringer and better testable theory, and similarly Fresnel's and Faraday's by Maxwell's. Newton's theory, and Maxwell's, in their turn, were unified and superseded by Einstein's" (Popper 1963, p. 220).

Here Feyerabend clearly acknowledged that the discussion containing the explicit introduction of 'incommensurable concepts' was motivated by Popper's discussion of the relationship between successive scientific theories. This was the staring point of Feyerabend's discussion. While these Popperian origins might seem somewhat surprising to readers who are only familiar with the more widely circulated 1981 reprint in Feyerabend's collected papers, it should not be surprising for readers who are familiar with the 1962 original text. An entire page was deleted from the introductory section of the 1981 reprint of the original 1962 text.[113] The deleted page contained three paragraphs all of which were lengthy and generous acknowledgements, and all three of which mentioned Popper (among others). Some of the others, such as Bohm, Feigl, Körner, Maxwell, Putnam and Tranekaer-Rasmussen, still appeared in the lengthened footnote added 1980, (1981a, p. 47, fn. 6), though with considerably fewer details about Feyerabend's debt to them (except in the case of Bohm, where Feyerabend specified his debt in more detail than in the original). The first deleted paragraph began by acknowledging Feyerabend's debt to Popper:[114]

> Popper's admirable *Logic of Scientific Discovery* and his paper "The Aim of Science" have been both the starting point and the motive force of the investigation to follow (1962a, pp. 32-33; rp 1981a, p. 47, removed).[115]

The second deleted paragraph was mainly an acknowledgement to discussions with Kuhn, but it also ended by citing Popper as also having pointed out the need for an alternative theory in the process of refutation (1962a, p. 32; rp 1981a, p. 47, removed). The third deleted paragraph was mainly devoted to Bohm and Vigier as the source of the idea that direct refutation may, in some cases, be impossible for empirical reasons (see chapter 7). However, Popper was again mentioned. He and John Watkins were thanked for useful constructive criticisms of earlier versions of the paper.

All of this would appear to lend support to the idea that Feyerabend, in 1962, can rightly said to have been a Popperian. Moreover, it would seem to suggest that Feyerabend's original explicit introduction of the

113 See and compare (1962a), pp. 31-32 with (1981a), p. 47. Instead of this page, Feyerabend lengthened a footnote (see below).

114 Along with Popper, Sellars and Vigier's acknowledgements also disappeared from the 1981 reprint. Originally, Sellars (along with Körner) had been cited as sharing similar views about the nature of observation languages. Feyerabend referred to Sellars' 'admirable' (1961). Originally, Vigier had been cited along with Bohm, as having emphasized what Feyerabend 'regarded as a most important feature of the situation . . . that direct refutation of a fairly complicated theory may be impossible for empirical reasons (1962a), p. 32, removed from the 1981 reprint.

115 Feyerabend cited the 1959 edition of Popper (1934) and Popper (1957b), and notes that the basic ideas of the latter had been published in German in Popper's (1948).

notion of incommensurable concepts belongs squarely within Popperian philosophy.[116]

Another compelling reason that Feyerabend's introduction of the notion of incommensurable concepts appears to be part of Popper's critical rationalist philosophy is that it was part of a criticism aimed at *exposing a dogmatic element within modern empiricism.* As we shall see in more detail in chapter 5, Feyerabend's attacks on the principle of deducibility and the principle of meaning invariance attempted to expose a dogmatic element in Nagel's theory of reduction and the Hempel-Oppenheim theory of explanation. According to Feyerabend, these modern empiricists share this dogmatic element with school philosophies such as Platonism and Cartesianism (1962a, pp. 30-31; rp 1981a, p. 46). In short, the dogmatic element is the meanings are not allowed to change. For example, just as Platonists require that words referring to Platonic ideas do not change their meaning, and the key terms in Cartesian Physics and metaphysics, such as 'matter', 'space', 'motion', 'god' and 'mind' are not allowed to change their meaning in any explanation involving them, logical empiricism demands that the meanings of observable terms remain unchanged. According to Feyerabend, such a demand is inconsistent with the existence of incommensurable concepts (1962a, p. 31; rp 1981a, p. 46). Feyerabend credited Popper with pointing out these similarities between logical empiricism and school philosophies,[117] and the very nature of the accusation that modern empiricism maintains a *dogmatic element* is indeed a critical rationalist complaint. For critical rationalists, no such dogmatic element is allowable in science. Feyerabend's use of the notion of incommensurable concepts to expose a dogmatic element that modern empiricists share with school philosophies is critical rationalist by nature.

Watkins has repeatedly criticized Feyerabend for not making the Popperian origins of Feyerabend's use of this Duhemian/Popperian argument against inductivism, and his indebtedness to Popper more generally, more explicit.[118] Watkins also suggested that the footnote citing Popper as the starting point of Feyerabend's discussion was the result of his own criticism of a draft of Feyerabend's essay (Watkins 2000, p. 48). It is somewhat odd that after criticizing Feyerabend yet again for not making the

116 Feyerabend also acknowledged that he used the term 'realism' in Popper's sense, which, at other times, he attributed to Viktor Kraft. For Popper's views on realism, see especially Popper (1963), "Three views Concerning Human Knowledge". Also see (1968c), p. 321, fn. 37, and especially (1964d); rp (1981a), pp. 161-175.

117 See (1962a), p. 31, fn. 5; rp (1981a), p. 46 fn. 5. Feyerabend referred the reader to Popper's discussion of essentialism in Popper (1954). This is an instance where Feyerabend did not delete his acknowledgement to Popper from the 1981 reprint.

118 See e.g. Watkins (1966), Watkins (1978), and most recently Watkins (2000), pp. 48-49.

Popperian nature of his ideas more clear, Watkins proceeds to argue that he had some "small influence on the published version" of (1962a), citing the above mentioned footnote in the 'Summary and Conclusion' section which acknowledged Popper (1957b) as the starting point for the essay, and claiming that Feyerabend "pooh-poohed his indebtedness" to Popper in the passage added to the footnote in 1981 (Watkins 2000, pp. 48-49, cf 1981a, p. 47, fn. 6). Watkins could have solidified his point by pointing out that in the original published version, there are many significant acknowledgments to Popper, and also several to Watkins himself. Watkins, it seems, has overlooked the major differences between the original version and the reprint. He devotes several pages of his article to bolstering his claims that Feyerabend was not forthcoming about his indebtedness to Popper, emphasizing a single footnote that he suggests was a result of his own critical comments on a draft version. He could simply have pointed out that Feyerabend removed an entire page filled with acknowledgments to Popper, including one thanking Watkins himself for comments on an earlier draft. Feyerabend was extremely forthcoming about his indebtedness to Popper and Watkins in the original 1962 version. He was equally extremely unforthcoming about his indebtedness to Popper and Watkins in the 1981 reprint.

In response to accusations that he had high jacked Popper' ideas, and that the idea of incommensurability is a critical rationalist thesis, Feyerabend amended a footnote to the 1981:

> Popperians [apparently he means John Watkins in particular] . . . have described it [(Feyerabend 1962a)] as a rehash of Popper's own ideas. Thus in his *Objective Knowledge* . . ., Karl Popper sadly notes that after 1962 people started referring to me instead of him when discussing matters of explanation and theory comparison while Imre Lakatos proclaimed, with his usual propagandistic flair, that Popper's theses 'were given wider circulation' by my paper which, he added, was 'not as good (and definitely not as clear) as Popper's and Duhem's original exposition' . . . the ideas which Popper and Lakatos claim for the Popperian school (1981a, p. 47, fn. 6, added to 1962a).

The relevant question is, what exactly was Feyerabend's debt to Popper? Does it concern the incommensurability thesis?

Feyerabend may have used Popper as a starting point for his discussion of the relationship between certain pairs of successive scientific theories, but even so, the incommensurability thesis should not be considered to be Popperian, nor part of Popper's critical rationalist philosophy.

In order to see this, let us begin by comparing Feyerabend's claims about the relationship of successive scientific theories to Popper's claims in a little more detail. According to Feyerabend, the *inconsistency* between Galilean Physics and Newton's theory arises at the *quantitative* level of empirical observation or measurement. To illustrate this with the example of the relationship of Kepler's first law and Newton's theory, the claim that they are *quantitatively inconsistent* boils down to the following: If one were to apply Newton's theory in order to understand the motion of the planets as they are actually observed through a telescope, then the mass of all of the planets would be given positive values, in which case Newton's theory (together with the appropriate correspondence rules identifying the mass and location of each of the planets and the sun) would give different *quantitative* results than Kepler's first law, i.e. a deformed, not a perfect ellipse. Similarly, Feyerabend argued that Galilean science and Newton's theory are *quantitatively inconsistent.* They give different results about the acceleration of a falling body near the surface of the earth (Galilean Physics suggests it is constant, while Newton's theory suggests that it is slightly decreasing).

However, while Feyerabend suggested that in these particular examples, we are confronted with *quantitative inconsistencies,* Feyerabend did not suggest that in these cases we are confronted with incommensurability (i.e. *qualitative inconsistencies*). While the argument that these particular examples are *quantitatively* inconsistent is enough to belie Nagel's theory of explanation and reduction, which takes exactly these examples as paradigmatic of successful reduction and explanation (making Feyerabend's an immanent criticism in this respect), the example of Galilean laws and Newton's theory is merely the *starting point* of Feyerabend's discussion of the relationship between successive scientific theories.

Feyerabend proceeds to *distinguish* between the case of the relationship between Galilean science and Newton' theory (his first example of the failure of Nagel's account that does not involve incommensurable concepts) and the case of 'the inertial law' of the impetus theory and Newtonian Physics, in which he argued that there is a *qualitative*, not a *quantitative*, reason for the failure of Nagel's account of reduction and explanation. Many of the confusions surrounding Feyerabend's claims about whether or not theories delineated in incommensurable concepts are 'inconsistent' can be remedied by applying Feyerabend's distinction between quantitative and qualitative inconsistency (see e.g. Preston 1997, pp. 104ff.). This is because many critics took 'inconsistent' always to mean quantitatively inconsistent, instead of sometimes qualitatively inconsistent. Confusion about Feyerabend's claims ensued.

Feyerabend called that *qualitative* reason *incommensurability* (cf. 1962a, pp. 46ff. to pp. 52ff.; rp 1981a, pp. 57ff. and pp. 62ff.), which he sharply distinguished from the quantitative reasons (logical inconsistency). The details of Feyerabend's claims about incommensurability will be the focus of chapter 5. Here the point is that Feyerabend's notion of incommensurable concepts does not at all cohere with Popper's critical rationalism.

As we have seen, according to Popper, successive scientific theories formally (or logically) contradict one another. By claiming that certain pairs of successive scientific theories are incommensurable, Feyerabend was explicitly rejecting the idea that these pairs of theories formally contradict one another. By calling particular pairs of scientific theories 'incommensurable', Feyerabend meant just exactly that they have *no logical relations* (1958b, p. 83, and 1961e, p. 388). As we have seen, Feyerabend's discussion of the relationship between successive scientific theories and his characterization of that relation with the notion of incommensurability does indeed have Popperian origins, and Feyerabend does accept and repeat the Duhemian/Popperian point that sometimes the two theories are indeed quantitatively inconsistent, and that this shows the poverty of inductivism and of the logical empiricists models of explanation and reduction. However, by calling other examples of successive theories incommensurable, he also explicitly denied the claim that all successive scientific theories actually logically contradict one another. Whereas Popper claimed that more recent scientific theories *contradict* and *correct* older theories, Feyerabend also argued that some pairs of scientific theories do not *contradict* and *correct* older ones. They *replace* them.[119]

It is these differences in their conceptions of the relationship between successive scientific theories (between *formal contradiction* and *correction* according to Popper, and *incommensurability* and *replacement* according to Feyerabend) that show that Feyerabend was not merely a Popperian until the late-1960s. These differences in their conceptions of the relation between successive scientific theories conflict with Popper, Lakatos, and Watkin's claims that (1962a) is merely a bad repetition of Popper's ideas. For by calling the concepts of successive scientific theories 'incommensurable', Feyerabend was explicitly denying the universal applicability of the argument Popper appropriated from Duhem against inductivism and in support of falsificationism.

Feyerabend himself argued that there is no room for incommensurability in Popper's falsificationist methodology (1988, p. 156ff.). In the passage added to a footnote to the 1981 reprint, Feyerabend reacted

119 Feyerabend emphasized his conception of scientific advance as one of replacement of older theories or concepts very often. See e.g. (1954c), pp. 470ff., (1958a), pp. 29ff.; rp (1981a), pp. 160ff., (1958b), pp. 81-83, (1958d), especially p. 50, p. 59, fn. 6; rp (1981a), p. 240, fn. 6, (1960e), pp. 321ff.; rp (1981a), pp. 219ff., (1961h), pp. 246-247, (1961i), p. 52; rp (1999a), p. 70, (1962a), p. 82 and p. 88; rp (1981a), p. 83 and p. 88, (1962b), (1963a), p. 13; rp (1989a), p. 108; rp (1999a), pp. 83ff., (1965b), p. 170, and p. 199, and (1965c), p. 227; rp (1981a), p. 110. Also see (1967c), p. 117, and (1970b), p. 291. For an excellent discussion of Feyerabend's notion of theory replacement, see Laudan (1976), pp. 588ff., and see Gaa (1975).

strongly against the suggestion that the idea of incommensurability should be seen as falling within the Popper's critical rationalism:

> *Added 1980* . . . Now the ideas which Popper and Lakatos claim for the Popperian school (Popper: 'the same idea of *mine* . . .' *Objective Knowledge*, my italics) are neither Popper's own (as Lakatos soon realized) nor are they the ideas I had in mind when writing my paper. Popper correctly remarks that his *Aim of Science* was a 'starting point' of my paper: he overlooks that it was a starting point of criticism, not of repetition: Popper, repeating Duhem, emphasizes that high level theories often conflict with established laws and thus *inspire* crucial experiments. I, on the other hand, was interested in cases where alternatives not only inspire such experiments but are necessary for producing the corresponding evidence (cf. the Brownian motion example in section 6 and the more detailed discussion of the case in *AM*; the Brownian motion example was suggested to me by David Bohm) (1981a, p. 47, fn. 6).

Such cases, according to Feyerabend, do not involve the kind of quantitative inconsistencies that are the basis of the standard conception of crucial experiments. Instead, they involve qualitative inconsistencies that also can be used to stage crucial experiments.[120] They are cases of incommensurability. Feyerabend continued:

> Lakatos (Philosophical Papers, I, 109, n.5) is aware of the purely psychological (or, as he calls it, 'catalytic') function of alternatives in Popper's methodology, regards it as insufficient, includes me in his criticism and ascribes to himself the view I actually hold in the present paper (and explain in greater detail in ch. 6.1) viz. that 'alternatives are *necessary* parts of the falsifying process'. He also overlooks that the paper introduces incommensurability and ends up by arguing that explanations inevitably contain 'subjective' elements. Small wonder that a paper with such unpopperian tendencies should look 'not so clear' to an author who reads it as a repetition of Popper's repetition of Duhem. Or do Popperians claim to have invented subjectivism and incommensurability as well?' (1981a, p. 47, fn. 6).

Feyerabend explicitly denied that the ideas he developed in his 1962a should be credited to Popper, and he rejected any suggestion that the notion of incommensurability belongs within the bounds of Popperian philosophy. As we have seen, Feyerabend's notion of incommensurability, as denying logical relations between particular pairs of successive scientific theories is not the same point that Popper and Duhem made about the logical inconsistency between some pairs of successive scientific theories. It is a qualitative. Not a quantitative reason for the failure of logical empiricist models of explanation and reduction, and it applies to different example.

---

120 For some revealing history on the notion of a 'crucial experiment', see Shapiro (1996), pp. 76ff., where he traced it through Newton's famous 'experimentum crucis' to Boyle and Hooke.

There are additional reasons supporting the claim that Feyerabend should not be placed solely or squarely in the Popperian camp. First of all, the notion that some pairs of successive scientific theories are incommensurable is also incompatible with Popper's *verisimilitude* conception of scientific progress.[121] As this has been convincingly argued by Kuhn, Feyerabend and Dilworth, the point will not be developed in any detail here.[122] It will suffice to point out that according to both Kuhn and Feyerabend, given incommensurability, both Popper's particular notion of nearness to the truth as well as the basic, general image of scientific progress as an advance on truth must be abandoned.

Second, in his 1962 development of incommensurable concepts, Feyerabend rejected the Popperian conception of a scientific theory in favor of the logical empiricist distinction between instantial empirical generalizations, such as Kepler's laws and non-instantial scientific theories, such as Newton's dynamics.[123] This is a further *ad hominem* aspect the criticism Feyerabend launched in (1962a), and another reason for some caution in characterizing Feyerabend's ideas in this piece as strictly Popperian.

A third additional point, which also supports the claim that Feyerabend's view of successive scientific theories as incommensurable does not fit within Popper's critical rationalism, is that Popper's own claim that Newton's theory *unifies* and *explains* Kepler's laws does not *itself* fit within Popper's critical rationalist philosophy.[124] This has been convincingly agued by Dilworth (1981) and Jarvie (1999). The reason is simply that according to Popper's critical rationalism, explanation consists of deductive subsumption. How can two theories formally contradict one another, if one supposedly deductively subsumes the other? As Dilworth puts it:

> we have an instance of Popper presenting a relatively acceptable description of what occurs in actual science [unification and explanation], but one which cannot

121 For an excellent and clear analysis of Popper's notion of verisimilitude, see Dilworth (1981), pp. 37ff. The point has been briefly discussed in Preston (1997), p. 61. Also see Oberheim and Hoyningen-Huene (1999), p. 230.

122 For Kuhn, see e.g. Kuhn (1970b), pp. 234-235. For Feyerabend, see e.g. (1970c); rp (1981b), pp. 131-161. For Dilworth, see Dilworth (1981), especially p. 50.

123 Popper did not distinguish between scientific theories and empirical laws, as did the logical empiricists. See e.g. Popper (1934), chapter 3, as well as Popper (1959), pp. 426ff. Note also that in all the citations offered, Popper speaks of Kepler and Galileo's theories, not laws. See (1962a), p. 28; rp (1981a), p. 44, slightly modified.

124 Perhaps one could argue that the fact that Popper blatantly contradicted himself in his discussion of the relationship between successive scientific theories (claiming both that they logically contradict one another and that one deductively subsumes the other) actually does suggest a connection between Popper and Feyerabend, as they were both notorious for promoting conflicting ideas. However, such an argument would seem to suggest that Popper, in this instance, was a Feyerabendian, and not the other way around.

be captured on his own conception . . . Popper wants to have his cake and eat it too (Dilworth 1981, p. 139).

In summary, there are several strong reasons for rejecting the idea that Feyerabend notion of incommensurable concepts, and his conception of the relation between successive scientific theories, should be placed within the critical rationalist camp. To begin with, although Feyerabend did repeat Popper's argument that some pairs of successive theories do *formally contradict* each other, this was merely the starting point for his discussion. Feyerabend proceeded to develop a second *qualitative* reason for their failure that applies to different cases of purported reductions and explanations. He called this reason incommensurability. Feyerabend himself explicitly argued that while he had taken Popper's claims as a starting point, it was a starting point for critical development, not for repetition. Moreover, Feyerabend's notion of incommensurable concepts does not fit within Popper's verisimilitude conception of scientific progress, which takes explanation to be deductive subsumption.[125]

Before concluding this discussion of the controversial nature of the supposedly Popperian roots of (1962a) and the explicit introduction of the notion of incommensurable concepts, it is important to make a more general point recommending caution in placing Feyerabend in the *Popperian school*. Part III argues that Feyerabend practiced pluralism in philosophy throughout his career. Although Feyerabend very often adopted, used and developed the tools of critical rationalism, his ideas were not confined to critical rationalism. Feyerabend did not belong to any particular philosophical school. This point has already been made as a response to Preston's presentation of Feyerabend as a strict adherent to Popper's conception of normative epistemology (Oberheim and Hoyningen-Huene 1999, p. 226), where it is argued that Feyerabend may best be understood as an "anti-isms" philosopher who used a pluralistic philosophical method aimed at critically investigating ideas, regardless to allegiance to philosophical *isms* like Popperianism. This suggestion met with the following response from Preston:

> Ironically, for philosophers well-acquainted with the views of Kuhn and Feyerabend, H-H & O's [Hoyningen-Huene and Oberheim's] account is so heavily slanted by awareness of some of Feyerabend's later work that their vision of his work as a whole borders on Whig history. What is more, they accept too much of his own self-mythologizing, little of which stands up to an examination of the ac-

125 This point is supported by Baum, who introduced Albert's *Briefwechsel.* He claims that Feyerabend accepted critical rationalism through 1965, but that this did not prevent him from criticizing certain theses of Popper's philosophy during this time and with developing thesis, such as incommensurability, which did not fit within Popper's critical rationalist conception. See Baum's introduction to (1997), p. 11.

> tual texts. No one who knows his work in depth can believe that the latter Feyerabend was a reliable guide to the productions of his earlier self. . . In suggesting, for example, that "even in Feyerabend's early work, it may be a mistake to ascribe any particular position to Feyerabend", H-H & O uncritically swallow his later line (see, for example, Feyerabend 1991a, pp. 127, 149, and 1991b, p. 489) that he did not have or need a philosophical 'position'. Although there are philosophers who officially eschew the production of philosophical theories, it is hard to imagine what philosophy without a position (which I take to mean 'view', and which might be held for no more than a fleeting moment) could possibly be. In earlier work, Feyerabend clearly and repeatedly identifies himself as a realist, a materialist, etc. (as detailed at length in my book). H-H & O make no mention of my demonstration that Feyerabend later doctored his earlier papers, excising many of the important references to Popper, and thereby giving the lie not just to Feyerabend's own claim that he was never a Popperian, but also to H-H & O's eagerness to believe it (Preston 1999, p. 234).

Preston, it would seem, interprets the claim that Feyerabend was not tied to any particular position as being based on Feyerabend's later 1991 pronouncements to that effect. He proceeds to accuse such interpretations of being based on Whig history, of projecting from the present into the past, and he finds it somewhat puzzling that such a conclusion could be reached by anyone familiar with Feyerabend's early publications. The basis for the accusation of Whig history is the projection of Feyerabend's later 1991 pronouncements about his former self into an interpretation of his earlier texts. However, the interpretation of Feyerabend as an anti-isms philosopher in question did not make use of, or refer to, any of Feyerabend's later pronouncements.

The claim about Feyerabend's pluralism, and the fact that he eschewed philosophical positions, will be explained, developed and supported in detail in Part III. The relevant point here is that the claims made in this section that incommensurability does not fit within the critical rationalist tradition does not rely on Feyerabend's later pronouncements about his former self. The claim that Feyerabend was an 'anti-isms' philosopher who eschewed allegiance to philosophical *schools* is not based on Whig history. Nor is downplaying Feyerabend's debt to Popper an exercise in Whiggism. The interpretation offered is based on an investigation of the differences between Popper and Feyerabend's conceptions of the relations between successive scientific theories in the early 1960s. It does mention the fact that Feyerabend later pointed these differences out himself, but that is not the basis of the claim. The historical evidence suggests that Feyerabend attempted to keep some critical distance to Popper, Popper's critical rationalism, and the Popperian school *from the early 1950s*. According to Popper:

> In all or almost all civilizations we find something like religious or cosmological teaching, and in many societies we find *schools*. Now *schools*, especially primitive

> *schools*, all have, it appears, a characteristic structure and function. Far from being places of critical discussion they make it their task to impart a definite doctrine, and to preserve it pure and unchanged. It is the task of a *school* to hand on the tradition, the doctrine of its founder, its first master, to the next generation, and to this end the most important thing is to keep the doctrine inviolate. A *school* of this kind never admits a new idea . . . [changes of doctrine] are all presented as re-statements of the true sayings of the master, of his own words, his own meaning, his own intentions (Popper 1972, p. 149, italics inserted).

Popper's critical rationalism was intended to contrast dramatically with such schools,[126] and this was the basis of Feyerabend's attraction to Popper in the first place. Feyerabend boasted in his autobiography that throughout the course of his career he only joined one professional organization, and emphasized how he originally admired Popper's disrespect for established traditions.[127] However from very early on, Feyerabend *began* to suspect that Popperianism itself was on its way to becoming just such a *school*, and that this was his main reason for trying, *even in the early to mid-1950s*, to keep some critical distance to Popper and Popper's *school*.[128] In fact, that is why Feyerabend rejected Popper's offer to become his assistant in 1953, although he had no other concrete alternative at the time:

> I declined [Popper's offer to become his assistant] despite the fact that I was broke and did not know where my next meal was going to come from. My decision was not based on any clearly recognizable train of thought but I guessed that *having no fixed philosophy* I preferred stumbling around in the world of ideas at my own speed to being guided by the ritual of a 'rational debate'" (1978a, p. 116, (1987), p. 312, 1993a, p. 261, italics inserted. Also see 1995a, p. 9 and pp. 96ff., as well as 1995c, Letter from 18 July 1971, p. 20).

According to Feyerabend's admittedly somewhat unreliable testimony about his relationship to Popper, *Feyerabend had no fixed philosophy* in the early 1950s. He preferred having no fixed philosophy to joining Popper's school. He preferred his intellectual freedom to being tied down to any one point of view.

---

126 Feyerabend criticized philosophical schools from a critical rationalist perspective in (1961g), p. 441, and especially (1961i), pp. 1ff.; rp (1999a), pp. 50ff.

127 See (1995a), pp. 71ff. Also see (1978a), p. 115, and similarly (1993a), p. 260. Also see (1995c), p. 218, Letter from 1982, where he also emphasized that his original attraction to Popper was not Popper's ideas, but his cheeky attitude.

128 Note that Preston paints a very different picture of Feyerabend's relationship to Popper in Preston (2002) than in Preston (1997). In Preston (2002), Preston emphasized that Feyerabend did have many reservations about the 'Popperian church' from very early on.

3.4. Casting off critical rationalism

We have seen that Feyerabend fell for falsificationism in Popper's seminars in 1952. When did Feyerabend begin to rethink falsificationism? According to Feyerabend, in 1955/56 in Bristol, he met David Bohm.

> While in Bristol I continued my studies of quantum theory. I found that important physical principles rested on methodological assumptions that are violated whenever physics advances: physics gets authority from ideas it propagates but never obeys in actual research . . . *That falsification is not a solution* became very clear in discussion with David Bohm who gave *a Hegelian account of the relation between theories, their evidence, and their successors* (1978a, p. 116, (1993a), pp. 261-262, italics inserted).

Thus, there appears to be some evidence that suggests that as early as 1955/56, Feyerabend began to doubt the efficacy of Popper's falsificationist model of the relation between theories, their evidence, and their successors. However, this particular evidence is rather dubious, as Feyerabend *rejected* Bohm's methodology in favor of Popper's falsificationism in his review of Bohm's book (1960e; rp 1981a, pp. 219-235).

However, Feyerabend described the source of his discomfort with the Popperian *school* in his autobiographical recollections of the early 1950s at some length (i.e. even before he turned down Popper's offer to become his assisitent):

> Joske [Agassi] urged me to become a faithful Popperian. I find this somewhat puzzling. As I have said, Popper's ideas were very seductive and I had fallen for them. So why did Joske try to push me further? My only explanation is that I had not fallen far enough and that my lack of commitment showed. Indeed, I always hesitated when it came to declarations of faith, in private, and even more so in public. Falsificationism, I seemed to say to myself, may be OK; but why should I act as if it were a sacrament? Why, for example, should I put Popper on every page and into every footnote of everything I wrote? . . . *I seemed to be entering into the domain of religious PR, group dynamics, or intellectual greed—none of which is my cup of tea* (1995a, p. 97, italics inserted).

This, I believe, represents the real source of tension between Feyerabend and the Popperian *school*. He felt as if it were infringing on his intellectual independence:

> He [Agassi] says I idolized Popper. Now it is quite true that I like to idolize people, I like to be able to look up to somebody, to admire her, to take her as an example – but Popper is not the stuff idols are made of. Agassi calls me a *disciple of Popper*. This is true in one sense, quite untrue in another. It is true that I listened to Popper's lectures, sat in his seminar, occasionally visited him and talked to his cat. This I did not do of my own free will but because Popper was my supervisor: working with him was a condition of my being paid by the British Council. I had not chosen Popper for this job, I had chosen Wittgenstein and Wittgenstein had accepted. But Wittgenstein died and Popper was the next candidate on my list.

> Also doesn't Agassi remember how often he begged me, on his knees, to give up my *reservatio mentalis*, fully to commit myself to Popper's 'philosophy' and, especially, to spread lots of Popper footnotes all over my essays? I did the latter – well, I am a nice guy and quite willing to help those who seem to live only when they see their names in print . . . (1987, p. 312).

Feyerabend recalled how Watkins (a Popperian at Bristol) chastised him for being a bad Popperian in 1957:

> Walking up and down with a stern face he would chastise me for having been a bad Popperian: too little Popper in the text of my papers, no Popper in the footnotes. Having explained where and in what manner Popper should have made an appearance, he would have a sigh of relief, lead me to the dining room, and allow me to eat. Imre Lakatos, whom I met much later [1968], attacked me in almost identical terms: "Why did you say X when Popper says Y, and why don't you mention Popper who, after all, also said X on a few occasions?" I hear that even now [1994] the sacred word POPPER continues to give strength to the faithful (1995a, p. 109).

There are also clear traces of Feyerabend's critical attitude toward Popper as early as 1961. For example, Feyerabend comments that Popper's philosophy is non-democratic in (1995b, written 1961, p. 375). However, Feyerabend clearly still followed Popper in many respects in these letters.[129] Moreover, after reading a draft of Kuhn's Structure of Scientific Revolutions, Feyerabend praised some similarities between Popper and Kuhn, as for example,

> 'culmination of unanticipated novelties proves to be an almost non existent exception to the rule of scientific development' {very similar in SSR, p. 96}—excellent! Popper says, in his lectures, that discovery is always intimately connected with refutation or, as you would say, with the changes of paradigms brought about by a crisis" (1995b, written 1961, p. 367).

; or,

> On p. 94 you are exactly Popper when you say 'Unanticipated novelty, the new discovery, can emerge only to the extent that his anticipations about nature and his instruments prove wrong' [SSR, p. 96, bottom}. This I do not say in order to make priority claims, but in order to show to you who your friends are (or, to be more precise, who the friends are of your better self) (1995b, written 1961, pp. 376-377).

But what was it about Popper that actually irritated Feyerabend so much? Feyerabend reported that he disliked Popper's growing tendency to preach authoritatively (Feyerabend 1995c, written 27 January 1970, p. 156). Moreover:

> And by the way, I have discovered why I react so angrily to the Popper issue (I am not sure that I am right, but I think that I am). *Personal friendliness is always confused with membership in a church*. I had conversations with Popperians, I addressed

129 See e.g. p. 360 as well as pp. 362-363.

> their problems and talked in their language because they were my friends. But in fact, they were not really my friends, as they did not acknowledge the friendship – they took it as a declaration of loyalty to the principles of their church and they publicized it that way. It took a long time for it to become clear to me that it was *that* that really got to me and it is because that is how they transformed me from a person and a friend into an object – a Popperian. Not to mention the fact that I see the Popperian Church as a miserable latrine with an awful stench (Letter from 1980 in Duerr 1995, p. 113, my translation).

Feyerabend often used the phrase "the Popperian Church" (e.g. 1995a, p. 73) referring to the Popperian school, and Feyerabend sardonically mocked the creation of schools (Letter from 15 December 1966 in 1997, p. 19). Feyerabend also warned of the dangers of Popperian peer pressure, and he advised Hans Albert to keep a critical distance to any school, even to his own (Letter from 1967 in 1997, p. 40).

While Feyerabend may have been uncomfortable with the Popperian school from very early on, he was still praising Popper's ideas and his contribution to the philosophy of science as late as 1965, when his glowing review of *Conjecture and Refutations* appeared (1965d). However by 1967, three things had happened. First, Feyerabend had become unhappy with his role as a teacher of critical rationalism, *because he saw himself as propagating the Popperian school*:

> If all goes well, I will arrive in Karlsruhe on the 26th of October to sew thee seeds of critical rationalism (Letter from 22 August 1967 in 1997, p. 44, my translation).[130]

Second, he came to perceive his London students as *blinded* by Popperianism. They were no longer thinking for themselves, but merely following critical rationalist rules (Letter from London 1967 in 1997, p. 53). And third, Feyerabend had decided that critical rationalism is not helpful in making science more humanitarian, which was becoming his main priority:

> My main interest is becoming the *humanitainization of the sciences*. Critical Rationalism is *not* the right solution and this is mainly because it does not leave room for laziness and calm reflection (Letter from London 1967 in 1997, pp. 52-53).

In (1968c), Feyerabend launched the first part of a long, pointed and unsympathetic attack on Popper's philosophy of Quantum Mechanics (see also 1969d). Feyerabend reported that Popper was not amused, and that he refused to read his 'diatribe' (Letters from 2 October 1967 and 16 March 1968 in 1997, p. 70 and p. 94 respectively). In a passage added to a footnote in the 1981 reprint, Feyerabend remarked:

> Lakatos accuses me of having 'completely forgotten that in some of (my) earlier papers (I was) more Popperian than Popper on these issues [Bohr and the Co-

130 In this letter, Feyerabend also described himself as a suffering critical rationalist, see p. 45.

> penhagen Interpretation]'. He seems to assume that there exists a Popperian 'position' in quantum mechanics. The above essay has shown that such is not the case: propensity was introduced by Bohr long before Popper started thinking about the matter, and for the rest we have pompous declarations and sophomoric mistakes which are elevated into a position by Popper's disciples. Unfortunately (for Lakatos) I never participated in these machinations and I criticized them when they came to my notice (1981a, p. 294, fn. 100).

Feyerabend referred the reader to a footnote and accompanying text of chapter 17 of (1981a, pp. 334-335 and fn. 83), where Feyerabend had criticized some aspects of Popper's interpretation of Bohr. As this particular passage originally appear in (1962b), it supports Feyerabend's claim that he was very critical of some aspects of Popper's work on Quantum Mechanics already in the early 1960s.

However, it was not Popper's ideas themselves that drove Feyerabend away from Popper. It was rather (a) the uncomfortable feeling that he was being pressured to join a school, (b) Popper's growing authoritative attitude, (c) the feeling that he himself was propagating a school mentality to his students, and (d) Popper's reaction to his criticism of him that drove Feyerabend from Popper and Popperianism.

Even so, in 1970, Feyerabend could still admit that he had learned a lot from Popper, and was personally thankful to him, and that Popper's book *The Logic of Scientific Discovery* is indeed a rich source for theoretical philosophy. What he really hated was the ideology. He somewhat ironically claimed to have felt the moral duty to free the world from Popperianism (Letter from 14 March 1970, in 1977 p. 181). By the end of 1967, Feyerabend had already begun to attack Popper's methodological model on the grounds that it does not reflect actual historical development. Moreover, according to Feyerabend, the attempt to squeeze the history of science into Popper's model makes the history of science uninteresting. In lectures at the London School of Economics in 1967, which were published in 1970 as "Problems of Empiricism, Part II", Feyerabend had begun explicitly to favor the historical approach to the philosophy of science, and to downplay Popperian methodological abstraction. According to Feyerabend, 'Problems of Empiricism, Part II' (1970b), continued the argument of "Problems of Empiricism" (1965b) that:

> it is not only possible but also desirable to introduce and elaborate hypotheses which are inconsistent with highly confirms theories and with the evidence. The reasons that I gave were partly abstract and partly historical. In the abstract part [of (Feyerabend 1965b)] I tried to show that evidence which refutes a theory can often be found only with the help of an alternative . . . the historical part introduced examples which explain how this works in practice" (1970b, pp. 275-276).

According to Feyerabend, his approach in "Part II" was significantly different from that of (1965b):

> It is mainly historical. Abstract considerations are used only sparingly, in the form of comments on the historical material . . . Contemporary philosophy of science not only fails to describe adequately some of the most exciting episodes in the history of thought; it would also have given some extremely bad advice to the participants. The change of style and the prevalence of history . . . (1970b, p. 276).

In these 1967 lectures, Feyerabend began to emphasize the historical approach to the philosophy of science *over* the normative approach, although as one can see, the normative aspect is still clearly apparent. In these 1967 lectures, Feyerabend first proposed the slogan "anything goes" (1970b), p. 278). Feyerabend began to argue for a Hegelian model of advance instead of Popper's.[131] He also began to criticize critical rationalism because it cannot help the sciences become more humanitarian, as could a more liberal epistemological anarchism. Or as Feyerabend put it in a 1967 letter:

> [At the London School of Economics] in the middle of Popper land, I show how little the Popperian model matches the historical facts (first point), and how uninteresting the history becomes when one tries to work through the history according to the rules of the model (second point). *Do what you want* – I shout at the students – *everything is legitimate.* Whatever makes you happy and does not hurt others, what makes a human science, and a cheerful poetry – that will definitely be profitable. Do not let anything that you like be forbidden to you by any methodology, and do not be convinced of anything you dislike by any methodology (Letter from London 1967 in 1997, p. 53).

In Berlin in 1968, in response to the student movement, Feyerabend finally casts off his ties to Popper completely. Feyerabend implied that as a consequence of the student movement, which had led to political polarization in Berlin, there was a slight split in Popper's circle. The right supported Popper, the Vienna Circle, and the idea that facts decide between theories. The left supported Feyerabend, proliferation, and incommensurability (Letter from 28 May 1968 in 1997, p. 76). Feyerabend had begun by working against the student movement, but gradually he moved farther and farther to the left (Letter from 28 May 1968 in 1997, p. 77). After 1968, Feyerabend may still have *sometimes* considered himself to be *some* sort of critical rationalist, but now in the spirit of Mill, no longer in the spirit of Popper.[132] For example:

> Popper's philosophy, which some people would like to lay on us as the one and only humanitarian rationalism in existence today, is but a pale reflection of Mill. It is specialized, formalistic and elitist, and devoid of the concern for individual happiness that is such a characteristic feature of Mill. We can understand its peculiarities when we consider (a) the background of logical positivism, which plays

131 Not in the late 1950s, as Feyerabend sometimes suggested. According to Feyerabend, Popper was very fashionable in London in the late 1960, see (1997), pp. 122-123.

132 See (1997), e.g. pp. 99, 100, 101, 115, and especially 166-167.

> an important role in the Logic of Scientific Discovery, (b) the unrelenting Puritanism of its author (and most of his followers), and when we remember the influence of Harriet Taylor on Mill's life and on his philosophy. There is no Harriet Taylor in Popper's life (1988, p. 34, fn. 2, 1993a, p. 34, fn. 2).

Feyerabend wrote:

> John Stuart Mill, for example, in his *On Liberty*. He says himself, he got it practically from Harriet Taylor, the woman he latter married. And he had a good relation with her. Sir Karl didn't have such a good relation with his wife. He ordered her around as if she were a maid. And shouted at her if she brought the wrong book. 'Here, you bring me so and so, and it is not the right one!'. It was very embarrassing to watch that (2000, p. 164).

By 1969, Feyerabend declared:

> With Popper it is finally over. Long live the flower power generation (Letter from September 1969 in 1995c, p. 14).

After 1969, Feyerabend made increasingly sarcastic and sometimes even personal criticism of Popper, who he felt had changed do to his success. Rather amusing examples include "Ayatollah Popper" (Letter June 1979 in 1995c, p. 64), and:

> Popper's but and the Pope's ring, that is no alternative, that is the same (Letter from 7 April 1972 in 1995c, p. 185).

However, *none of this prevented Feyerabend from temporarily explicitly adopting a critical rationalist standpoint as a useful tool*, and for all of Feyerabend's criticism of Popper and his critical rationalist philosophy, even as late as 1970, Feyerabend still acknowledged that the book to be a rich source of interesting ideas (Letter from 14 March 1970, in 1977 p. 181). For example, in his 1970 critique of Kuhn's views, Feyerabend adopted critical rationalist standpoint in order to assessing the strengths and weakness of Kuhn's competing accounts (1970c; rp 1981b, pp. 131-161).

In the late 1970s, Feyerabend was very explicitly critical of Popper in print. He traced "the deterioration of philosophy of science from Mach to the Vienna Circle to Popper and his followers" (1978a, p. 10), doubted Popper's intelligence (1978a, p. 26, fn. 11), implied that Popper stole ideas from Aristotle (1978a, p. 60, fn. 98) and called Popper a "slavish imitator" of the true form of enlightenment ideas (1978a, p. 199, fn. 12). He was also increasingly critical of Popper in light of Duhem: "Popper repeats what others said before him, but he repeats it badly and without the historical perspective of his forerunners" to which a footnote is added where Feyerabend included similar criticisms from some of "Popper's fans", such as P. Medawar, O. Neurath, and I. Lakatos:

> . . . 'I [Neurath] read Popper again. I hope that after so many years you will see, how empty all that stuff is . . . What a decrease after Duhem, Mach, etc. No feel-

> ing for scientific research.' Lakatos, late in his life, came to the same conclusion" (1987, pp. 190-191, fn. 2).

Or similarly:

> He [Lakatos] genuinely admired Popper and wanted to form a movement around Popper's philosophy. Eventually he became disillusioned. 'What has Popper done over and above Duhem?' he wrote on one of his last postcards. 'Nothing' (1995a, p. 130).

Feyerabend's 1979 "Dialogue on Method" began with a mock discussion of his relation to Popper:[133]

> A: What have you got against critical rationalism?
> B: Critical rationalism?
> A: Yes, critical rationalism; Popper's philosophy.
> B: I didn't know Popper had a philosophy.
> A: You cannot be serious. You were his student . . .
> B: I listened to some of his lectures . . . .
> A: And became his pupil . . . .
> B: I know this is what Popperians say . . . .
> A: You translated Popper's Open Society . . . .
> B: I needed the money . . . .
> A: You mentioned Popper in footnotes, and quite frequently . . . .
> B: Because he, and his pupils begged me to do so and I am kind-hearted. Little did I know that one fine day such friendly gestures would give rise to serious dissertations about 'influences'.
> A: But you were a 'Popperian' – all your arguments were in the Popperian style.
> B: That is where you are mistaken. It is quite true that some of my discussions with Popper are reflected in my earlier writings – but so are my discussions with Anscombe, Wittgenstein, Hollitscher, Bohr, and even my reading of Dadaism, Expressionism, Nazi-authorities has left a trace here and there. You see, when I come across some unusual ideas I try them out. And my way of trying them out is to push them to the extreme. There is not a single idea, however absurd and repulsive that has not a sensible aspect and there is not a single view, however plausible and humanitarian that does not encourage and then conceal our stupidity and our criminal tendencies. There is much Wittgenstein in all my papers – but Wittgensteinians neither seek, nor are in need of great numbers of followers and so they not claim me as one of their own. Besides they understand that while I regard Wittgenstein as one of the great philosophers of the 20th century . . . .
> A: Greater than Popper?
> B: Popper is not a philosopher, he is a schoolmaster – this is why the Germans love him so. At any rate – the Wittgensteinians realize that my admiration for Wittgenstein does not yet make me a Wittgensteinian. But this is all beside the point . . . .
> A: Not quite. For what you claim is that while you may use v[c]ertain ideas you need not accept them.
> B: Yes.

133 (1979), pp. 63-64. Also see (1976b) and (1981d).

Sometime thereafter, Feyerabend proceeded systematically to remove acknowledgements to, and praise of, Popper in his early papers when they were reprinted in two volumes of collected papers in 1981.[134] By the mid- to late 1980s, Feyerabend was highly critical and dismissive of Popper's ideas, and the philosophy of science more generally. For example, Chapter 6 of *Farewell to Reason* "Trivializing Knowledge: Comments on Popper's Excursions into Philosophy" (1986, pp. 161-191) is, as the title suggests, a sustained, pointed attack on Popper and his philosophy, the gist of which can be summarized in a single line: "Popper's 'ideas' are best forgotten as quickly as possible" (1987, p. 185).[135] In the same book, Feyerabend also called Popper, "just a tiny puff of hot air in the positivistic tea cup" (1987, p. 282). He summarized his critique of Popper's view on Quantum Mechanics as follows:

> Popper's criticism of the Copenhagen Interpretation, and especially of Bohr's ideas, is irrelevant, and his own interpretation is inadequate . . . Altogether it [Popper's own positive view] is a big and unfortunate step back from what had already been achieved in 1927 (1981a, p. 293, also see 1987, p. 182).

In the 1990s, Feyerabend continued to downplay his Popperian past. For example:

> True, there are some acknowledgments to Popper in these papers [his early work], but they are friendly gestures, not historical statements (I also mention my girlfriends). True, some points sound quite Popperian to somebody who has only read Popper, but they stem from Mill, Mach, Boltzmann, Duhem and above all Wittgenstein (1993a, p. 261).

Finally in 1991, Feyerabend denied that he had a philosophical position, and in a mock dialogue, he responded to the statement "A: Popper introduced falsification by negative instances . . .," with, ". . . you've got to be kidding. Falsification by counter examples is as old as the hills . . . making Popper the inventor of falsificationism! You might as well make Ronald Regan the inventor of rhetoric!" (1991b, p. 490).

Having documented and explained Feyerabend's acceptance and then increasingly vocal rejection of Popper's critical rationalist philosophy in some detail, I will now summarize the point most relevant to my interpretation of Feyerabend's philosophy. On my interpretation, Feyerabend was an epistemological opportunist and a philosophical pluralist, who developed ideas regardless to philosophical *isms* such as Popperianism. This

---

134 As the footnotes added sometimes begin "Added 1980", perhaps he also removed the footnotes to Popper in 1980.

135 Feyerabend claimed that his scathing attacks on Popper's work on Quantum Mechanics were a belated effect of a discussion he had in the autumn of 1965 with C. F. von Weizsäcker. See (1981a), p. 247, fn. *, as well as (1968c) and (1969d) of which chapter 17 of (1981a) "Niels Bohr's world view", pp. 247-297 is a compilation.

interpretation clashes with the view that Feyerabend was a Popperian until the late 1960s. My question is: Did Feyerabend (a) belong to the Popperian *school*, or was he rather (b) a pluralist who tried out ideas regardless of allegiance to philosophical isms, such as Popper's Popperianism? Here we have seen ample evidence against (a). I would like to re-emphasize, once again, that this does *not* imply that Feyerabend did not adopt, support, and use many Popperian ideas. *He did. Karl R. Popper was the single most significant source of influence of Feyerabend's philosophical development.* However, from as early as 1952, Feyerabend was uncomfortable with Popperianism as a school. In the early 1960s, he explicitly developed ideas that no longer cohered with critical rationalism, and by 1967, the feeling that he had indeed contributed to the Popperian school and its propagation contributed to the complete destruction of Feyerabend's positive relation to Popper. In the Part III, I will argue for (b). Here I will conclude by pointing out that according to Feyerabend:

> They [Curthoys and Suchting: authors of a Marxist critique of *Against Method*] cannot imagine that a writer might write an account of one kind of life, live another, belong to a group that is connected with still another and make propaganda for a style that is different from all three. For them a human being is like a statue made of sand and dried: touch it at one point, and it disintegrates entirely (1978a, p. 179).

To which a footnote is attached: "A similar mistake is committed by those who regard me as a Popperian, or a former Popperian" (1978a, p. 179, fn. 47). Feyerabend's ideas were not constrained to Popper's critical rationalist philosophy. Feyerabend was not a member of the Popperian school, and he often said as much himself (e.g. Letter from 1993 to Baum in 1997, p. 16).

While Feyerabend's criticisms of Popper clearly become increasingly pointed and occur with greater frequency at the end of the 1960s and in the 1970s and 1980s, Feyerabend's views in *Against Method* still clearly owe much to Popper. For although Popper was a main target of Feyerabend's attack in *Against Method*, Feyerabend himself, in a letter to Lakatos dated 17 December 1967, pointed out:

> It is settled: the best medicine for my present anti-theoretical mood is history; and so I have decided, as you have decided to do so, to start work on the two historical books I am supposed to write . . . The new version . . . will criticize *all* theoreticians of science . . . My "position" is thoroughly Popperian: Karl always starts his lectures on scientific method (he did so in 1952, when I listened to him in London; he did so again in Berkeley in 1969) with the remark that "there is no scientific method." It is this remark that I accept . . . The rest, however, is open to attack (Motterlini 2000, pp. 120-121).

Chapter 4
Felix Ehrenhaft
The impotence of experiment

It is well-known that Wittgenstein and Popper were main influences on Feyerabend's philosophical development. By contrast, the decisive episode that shaped Feyerabend's philosophical development is little known, and has gone almost completely untold in the secondary literature on Feyerabend. It happened while Feyerabend was still at university in Vienna in the mid to late 1940s. The events that unfolded taught Feyerabend a lesson about experience and its relation to theory that grounded his entire philosophical outlook: Because established theories can be insulated from recalcitrant experience through *ad hoc* means, alternative theoretical perspectives are need to promote progress.

4.1. Ehrenhaft, oil drops and experimental effects

While at the University of Vienna, but before beginning his Ph.D. in philosophy, Feyerabend studied physics under Hans Thirring, Karl Przibram, and Felix Ehrenhaft. In his autobiography and in (1978a, pp. 109-111), Feyerabend discussed a major lesson that he learned from Ehrenhaft's failure to convince theoreticians of the significance of his experimental effects.[136] Feyerabend reported that strange stories preceded Ehrenhaft's return to Vienna in 1947.[137] Ehrenhaft had fought, and lost to, Millikan on the charge of an electron, based on the oil drop experiment. Today, the oil

136 It is clear that in writing these two pieces, Feyerabend drew heavily from an unpublished appendix that he wrote to his 1967 translation of the short-hand transcripts of Ehrenhaft's 1947 lectures: ("Single Magnetic Northpoles and Southpoles and Their Importance for Science" – Ten Lectures delivered at the University of Vienna during the summer semester of 1947 by Dr. Felix Ehrenhaft, U.S. Visiting Professor", compiled with the assistance of Professor Ehrenhaft, and Dr. Schedling, by J. Ferber and P. K. Feyerabend.), as many passages are identical. The translated lectures, 27 pages complete with hand-drawn figures of the experimental set ups, include two appendices (3 pages and 7 pages) and an introduction (1 and a half pages) written by Feyerabend, hereafter cited as (1947/1967).

137 (1947/1967) were the first lectures Ehrenhaft performed on his return from London in 1947, see appendix 2, p. 1.

drop experiment is generally recognized as crucial for providing confirmation of the elementary electrical charge. The acceptance of the elementary electrical charge was preceded by a bitter dispute between Millikan and Ehrenhaft that began in 1910.[138] Although both Millikan and Ehrenhaft obtained very similar experimental results, Millikan had used them to defend the elementary electrical charge (electrons), while Ehrenhaft had used them to defend fractional charges (sub-electrons). In his autobiography, Feyerabend referred to Holton (1978), which presents a detailed reconstruction of the research methodologies of Millikan and Ehrenhaft (see 1995a, p. 65).

As Holton tells the story, Ehrenhaft had followed the *traditional* scientific method in that he allowed his theory to be dictated by experimental data, while Millikan was guided by the presuppositions of his theoretical framework. (According to Holton, Millikan's hand-written notebooks show that Millikan discarded 59% of the drops, as they did not provide support for his hypothesis of the elementary electrical charge). Even though Ehrenhaft purportedly stuck to the scientific method as presented in scientific textbooks, while Millikan allegedly did not, Ehrenhaft lost the dispute to Millikan. Interestingly, the entire episode is not usually reported in Chemistry textbook in the United States. In Niaz's (2000), "The oil drop experiment: A rational reconstruction of the Millikan-Ehrenhaft controversy and its implications for chemistry textbooks", it is reported that none of the 31 general chemistry textbooks published in the U.S. that were analyzed mentioned the Millikan-Ehrenhaft controversy. The implication is that while chemistry textbooks pontificate about how the scientific method subordinates theory to experiment, they disregard controversies in which scientists who based their findings strictly on the experimental results lost to scientists who did not strictly follow the scientific method as the textbooks present it (see Niaz and Rodriguez 2000, pp. 160-161). In any case, the episode had damaged Ehrenhaft's reputation, who thereafter was sometimes considered to be a charlatan.[139]

Ehrenhaft also claimed to have discovered 'magnetolysis' (the dissociation of water in strong magnetic fields) and other effects that conflicted with orthodox theory.[140] According to Feyerabend, "Many scientists re-

138 For some background, see Hacking (1983), pp. 235ff.

139 See (1947/1967), appendix 2, p. 1, fn. 1: "The main element of these rumours ["which denounced him as a charlatan", inserted from main text] was his alleged discovery of subelectrons. On this point Ehrenhaft never admitted to have been wrong. His main counterargument was the utter unreliability of Stoke's law."

140 "The Magnetic Current", supposedly published in (*Science* 94, No. 2436). A number of popular versions can be found on the net, especially on 'Extraordinary Science' and 'Anomalous Phenomena' cites, one of which even cite (1946/1967). My copy is from the Feyerabend Archive: PF 5-33. Feyerabend claimed that it is the only existing record of

garded him [Ehrenhaft] as a charlatan, and we, the physics and math students set out to expose him" (1995a, p. 65, rp 1987a, pp. 109-111). However, as it turned out, "Ehrenhaft [who] exposed us instead" (1995a, p. 65).

Feyerabend had been quite excited by the prospect of Ehrenhaft's return to Vienna:

> We knew that he had taught theoretical physics which was then (as it is now) a quite exceptional feat for an experimentalist. But we were also familiar with the persistent rumors which denounced him as a charlatan who had been, and still was retarding the progress of science. It was the last item which made us quite curious. How was it possible to be a physicist, a professor of physics, and a charlatan at the same time? And if such a combination was possible, did this not reveal a very fascinating aspect of physics? We could hardly wait to see the phenomenon in the flesh and to listen to his theories (1947/1981, appendix 2, p. 1).

Ehrenhaft performed what Feyerabend described as 'simple' and 'straightforward' experiments, which left him, and his fellow students, speechless. Some of the experiments purportedly demonstrated completely novel phenomena: "We realize, then, that we are dealing with an entirely new phenomenon, viz. a magnetic action upon static charge" (Ehrenhaft in 1947/1967, p. 12). Moreover, "we have now at our disposal observations which cannot be interpreted at all in the customary manner. An example is the rotation of the droplet of FeCl between the poles of a magnet. These observations seem to force a new way of describing electromagnetic phenomena" (1947/1967, p. 24). It certainly *appeared* as if the experiments undermined orthodox theory (and it seems as though at some later point Feyerabend was indeed willing to believe that they did).[141] Even so, at first, he was not prepared to give up established theory or to deny its excellence (1987a, p. 111). And according to Feyerabend, Ehrenhaft certainly did not convince the theoretical physicists either (though "Paul Dirac wanted information about the strength of his magnetic poles)" (1995a, p, 66). Ehrenhaft was quite critical of theoretical physics, preferring experimental approach:

Ehrenhaft's views in 1947, see (1995a), p. 65. In several footnotes to his 1960s and 1970s publications, he mentioned that copies could be obtained by dropping him a postcard. Ehrenhaft's experimental effects continue to be discussed by some physicists on the fringes.

141 See (1947/1967), appendix 2, also see (1963a), p. 23; rp (1999a), p. 92.

> Each demonstration was accompanied by a few gentle remarks on 'school physics' and ' the theoreticians' who built theories upon theories *but refused to consider the decisive experiments* which Ehrenhaft devised and continued devising in all fields of physics, and which constantly produced inexplicable results (1947/1967, appendix 2, p. 2, italics inserted).

In Alpbach, Feyerabend witnessed turbulent discussions, between Thirring, Popper, Rosenfeld, Pryce and von Hayek. The heated discussions were about the significance of Ehrenhaft's effects. They had just witnessed graphite dust particles that described circles, ellipses and more complex figures in the air, in the focal point of a large lens collecting sun light, for over an hour (see 1947/1967, appendix 2, pp. 3-4).[142] These effects appeared to challenge established theory. However the theoretical physicians discounted the effects as too complicated. A *direct* relationship to the theories on question could not be demonstrated. For this reason, they were dismissed as mere *Dreckeffekt.*

4.2. The impotence of Ehrenhaft's experiments

Back in Vienna, Feyerabend's reaction to Ehrenhaft's experimental effects was to organizing a 'seminar' or 'club' or 'student run classes in the evenings' (see and compare (1995a), pp. 66-67 to (1987a), p. 111, to (1947/1967), p. 5, respectively), called the "Club for the Salvation of *Theoretical* Physics" (italics inserted).

> having overcome the first shock we were not at all prepared to reject theory and we tried to criticize Ehrenhaft wherever we could. His attack . . . made us study . . . even more . . . and from a quite different point of view (1947/1967, appendix 2, p. 5).

The group first tried to explain Ehrenhaft's experiments and the unusual phenomena that they produced according to orthodox theory: "We did not succeed. We were not converted either—we simply believed that a better and more sophisticated approach would do the trick" (1995a, p. 67). In other words, Feyerabend *himself* believed established theory, even though he could not explain the effects of the experiments he himself was witnessing according to them. He reported that, at the same time, nobody

142 Apparently some participants suggested that they were quantum effects, to which Popper replied that they could be used as an argument against Quantum Theory (1947/1967), appendix 2, p. 4, fn. 2.

"ever doubted that science had to be adopted to the facts". This led Feyerabend to the conclusion that normal science (in Kuhn's sense) requires a split consciousness (1995a, p. 67). On the one hand, Feyerabend was convinced that scientific knowledge and scientific theory are firmly based on the observable facts. On the other hand, the recalcitrant evidence did not have the power to undermine his belief in the excellence of a well-entrenched theory. He simply assumed that "a better and more sophisticated approach [to explaining the phenomena with orthodox theory] would do the trick" (1995a,. p. 67). Thus, even for Feyerabend, Ehrenhaft's phenomena were initially considered to be *Dreckeffekt*- the result of an as yet unknown disturbance (1995a, p. 66). Feyerabend the impotence of Ehrenhaft's experiments first-hand

The effects that this episode had on Feyerabend's conception of the relationship between theory and experiment should not be underestimated. This was no minor event, or aside, in Feyerabend's philosophical development. Feyerabend was directly involved: Not only did he organize the seminar mentioned above, he also even wrote up Ehrenhaft's lectures for publication, and sold them as gray literature (1995a, p. 56).[143] Feyerabend referred to the lessons he learned from this episode in subsequent publications in the 1960s and 1970s, not to mention his autobiography, and even in 1967, twenty years after the events took place, Feyerabend took the time to translate Ehrenhaft's lectures into English (1947/1967).

Feyerabend reported that these experiences taught him the lessons that, "The relation between theory and experiment was much more complex than is shown in textbooks and even in research papers" (1987a, p. 111), and that, "the connexion between theory and experiment [is] quite tenuous" (1947/1967, p. 6). According to Feyerabend, the fact that Ehrenhaft's experimental effects failed of convince the theoretical physicists exposed,

> "an iron curtain surrounding Maxwell's equations [that], protect[ed] physics from Ehrenhaft—an iron curtain of *precisely the kind that had protected Galileo's opponents*" (1995a, p. 66, italics inserted).

After Erhenhaft died, the planned publication of his work was cancelled. It was simply more convenient to forget about his disturbing effects than to confront them:

> (Now that Erhenhaft is dead] I am sure that many people are breathing a sigh of relief when seeing that things, at last, are "back to normal" [Feyerabend specifies

143 See also (1963); rp (1999a), p. 92, fn. 27 and (1970d), p. 40 and p. 117, fn. 122. The fact that they are sometimes cited appears to confirm this.

that he means 'normal' in Kuhn's sense]:[144] and so they are, considering that physics is often governed by the same urge for unity which St. Irenaues has described so well when speaking about Catholic Church (adversus haereses I/x/2): 'This is the preaching which the Church has received, this is her faith, as we have said. And although she is dispersed throughout the whole world, she keeps it carefully, as if she dwelt in one single house, and believes in it unanimously, as if she possessed but one soul and one heart, and in complete agreement she preaches, teaches, and transmits it, as if she had but one mouth. Certainly the languages on the surface of the earth are different, but the force of tradition is one and the same (1947/1967, appendix 2, p. 7).

4.3. Conclusions

Feyerabend's repeated 1960s attacks on empiricism, his criticisms of the role of observation in theory testing, and the development of his pluralistic test-model and arguments for pluralism (see chapter 8) should be understood in light of the lessons Feyerabend learned from this episode.[145] Although it would take more than two decades for those lessons to mature into *Against Method*, his views on the limited role of observation in theory testing are best understood in the context of the lessons Feyerabend drew first-hand from the impotence of Ehrenhaft's experiments. For Feyerabend, *the impotence of Ehrenhaft's experiments did not simply show that experience is too weak to challenge established theories, it exposed the necessary role alternative theoretical perspectives play in promoting progress.* Feyerabend realized that an unexplainable experimental effect by itself lacks the epistemological power to challenge a well-established fundamental theory. Without an alternative theoretical perspective that can explain the recalcitrant observations, they will simply be pushed aside as *Dreckeffekt.* However, against the background of a competing theory, an experimental effect can launch such a challenge.

We will postpone a detailed reconstruction of Feyerabend's pluralistic test-model and his arguments for the benefits and even necessity of theoretical alternatives in promoting progress until chapter 8, where we will see just how deeply the lessons Feyerabend learned from this episode influenced the development of Feyerabend's pluralistic philosophy. For now the following will suffice. According to Feyerabend, without alternatives:

144 This suggests that appendix 2 was written after 1961. Feyerabend also referred to a 1966 article by Hempel that mentions Ehrenhaft's experiments. Perhaps Feyerabend added the second appendix when he translated the lectures from German into English in 1967.

145 Feyerabend collected these attacks in (1965b). Also see (1970d), pp. 40ff.

> … the gradual acceptance of a theory by an ever-increasing number of people must finally bring about a transformation even of the most common idiom that is taught in very early youth. In the end, all the key terms will be fixed in an unambiguous manner, . . . and the idea that they are copies of unchanging entities and that change of meaning, if it should happen, is due to human mistakes—this will now be plausible. Such plausibility reinforces all the maneuvers which may be used for the preservation of the theory (elimination of opponents included) (1962a, p. 69).

The suggestion that proponents of alternatives will be eliminated was not merely a sarcastic sneer, or some science fiction fantasy. Feyerabend was serious:

> Today, of course, the 'elimination' takes the more 'refined' form of a refusal to publish (or to read) what is not in agreement with the accepted doctrine (1962a, p. 69, fn. 85; rp Feyerabend 1981a, p. 75, removed).

This is just what Feyerabend experienced concerning Ehrenhaft's effects. Long after Ehrenhaft's death, Feyerabend remained bitter that Springer-Vienna had reneged on their agreement to publish Ehrenhaft's research:

> While Ehrenhaft was teaching there was a great amount of research in the phenomena explored by him. Ehrenhaft himself prepared a comprehensive account of his ~~results~~ research to be published by Springer-Vienna. The agreement was sealed with a handshake, for Ehrenhaft found it unnecessary to have a written agreement drawn up. Then Ehrenhaft fell ill . . . and died. As far as I am aware, Springer has never honored the agreement. Ehrenhaft's collaborators went their own ways, his pupils dispersed, and his research is not almost forgotten – except for occasionally serving as an ~~bad~~ example in a philosophy text (1947/1961, appendix 2, pp. 6-7).

Ehrenhaft's effects had effectively been shut out of the scientific community by an iron curtain just much like the one that protected Galileo's opponents.

Part II. Feyerabend's Assault on Conceptual Conservativism

Chapter 5
Incommensurability as attack on conceptual conservativism

This chapter begins with a brief examination of the origins of the notion of incommensurability in Greek geometry. Then it investigates the origins of Kuhn's infamous notion of incommensurability. The rest of the chapter chronicles the less well-known story of Feyerabend's development of his notion of incommensurability. The aim is not to establish claims of priority, but to enhance our understanding of the notion and the central role it plays in Feyerabend's philosophy. Focusing on the *function* of the notion of incommensurability *common* to Feyerabend's various critiques explicates Feyerabend's philosophy as a series of challenges to forms of *conceptual conservativism* based on the idea that in the course of scientific advance, when fundamental theories change, meanings change, which can result in a new conception of the nature of reality.[146]

5.1. Preliminaries

The notion of incommensurability has undoubtedly been one of the major themes in the history and philosophy of science in the last half-century.[147] As Thomas Kuhn and Paul Feyerabend are commonly credited with introducing this provocative idea in independent publications in 1962, it may come as a surprise to learn that Feyerabend had developed his notion of incommensurability more than a decade prior to the publication of

146 Significant portions of this chapter are rreprinted from Oberheim (2005): "On the historical origins of the contemporary notion of incommensurability: Paul Feyerabend's assault on conceptual conservativism" *Studies in the History and Philosophy of Science* 36: 363-390 with permission from Elsevier.

147 See the incommensurability bibliography in Hoyningen-Huene and Sankey (2001).

Kuhn's *The Structure of Scientific Revolutions*. However, the idea is even older than that.

According to the dictionary, the term 'incommensurable' means:

> (Of magnitudes) having no common measure integral or fractional (*with* another); irrational, surd; not comparable with respect of magnitude; not worthy to be measured *with* (Fowler and Fowler 1934, p. 576).

This contemporary use of the term 'incommensurable' originated in Greek mathematics, where it meant having no common measure. For example, two lengths *a* and *b* have a common measure just in case some whole number *x* of lengths of *a* can be set against some whole number *y* of lengths *b*, and thus one set of lengths can be used to measure the other exactly. For example if length *a* is three centimeters and length *b* is nine centimeters, then the two lengths share a greatest common measure (3 centimeters), and length *b* is exactly three *as* long.

The sides and the diagonal of any square, however, do not have this property (neither do the sides and hypotenuse of any right, isosceles triangle). In such special cases, no number of sides of *s* can be used to measure exactly any number of diagonals *d*. More specifically, a procedure for finding the greatest common measure for two magnitudes which was widely used by mathematicians, carpenters, architects, and geographers was the method of *antanairesis*: subtract the smaller length from the larger length, the difference from the smaller length, and so on until zero is reached (1983b, pp. 75, and 1987, p. 230). The last number in the sequence before zero is the sought after greatest common measure of the two lengths. For incommensurable line segments, such as the diagonal and side of a square, *antanairesis has no end*. Feyerabend claimed that according to some authors, the Greeks discovered incommensurability by noticing this property.[148] Today, the idea can be expressed by pointing out that if *s* and *d* are incommensurable, then *d* divided by *s* is an irrational number. For example, in a square with sides of 3 centimeters, the diagonal is 4.2426406 centimeters. 4.2426406 divided by 3 equals 1.41421353333.... In this case, *s* and *d* have no common measure. One cannot be precisely measured in terms of the other.

The next, and much more interesting, question is how this geometrical notion of incommensurability found its way into the philosophy of science. But before we consider Feyerabend's development of the notion of incommensurable concepts and how it was heavily influenced by a num-

148 Feyerabend mentioned Kurt von Fritz as an example, but he provided no specific reference. In the surrounding discussion, Feyerabend touched on the rich history of the idea, considering the suggestion that the Pythagoreans founded geometric algebra specifically in order rigorously to treat such incommensurables. See (1983b), pp. 75-76; rp (1987), pp. 230-231.

ber of individuals who were discussing a wide range of different topics, let us investigate how Kuhn developed his notion of incommensurability. This serves two purposes. First, it emphasizes some key differences between Kuhn and Feyerabend's notions of incommensurability[149] and second it qualifies the popular myth that the modern use of the term 'incommensurable' is the product of conversations between Kuhn and Feyerabend that took place in Berkeley, California, on Telegraph Avenue around 1960 (Hacking 1983, p. 67). This myth is based on the fact that Feyerabend had read and extensively commented on a draft of *The Structure of Scientific Revolutions*, in which Kuhn introduced his notion of incommensurability ([1960/61] 1995). However, as wee shall see, Feyerabend had independently developed his notion of incommensurability, and had even independently been using the term 'incommensurable'.[150]

Kuhn initially used the term to capture methodological, observational, and conceptual disparities between successive scientific paradigms that he had encountered in his historical investigations of the natural sciences (Kuhn [1962] 1970, pp. 148-150).[151] He originally used the model of a Gestalt switch to illustrate some limited aspects of his point.[152] Dilworth has argued that Feyerabend's notion of incommensurability actually fits better to the Gestalt switch model than Kuhn's. This is because Kuhn claimed that individuals thinking in terms of different theories looking in the same direction actually see different (phenomenal) objects. In contrast, according to Feyerabend the objects remain the same. It is only our beliefs about them that change (Dilworth 1981, pp. 90ff.).

Dilworth is correct in that the Gestalt switch model fits Feyerabend's notion of incommensurable concepts somewhat better than Kuhn's. However, Dilworth has not identified the real reason that the Gestalt switch model fits Feyerabend's notion so well. The main point is not that for Feyerabend only our beliefs change, while the objects themselves do not change. Feyerabend did not usually express himself in such a scientific realist fashion, although in a discussion at a conference in 1970, Feyerabend did say: "When we go from classical physics to relativity, what remains the same are the objects. The objects are what they are, only we

149 There have always been several significant differences in their conceptions of incommensurability. Feyerabend discusses some in (1977) and (1978), pp. 66-67. Also see Hoyningen-Huene (2000a), Hoyningen-Huene (2001), and Hoyningen-Huene (2005).

150 Feyerabend claimed that his notion of incommensurability pre-dated the 1962 introduction of the term by about a decade. See (1988), pp. 228ff. and (1993), p. 211. Also see (1970b), p. 219, Kuhn (1983), pp. 669, 684, n. 2, Kuhn (2000), pp. 297-298. For discussions, see Hoyningen-Huene (1993), p. 207; Preston (1997a), pp. 102-104, 216-217; and Hoyningen-Huene (2005).

151 See Hoyningen-Huene (1993), p. 206 and Sankey (1993).

152 For a discussion, see Hoyningen-Huene (1993), pp. 38ff.

think different things about them. To a large extent, our operations for getting certain numbers remain the same. The functors which we use and much of the syntactical apparatus may also remain the same. What is different are all the concepts connected with the functors" (1970e, p. 247). Usually, he was very reluctant to talk about the objects themselves (as distinct from our various experiences of them). As we shall see in chapter 6, on Feyerabend's view, the independent existence of the external objects is a hypothetical postulate that is justifiable on the basis of normative, methodological considerations. It is not to be interpreted as a factual claim based on inductive or abductive reasoning. Moreover, Feyerabend rejected scientific realist metaphysics and a scientific realist conception of scientific advance.

Rather, the real reason that the Gestalt switch model fits Feyerabend's notion of incommensurability very well is that just as the Gestalt switch model involves one physical image that can be seen as two mutually exclusive pictures (e.g. Wittgenstein's infamous duck-rabbit picture), for Feyerabend, one *sentence* (as a physical event or object) can be interpreted as two competing, mutually exclusive *statements*. A single sentence can be interpreted into two statements according to the two competing theories delineated in mutually exclusive incommensurable concepts. As we have seen in chapter 2, the basis for this distinction between a sentence (as a physical object or event) and the different mutually exclusive statements that it can express is Feyerabend's distinction between a language's physical properties (its characteristic), and its semantic properties (its interpretation). This holds measurements as physical events as well. The point of incommensurability is that the observable results of one physical measurement can be interpreted on the basis of two mutually exclusive sets of concepts. To use an example from 'Explanation, Reduction and Empiricism' (1962a), a single measurement of how deeply a projectile penetrates a block of wax can be used to determine the value of two incommensurable concepts (the projectile's impetus and momentum). One sentence can even be used to represent two different interpretations of a single observation or measurement. For example, the sentence "The planet has disappeared behind the horizon" can be used to make two very different statements depending on the theory used to give an interpretation to the term 'planet' (Ptolemaic or Copernican). While it has never been explicitly stated by Feyerabend or in the secondary literature, this point exposes the essence of incommensurability as Feyerabend conceived of it. Forty years of confusion concerning Feyerabend's views on the possibility of comparing the content of incommensurable theories through crucial experiments is due to a lack of appreciation of exactly this point.[153] Feyerabend did not initially use the Gestalt switch metaphor to illustrate incommensurability. The first time that he did so was in (1975a, chapter 17). Kuhn, by contrast, eventually retracted the Gestalt switch as a model for the kind of transition involved in paradigm change because it had led to so many misunderstanding.[154]

Later, Kuhn refined the idea of incommensurability and began to argue that there is translation failure due to differences in the taxonomic structures of successive scientific theories.[155] Kuhn's developing notion of incommensurability has received much attention, and it continues to provoke plenty of discussion.[156]

According to Kuhn, he arrived at the notion of incommensurability sometime in the mid to late 1940s as a graduate student, while attempting to understand what at first appeared to be nonsensical passages that he encountered in Aristotle's Physics.[157] However, the notion of incommensurability is conspicuously lacking from Kuhn (1957), and Kuhn has been accused (by various sources) of having borrowed the idea (from various sources) without having given them their due credit.

For example, Agassi recently claimed that Kuhn "borrowed" the idea from Duhem, and that "Kuhn ignored his debt to Duhem" (Agassi 2002, p. 409). In addition, according to Agassi, Kuhn denied (in a personal communication) ever having read Duhem. Agassi argues that this is highly implausible, as all members of Conant's circle, such as Kuhn, were familiar with Duhem's work. Thus, Agassi is "puzzled" by Kuhn's claim that incommensurability was the central innovation of his famous 1962 book, as "the word [incommensurability] denotes an important idea that Duhem explained in some detail. *It is that we do not forget old theories even after they are dated*" (Agassi 2002, p. 410, italics inserted). This idea, however, does not at all capture Kuhn's notion of incommensurability. Moreover, Agassi

153 Feyerabend later used Gestalt switch figures to illustrate incommensurability in chapter 17 of (1975a), but even there, he never really nailed down the point.

154 See Hoyningen-Huene (1993), p. 205 for a discussion with references.

155 According to Sankey, Kuhn's notion of incommensurability underwent a major transformation to the point that his later notion has little in common with his original account, see Sankey (1993) and Sankey (1998). Feyerabend can be interpreted as having anticipating Kuhn's 1990s notion of taxonomic incommensurability, with his 1965 claim that two theories are incommensurable "if a new theory entails that all the concepts of the preceding theory have zero extension or *if it introduces rules. . . which change the system of classes itself*" (1965a, p. 268). Also see Carrier (2001), p. 88.

156 For example, see Hoyningen-Huene (1993); Sankey (1993); and Bird (2003).

157 Kuhn (1991), p. 4. For more on Kuhn's account of his discovery of incommensurability, see Kuhn (1977), pp. xi-xiii; Kuhn (1981), pp. 3-6; Kuhn (1989), pp. 49-50; Kuhn (1991), p. 4; and Kuhn (1999), p. 33.

offers no evidence suggesting either that Duhem had actually developed a notion of incommensurability, or that Kuhn had borrowed the idea from Duhem.

Struan Jacobs has also recently rejected Agassi's claim about Kuhn's unpaid debt to Duhem. However, Jacobs proceeds to accuse both Kuhn and Feyerabend of having appropriated the idea of incommensurability from Michael Polanyi, without having given Polanyi his due credit (Jacobs 2002, p. 106 and 2003, pp. 58, 69, 70. By 1958, Polanyi had indeed developed a notion of incommensurability that was similar to Kuhn's in some significant respects,[158] and Jacobs presents some strong evidence in support of Kuhn's unpaid debt to Polanyi.[159] However, Jacobs' account of Polanyi as the source of Kuhn's notion of incommensurability is incomplete, as it does not temper Kuhn's debt to Polanyi in light of Kuhn's debt to Ludwik Fleck.

In the forward of Kuhn ([1962] 1970), Kuhn acknowledged Fleck's work as anticipating many of his ideas, and he candidly admitted that although readers will find few references to Fleck, he is indebted to Fleck "in more ways than [he] can now reconstruct or evaluate" (Kuhn [1962] 1970, p. vii). Fleck's concept of a 'thought-style' clearly anticipates several senses of Kuhn's central notion of a 'paradigm',[160] and Fleck had presented a phase-model of scientific development covering pre-normal and normal science. He even illustrated the idea with some of the same examples Kuhn later adopted (cf. Fleck [1935] 1979, pp. 83ff).

Kuhn's debt to Fleck also directly concerns the notion of incommensurability. Fleck's book, *Genesis and Development of a Scientific Fact*, discusses conceptual change at length.[161] It emphasizes that terms acquire their meanings from their theoretical context, and that those meanings change through theoretical transitions (Fleck [1935] 1979, pp. 25, 39, 40, and especially 53-54). Kuhn even used some of the same examples as Fleck did to explain meaning change, such as the concepts of chemical 'elements' and 'compounds' (Fleck [1935] 1979, pp. 122-123, 128 cf. Kuhn [1962] 1970, pp. 130-134). Moreover, they both emphasized the theory-ladeness of observation, and they both drew explicitly from Gestalt

158 For a comparison, see Jacobs (2002) and (2003).

159 Polanyi had also used the *term* 'incommensurable' in 1951 and 1958, but in contexts clearly different than Kuhn and Feyerabend's. See Polanyi (1958), p. 174 and (1951), p. 100. For a discussion of these and other uses of the term outside the context of philosophy of science, see Jacobs (2002), p. 111 and Jacobs (2003), pp. 59-60.

160 There are also some significant differences between Flecks' thought-styles and Kuhn's paradigms. For a discussion, see Babich (2003).

161 See Fleck ([1935] 1979), pp. 9, 10, 19, 20-21, 25-28, 53, 36, 38-41, 53-54, 62, 64, 92-95, 98, 110, 122-123, and 128.

psychology, stressing that theory can determine the perception of meaning. They both even acknowledged Wolfgang Köhler in this regard.[162]

In 1927, Fleck had used the term 'incommensurable' to describe the *in principle* differences between '*medical thinking*', which attempts to understand irregular, temporally dynamic phenomena such as an illness, and '*scientific thinking*', which attempts to understand uniform phenomena such as atomism in chemistry and energetics in physics (Fleck [1927] 1986, pp. 44-45). And in 1935, Fleck had used the term 'incommensurable' to describe *conceptual replacements* in theoretical transitions in science: "The old concept of disease thus becomes quite incommensurable with the new concepts and is not *replaced* by a completely adequate substitute" ([1935] 1979, p. 62).[163]

Kuhn, who was instrumental in having Fleck's (1935) translated and then published in English, emphasized that his "revelation" (his discovery of incommensurability) had occurred several years before he had read Fleck (in 1949 or 1950) (Kuhn 1979a, p. vii). In fact, it was this revelation that had led him to Fleck in the first place (*Ibid*). Kuhn also explicitly emphasized that, at that time, he was "much engaged in exploring Köhler, Koffka, and other gestalt psychologists" (Kuhn 1979a, p. ix). As we shall see, Gestalt psychology's stress on the theory-ladeness of observation, and on our ability to perceive certain meanings only after we are in the possession of a particular theory is the main similarity in Kuhn and Feyerabend's developments of their notions of incommensurability.

Thus, Fleck had developed a notion of incommensurability that anticipates Kuhn's, and Kuhn explicitly acknowledged his debt to Fleck. While this does not *exclude* Jacobs' claims about Kuhn's unpaid debt to Polanyi, Fleck was arguably the main influence on Kuhn's development of his notion of incommensurability.

In contrast to Kuhn, and Hacking myth about Telegraph Avenue, Feyerabend traced his notion of incommensurability back to his '*thesis I*' in "An attempt at a realistic interpretation of experience" (1958a).[164] As we have seen, in that paper Feyerabend developed '*Thesis I*': *The interpretation of an observation language is determined by the theories which we use to explain what we*

162 See Fleck ([1935] 1979), pp. 16, 44, 90-93, 120, 133, 142, (1935), pp. 76-78, and (1936), pp. 84-85, and compare Kuhn ([1862] 1970), e.g. p. 204. Kuhn acknowledged his debt specifically to Gestalt psychology in ([1962] 1970), p. vi, and to Köhler in (1979), p. ix. For an excellent discussion of Kuhn's use, and subsequent retraction, of the gestalt metaphor to characterize theoretical advance, see Hoyningen-Huene (1993), pp. 38-41, 204-205. For an introduction into Gestalt psychology including some aspects of its philosophical significance, see Köhler (1947).

163 This sentence and the term 'inkommensurabel' also appear in original German version; see Fleck (1935) on p. 82.

164 See (1978), pp. 67-78 and (1977), pp. 363ff.

observe, and it changes as soon as those theories change (1958a, p. 163), and this idea is based on a contextual theory of meaning according to which the meaning of a term is determined by the theories whose truth is presupposed when it is used to describe some feature of the world. In his (1958a), Feyerabend also argued that when older theories are replaced, the meanings of the observational terms used to test the theories change. The result is incommensurability: The idea that successive scientific theories are conceptually incompatible. *Thesis I* is unmistakably an early version of the incommensurability thesis. Moreover, as we shall see, Feyerabend used the notion of incommensurability in 1958 much in the same way he used in 1962. In 1958, he used it to challenge an implicit *conceptual conservativism* in logical positivism. He challenged the assumptions that theoretical terms derive their meaning *solely* through their connection with experience, and that experience itself is a *stable* (or unchanging) foundation on which theoretical meaning can be based. Clearly, then, Feyerabend had developed a notion of incommensurability by 1958, (before he had met Kuhn in Berkeley in 1960), and if that is the case, then Feyerabend and Kuhn developed the notions independently. Kuhn had it mainly from Fleck, where as Feyerabend had it before he met Kuhn in 1960. However, the notion of incommensurability can be found in Feyerabend's work even earlier than that, and this should come as no surprise as (1958a) is supposed to be a condensed version of Feyerabend's 1951 unpublished doctoral thesis.

Feyerabend's use and development of the notion of incommensurability has received less attention than has Kuhn's. The natural place to begin our investigation of Feyerabend's notion of incommensurability is his landmark essay "Explanation, Reduction and Empiricism" (1962a), where he first introduced the term 'incommensurable' to apply to the concepts of different universal theories. Since then, in the vast secondary literature on the subject, it has become customary to apply the term 'incommensurability' to a purported relationship directly between *scientific theories*, as well as many other kinds of things.[165] For instance, in addition to incommensurable concepts and theories, there have been discussions of incommensurable worldviews, traditions, cultures, communities, paradigms, values, practices, actions, perceptions, languages, terms, questions, problems, and so on. In the 1970s, Feyerabend also extended his use of the term 'incommensurable' from applying to just concepts and theories to a variety of things, such as incommensurable actions, (1975a, e.g. p. 271), incommensurable perceptions, (1975a, p. 225), incommensurable discoveries (1975a, p. 269), incommensurable statements, (1975a, e.g. p. 26),

165 For a long list of literature including discussions of incommensurability from many different fields, see Hoyningen-Huene and Sankey (2001c).

incommensurable attitudes, (1975a e.g. p. 269), and incommensurable paradigms, (1970c; rp 1981b, pp. 131-161).

Developing Feyerabend's notion of incommensurable concepts by focusing almost exclusively on the 1962 text avoids the problem that Feyerabend made a number of inconsistent claims about it.[166] The main confusion concerns the relationship between meaning (or semantic theory) and incommensurability. As we shall see, incommensurability was widely mistakenly taken to imply the rational incomparability of theories due to radical changes in the meanings of scientific terms. Incommensurability was understood as the claim that *scientists* cannot rationally compare their theories, because through scientific revolutions, the meanings of their terms change. However, Feyerabend repeatedly emphasized that when scientists decide between competing theories, meanings do not play a major role:[167]

> [I]n the decision between competing theories 'meanings' play a negligible part (1965a, p. 267, cf. p. 272 and p. 273; rp 1981a, p. 97, cf. p. 102 and p. 103).

This, by itself, does not conflict with Feyerabend's use of semantic considerations to develop the notion of incommensurable concepts. While semantic considerations may be used to show that the logical empiricist accounts of theory comparison fail because of the existence of incommensurable concepts, it by no means follows that scientists consider meanings in their decisions between competing theories. In 1993, Feyerabend put the point succinctly:

> I think that incommensurability turns up when we sharpen our concepts in a manner demanded by the logical positivists and their offspring and that it undermines their ideas on explanation, reduction and progress. Incommensurability disappears when we use concepts as scientists use them, in an open, ambiguous and often counterintuitive manner. Incommensurability is a problem for philosophers not for scientists, though the latter may become psychologically confused by unusual things (1993a, p. 211).[168]

166 In his published work, Feyerabend began to apply the term incommensurable to traditions in 1975. See e.g. (1975a), chapter 17ff. This later notion of incommensurability differs significantly from his notion of incommensurable concepts in the 50s and 60s. For a discussion, see (1997), pp. 185ff. In (1975a), p. 225, Feyerabend claimed that incommensurability cannot be defined, and is not susceptible to explication, but can only be shown.

167 Note, however, that Feyerabend was critical of the use of the notion of the 'meaning' of term in discussion about theory testing already in 1961, i.e. before critics had begun to criticize him for failing to specify a theory of meaning. See e.g. Achinstein (1964). Also see "ARGUMENTS FROM MEANING NEVER LEAD ANYWHERE" (1961c, p. 158 and p. 159 and 1995b, p. 386, written 1961).

168 Torretti launched a similar objection to the idea of incommensurability in Torretti (1990), p. 83. For a discussion, see Maudlin (1995), pp. 1009ff.

In other words, the fact that there is incommensurability is a serious (logical) problem for some philosophical accounts of science and its advance. Scientists, on the other hand, only encounter a kind of psychological confusion, for example, in situations where they are discussing the merits of two conceptually incompatible theories, they may misunderstand each other because they use the same terms to mean something different. However, the kind of temporary psychological confusion that *may* arise because one scientist mistakenly attaches the wrong meaning to a term is by no means an 'in principle' hindrance for the rational comparison of the two theories. Scientists can handle any incommensurability that may arise. They must merely specify how they are using the terms. For example if they use the term 'mass', they must specify 'mass' according to either Newtonian or Relativity theory, if they want to avoid potential misunderstandings. In fact, as we shall see, Feyerabend borrowed the notion of incommensurability from scientists who were having no trouble understanding the relationship between competing mutually exclusive conceptual accounts of the same phenomena that Feyerabend used the notion of incommensurability to describe.

Because the notion of incommensurability became bogged down in discussions of meaning, Feyerabend became increasingly keen to avoid the impression that his views on the relationships between successive scientific theories were based on controversial claims about meaning. This led to some highly misleading claims made by Feyerabend. For example:

> In his paper Putnam creates the impression that I am mainly interested in meanings and that I am eager to find change where others see stability. This is not so. *As far as I am concerned even the most detailed considerations about meaning belong in the gossip columns and have no place in the theory of knowledge. This is true even in those cases where meanings are invoked to force a decision about some different matter.* For even here their only function is to conceal some dogmatic statement which would not be accepted, if presented by itself, and without the chatter of semantic discussion (1965c, p. 230; rp 1981a, p. 113, italics inserted).[169]

If this statement is taken at face value, then the large majority of Feyerabend's early major publications (from his 1951 Ph.D. thesis to his 1962a) all belong in the gossip columns.[170] In fact, in the same year that Feyera-

---

169 Such claims have led some critics to suggest that Feyerabend did not have, and did not need a theory of meaning. See e.g. Hacking (1975) and Rorty (1979). Other critics, who highly value and contribute to semantic speculation in philosophy of science, have criticized Feyerabend for having a "half-baked idea of what meaning is." Preston (1997), p. 25. For similar criticisms of Feyerabend's purported theory of meaning, also see Achinstein (1964), Achinstein (1968), Putnam (1965) reprinted in Putnam (1975a), and Shapere (1966).

170 Feyerabend had written many pieces concerned with meaning and interpretation from the 1950s to the mid-sixties. See e.g. (1951), (1954a), (1954b), (1955a); rp (1981b), pp. 99-130,

bend made this statement, he also published (1965b) in which the contextual theory of meaning is used to support the idea of incommensurable concepts. In this paper, just as in his Ph.D. thesis and in (1962a), differences in *meanings are invoked to force a decision about some other matters*: be it the logical positivist account of the role of basic statements, the logical empiricist accounts of explanation and reduction, or Bohr's principle of complementarity.

By 1981, some of Feyerabend's claims about his own former publications on incommensurable concepts are patently false:

> My discussion [in (1962a)] of the relation between impetus and momentum [the first example of incommensurable concepts he develops] . . . is not an attempt to draw consequences from a contextual theory of meaning – *theories of meaning play no role in this discussion* – it simply shows that both the facts and the laws of Newtonian mechanics prevents us from using the concept of impetus as part of Newton's theory of motion (1981a, p. x, italics inserted).

As we shall see, a contextual theory of meaning plays a major role in Feyerabend's 1962 discussion of the relationship between impetus and momentum.

Another reason for restricting our initial analysis of Feyerabend's notion of incommensurability to the 1962 paper in which he introduced the term 'incommensurable' is that Feyerabend's notion of incommensurability is not something stable. It changes as Feyerabend's philosophy develops. For example, in 1962, Feyerabend talks about the impossibility of inter-defining the concepts of incommensurable theories. In 1965, he characterizes his notion of incommensurability by claiming that two theories are incommensurable when the meanings of their main descriptive terms depend on mutually inconsistent principles.[171] Later, he claimed that "[w]hen using the term "incommensurability" [he] always meant deductive disjointedness, and nothing else" (1977, p. 365). In the mid-1970s, Feyerabend presented a very different conception of incommensurability,[172] which he applied to the transition from the Greek archaic, aggregate worldview of Homer to the substance world-view of Aristotle.[173] In short,

(1958a); rp (1981a), pp. 17-36, (1960a); rp (1981a), pp. 37-43, (1962a); rp (1981a), pp. 44-96, and (1965b).

171 Feyerabend (1965b), p. 227. Also see Feyerabend (1975), pp. 269-270, 276. For a discussion, see Hoyningen-Huene (2005), p. 106.

172 See also the companion to *Against Method*: The unpublished German monograph, *The Introduction of Natural Philosophy into Pre-Greek and Greek Thought* (my translation), which was written from the perspective of epistemological anarchism in 1971, and revised in 1976. It will appear edited by Heit and Oberheim, as Feyerabend (forthcoming [1971/1976]).

173 Neto (1991) claims that Feyerabend was lead to incommensurability through his study of Greek philosophy and its relationship to Western Rationalism. This puts the cart before the horse. While this affected Feyerabend's development of the notion of income-

Feyerabend both characterized incommensurability, and applied the term in quite different ways in the course of using and developing the idea. These changes do not represent a development in positive sense, as is the case with Kuhn who improved his notion of incommensurability by clarifying and then distinguishing issues he had original conflated.[174] Rather, the changes in Feyerabend's notion of incommensurability reflect his changing interests, more than some systematic attempt to develop the notion.

For these kinds of reasons, a piecemeal approach to Feyerabend's notion of incommensurability recommends itself. Rather than trying to synthesize a single notion of incommensurability from all the different kinds of things Feyerabend said with the term and about the idea, we will investigate each instance of the notion of incommensurability separately, and examine how the notion developed in the context of Feyerabend's philosophy. What holds these notions of incommensurability together is how they were used, i.e. the function the play in Feyerabend's philosophy.

5.2. Formal accounts of reduction and explanation (1962)

According to the first sentence of "Explanation, Reduction and Empiricism", where Feyerabend original introduced the term 'incommensurable':

> The main contention of the present paper is that a formal account of reduction and explanation is impossible for general theories (1962a, p. 28; rp 1981a, p. 44).

In the following sentence, he specifically named Nagel's theory of reduction and the theory of explanation associated with Hempel and Oppenheim. These two theories, together with the hypothetical-deductive account of confirmation, represent the core consensus of the philosophical movement known as *logical empiricism*. Consequently, it will be useful to begin with some general background into the logical empiricists conception of scientific knowledge, and then turn to the formal accounts of reduction and explanation that Feyerabend explicitly challenged on the basis that they cannot take into account the use of incommensurable concepts.[175]

---

mensurability in the 1970s, originally incommensurability was not a consequence of such a comparison.

174 For a discussion, see Sankey (1993).

175 For earlier discussions of explanation by Feyerabend, see e.g. (1961a), pp. 35ff., (1961i), pp. 45ff.; rp (1999a), pp. 66, and (1961d). For a more general critique of Nagel's views by Feyerabend, see (1966c); rp (1981b), pp. 52-64.

The *logical empiricists* characterized scientific knowledge as a set of universal descriptive statements that we can make about the world.[176] The leading idea is that scientific knowledge is just the set of those statements about the world that are ultimately justified empirically, that is, on the basis of systematic observations and measurements.[177] Their general goals were to understand "how scientific knowledge is arrived at, how it is supported, and how it changes. . .", and their strategy was to approach these questions by examining, "how science explains empirical facts, and what kind of understanding its explanations can give us" (Hempel 1966, p. 2). The basic assumptions included the idea that science aims at "a comprehensive body of sound empirical knowledge, represented by a well confirmed system of empirical statements" (Hempel 1966, p. 40), and that all explanations "fit the phenomena to be explained into a pattern of uniformities and show that its occurrence was to be expected, given the specified laws and the pertinent particular circumstances" (Hempel 1966, pp. 50, also see pp. 50-53). According to both Hempel and Nagel, explanations of particular facts can be given on the basis of general statements, which state empirical laws about observable regularities. These, in turn, can be explained on the basis of theories.[178]

Theories were characterized as sets of general statements, which make reference to unobservable entities such as fields and charges. Theories are linguistic entities that can be distinguished from each other according to their syntax. In other words, each theory has its own syntax, and a change in the syntax of a theory would result in a different theory (Carnap 1966, sec. 27 and 28). Both Hempel and Nagel shared the basic logical empiricist assumption that theories are formal languages.

According to Hempel, theories systematically unify diverse phenomena on the basis of some shared, underlying process. They explain how various uniform regularities are manifestations of one common set of basic laws (see Hempel 1966, p. 75). The logical empiricists developed a precise account of how discovering these underlying processes can be used to explain the regularities in how the world behaves. They developed the idea by characterizing *scientific theories as explanations of the regularities in the world's behavior on the basis of general laws that function axiomatically in an interpreted, deductive system.*[179]

176 For an informative critical discussion, see Giere (1999), pp. 32ff. For an excellent introductory account, see Salmon, Earmon et al. (1992), pp. 116ff. and 135ff.

177 For a discussion, see the introduction of Giere and Richardson (1966), pp. 1-13.

178 For critical discussions of these ideas see, for example, Suppe (1972), Giere (1999), especially pp. 97ff., and Salmon (1989).

179 See e.g. Hempel (1966), Hempel (1958), Nagel (1949), and Nagel (1961).

The account was developed from the idea that in order to explain regularities in how things behave, scientists use general statements that express laws as axioms in order to construct explanatory arguments. By using laws as axioms in a deductive, explanatory argument, scientists can use theories to explain a great many diverse facts on the basis of just a few laws. For example, the basic principles of Newton's theory of Gravitation and of motion can be used to explain diverse empirical regularities such as the motion of free falling objects, simple pendulums, the tides, the satellites, the moon, the planets, comets, and double stars.[180]

Hempel and Oppenheim attempted to capture this insight with the development of a formal account of deductive-nomological explanation, or the *D-N model of explanation.* The model was introduced in the classic article "Studies in the Logic of Explanation" (Hempel and Oppenheim 1948), which has served as a starting point for much of the subsequent work on scientific explanation in the last half-century or so. The basic idea of the D-N model of explanation is that scientific explanations provide understanding by demonstrating that the phenomena to be explained should be expected as a consequence of the general laws of nature.[181] By 1965, Hempel had refined and modified some aspects of the model,[182] which he set out in (Hempel 1965). It became known as the 'received view' of scientific explanation (Suppe 1972), or the 'covering-model', or what has been alternatively called just 'the standard view' of scientific explanation.[183]

According to the logical empiricists, scientific knowledge is arrived at by the method of hypothesis according to which tentative answers (hypotheses) to a problem under investigation are proposed and then subject to empirical test. Hypotheses can come in the form of descriptions of particular facts or events, general laws about the regularities in observable events, or even more inclusive sets of statements, such as theories, which bind diverse empirical laws together by referring to their common, unobservable, underlying processes.

The model of explanatory arguments is *deductive* because, the statements used to describe the phenomenon to be explained are deductive consequences of the premises used in the explanatory argument: Explana-

180 See Hempel (1948) who adds the Kinetic Theory of Gasses and Bohr's theory of the hydrogen atom to further illustrate the point.

181 For an overview, see Kitcher (1999).

182 For a detailed and informative discussion of the refinements and modifications see Salmon (1989).

183 See, for example, Hempel (1966), especially pp. 49-54 on "Deductive-nomological explanation" and chapter 6 on theories and theoretical explanation, pp. 70ff. For a synapse of the model and its history, see Kitcher (1999).

tory arguments make a deductive inference. The model of explanation is *nomological* because the explanations are based on laws. The form of the explanatory arguments was captured in a simple schema:[184]

$L_1, L_2, \ldots, L_I$
$C_1, C_2, \ldots, L_K$

E

Hempel called "E" the "explanandum sentence". The explanandum sentence describes the phenomena to be explained. He called "$L_1 \ldots L_r$", and "$C_1 \ldots C_k$" the "explanans sentences", or "explanatory facts" (p. 50). According to Hempel's model, all of these sentences state facts about the world. They are all true statements. The Ls state *general* laws, the Cs and E make particular assertions about actual circumstances. The C's make assertions about specific initial conditions, and Es are statements about some observable or measurable circumstances. In order to help clarify his model of explanatory arguments; Hempel called the laws used to explain "*covering laws*", and his account of explanation "deductive subsumption under these general laws" (Hempel 1966, p. 51).

Successful explanatory applications of this schema appeared readily apparent in the derivations in classical Newtonian dynamics. For example, the trajectories of bodies are deduced from force laws and initial conditions, and thereby their motion is explained.

Hempel also developed a model of *probabilistic explanations* (Hempel 1966, pp. 58-59). For example, Hempel attempted to show how we explain why *some* individual contracted measles, by observing that they had contact with someone who had measles, in conjunction with the observation that a large percentage of individuals contract measles, if they have contact with individuals who have measles. His schema of an explanatory statistical argument mirrors that of the D-N model. He illustrated it as follows (Hempel 1966, p. 59):

The probability for persons exposed to measles to catch the disease is high.
Jim was exposed to the measles.
=================== [makes highly probable]
Jim caught the measles.

184 See Hempel (1966), p. 51, as well as Hempel (1958), p. 37. For a concise synapses of the basic ideas underlying this model by Hempel, see Hempel (1958), section 4, 'Structure and Interpretation of a Theory', pp. 46-48.

A probabilistic law states the probability that a predicate is found among the members of a specific class. The general statements used as premises are probabilistic laws. Hempel's D-N model illustrates how scientific explanations meet two systematic requirements: explanatory relevance and the requirement of testability (Hempel 1966, pp. 48ff.). In order to meet the requirement of explanatory relevance, the explanatory information must provide good grounds for believing that the phenomenon to be explained does actually occur (Hempel 1966, p. 48). In order to meet the requirement of testability, the statements constituting a scientific explanation must be capable of empirical test (Hempel 1966, p. 49). A deductive explanation of an event is a logical deduction of an 'explanandum' sentence, which asserts that an event happens, from premises, the 'explanans' sentences, which are either general laws or statements of antecedent conditions (Hempel, 1965, pp. 249ff.). Specifying an explanandum event can serve both to identify a phenomenon to be explained, as well as to make predictions with which the potential explanation can be tested.

By 'reduction' the logical empiricists meant an explanatory relation between two theories.[185] Nagel's theory of reduction, like the Hempel-Oppenheim theory of explanation, takes scientific theories to be deductive systems. Moreover, Nagel's theory of reduction, like the Hempel-Oppenheim theory of explanation, conceives of explanation of the statements of one theory by the statements of another as the deduction of one from the other: "a reduction is effected when the experimental laws of the secondary science . . . are shown to be *logical consequences* of the theoretical assumptions . . . of the primary science" (Nagel 1961, p. 352, italics inserted).

Nagel discusses three formal *considerations* on reduction and he delineates two formal *conditions* on reduction (Nagel 1961, pp. 345-358, see especially p. 358). According to Nagel, the third formal consideration on reduction is that:

> The primary and secondary sciences involved in a reduction generally have in common a large number of expressions (including statements) that are associated with *the same meanings in both sciences* (Nagel 1961, p. 351, italics inserted).[186]

185 The term "reduction" is used in so many different ways in the philosophy of science that it is important to specify the sense in which it is being used. Here, I will be talking only about theoretical reduction, and only a particular model (Nagel's) of it. See Nagel (1961), pp. 336ff.

186 This is the third formal consideration. The first two are "the axioms, special hypotheses, and experimental laws of the science involved in a reduction must be available as explicitly formulated statements, whose various constituent terms have *meanings unambiguously fixed by codified rules of usage or by established procedures appropriate to each discipline*" (Nagel 1961, p. 345, italics inserted). The second formal consideration is "Every statement of a science S can be

The two formal requirements on reduction are: (1) *The condition of connectability*: "Assumptions of some kind must be introduced which postulate suitable relations between whatever is signified by 'A' and traits represents by theoretical terms already present in the primary science." (Nagel 1961, pp. 353-354); and (2) *The condition of derivability*: "With the help of these additional assumptions [see *the condition of connectability*] all the laws of the secondary science, including those containing the term 'A' must be *logically derivable* from the theoretical premises and their associated coordinating definitions in the primary discipline" (Nagel 1961, p. 354, italics inserted).

Nagel was primarily interested in cases of synchronic reduction in which one subject is purportedly reduced to another (such as Thermodynamics to Statistical Mechanics). However, he did also discuss examples of diachronic reduction (such as *successive* scientific theories in a field as for example Galileo to Newton), and he did claim that his account was supposed to hold for such theoretical transitions.[187]

It follows from these two formal conditions on reduction that Nagel's account of theoretical transitions is not one of *replacement* in which the older theory is replaced by the newer, but as *absorption* in which the older theory is shown to be a deductive consequence of the newer theory: "the phenomena of a relatively autonomous theory becoming *absorbed* by, or *reduced* to, some more inclusive theory is an undeniable and recurrent feature of the history of modern science" (Nagel 1961, pp. 336-337, italics inserted). Thus, Nagelean reduction is conceptually conservative or retentive: Through the reduction of an older theory to a newer theory, the newer theory conserves and legitimizes the older theory. As we shall see, it is to exactly this aspect of Nagel's theory of reduction to which Feyerabend objected.

To summarize, the logical empiricists characterized scientific knowledge as *a set of true statements* including theoretical explanations of the regularities in how objects behave. On this conception of scientific knowledge, the units of scientific knowledge are true statements which come in two forms: general statements and statements of particular facts. The structure of scientific knowledge is exemplified in the structure of explanatory arguments: From general premises about law-like regularities in conjunction with sets of particular known facts (assumed to be true for the purpose of the inference), we deductively infer the phenomena to be explained. Na-

analyzed as a linguistic structure, compounded out of more elementary expressions in accordance with tacit or explicit rules of construction" (Nagel 1961, p. 349).

187 See e.g. Nagel (1961), p. 291 and p. 362. Feyerabend has been criticized for interpreting Nagel's theory as applying to diachronic theory transition, when in fact it was primarily aimed at synchronic reduction. See e.g. Preston (1997), p. 61. For a short defense of Nagel from Feyerabend's attack, see Coffa (1967).

gel's theory of reduction characterized the relationship between successive scientific theories as an explanatory relationship according to which the older theories can be derived from their successors. In this way, allegiance to the older theories is vindicated, as they are conserved within their successors. For all of their promise, both the Hempel-Oppenheim theory of explanation and Nagel's theory of reduction have faced, and continue to face, a growing number of diverse criticisms of many of their aspects, as well as challenges to the legitimacy of the entire approach on which it is based (see Salmon 1989). The sketch that I have presented is by no means comprehensive. It is merely intended to provide a basis with which to facilitate an understanding of the challenge that Feyerabend set out to it with the notion of incommensurable concepts.

Feyerabend's attack on the Hempel-Oppenheim theory of explanation and Nagel's theory of reduction began by arguing that both theories implicitly contain the following two principles, "two principles which underlie the orthodox approach: (A) The principle of deducibility; and (B) the principle of meaning invariance" (1962a, p. 30; rp 1981a, p. 46). Feyerabend argued that according to (A) "explanation is achieved by deduction in the strict logical sense," which leads to the demand that "all successful theories in a given domain must be mutually consistent" (1962a, p. 30; rp 1981a, p. 46). According to (B), "an explanation must not change the meaning of the main descriptive terms of the explanandum" (1962a, p. 30; rp 1981a, p. 46). Feyerabend's main goal was to show that both principles are inconsistent with empiricism. In other words, Feyerabend intended to attack contemporary empiricism by showing that its theories of explanation and reduction are inconsistent with its own *methodological* views.[188] In *this* respect, it is meant to be an immanent criticism. It is *not* an immanent criticism with respect to the issue of 'realism', as the term was used by Popper and Feyerabend in the early 1960s.[189] As Feyerabend pointed out, the logical empiricists did not agree with the realist assumptions he used in making his argument. For this reason, Feyerabend identified these *unshared* assumptions as unshared, and argued for them at length on independent grounds.[190]

188 For an excellent reconstruction of Kuhn and Feyerabend's attacks on reduction, see Spinner (1973) pp. 43ff. See also Hovard (1971), pp. 89ff.

189 See Popper (1963), pp. 97ff. Feyerabend argued that sentences are supposed to be conjectured propositions; i.e. they are supposed to be about causal events in the world. We *ought* to try to interpret them realistically.

190 See (1962a), pp. 34-43, especially p. 34, and pp. 42-43; rp (1981a), pp. 49-55, especially p. 49 and pp. 54-55.

It is also supposed to be an immanent criticism in a second important respect. Feyerabend's introduction and use of the contextual theory of meaning is also supposed to be acceptable to Nagel, who claimed that:

> It is, however, of utmost importance to note that the expressions belonging to a science possess meanings that are fixed by its own procedures of explications. In particular, expressions distinctive of a given science (such as the word 'temperature' as employed by the science of heat) are intelligible in terms of the rules or habits of usage of that branch of inquiry; and when those expressions are used in that branch of study, they must be understood in the senses associated with that branch (Nagel 1961, p. 352).

And that:

> it must be admitted that in an earlier use the word 'temperature' had a meaning specified exclusively by the rules and procedures of thermometry and classical thermodynamics (Nagel 1961, p. 357).

In fact, Feyerabend's discussion of the "rules of usage" of a theory that determine the meaning of its terms in his development of the notion of incommensurable concepts (see below) is taken directly from Nagel himself.[191] Feyerabend made this point explicit:

> *According to Nagel* the meaning of these terms is to be regarded as "fixed" by the procedures and assumptions of the impetus theory (1962a, p. 55; rp 1981a, pp. 64-65, italics inserted. Feyerabend referred to Nagel 1961, p. 301)

Nagel also explicitly says as much when considering a possible objection to his "essential point" that:

> in the reduction of thermodynamics to mechanics a postulate connecting temperature and meaning kinetic energy of gas molecules must be introduced, and *that this postulate cannot simply be warranted by simply explicating the meanings of the expressions contained in it* (Nagel 1961, p. 357, italics inserted).

The reason that such a postulate cannot be warranted by explicating the meanings of the terms in it, is that the meanings of the terms, *according to Nagel,* are determined by the theories to which they belong. This has apparently gone unnoticed by commentators who reject Feyerabend's argument as being based on an acceptable theory of meaning.[192]

In order to show that Nagel's account of reduction implicitly assumes (B), Feyerabend stated the principle of deducibility and argued that it follows from it that *meanings are invariant with respect to the process of reduction* (1962a, p. 33; rp 1981a, p. 48). That Nagel does accept this principle fol-

---

191 Feyerabend cited several parts of this passage on p. 80 without providing any references. See also footnote 373 above, which states the third formal condition on reduction, which also implies that meanings are fixed by the rules of usage appropriate to *each* discipline.

192 See e.g. Shapere (1966) and Sankey (1993).

lows from (A) Meanings are not allowed to change in logical deductions.[193]

Feyerabend next task is to show that (A) and (B) are also implicit assumptions of the orthodox theory of explanation. Feyerabend citied Hempel: "The explanandum must be a logical consequence of the explanans; in other words, the explanadum must be logically deducible from the information contained in the explanans, for otherwise the explanans would not constitute adequate grounds for explanation" (Hempel 1948, p. 321, cited by Feyerabend, 1962a, p. 34; rp 1981a, p. 48).

Feyerabend pointed out that (A) and (B), just as any theory of reduction and explanation, can be interpreted in two different ways: As either a *description* of actual scientific practice, or as a *prescription* which must be followed to guarantee the scientific character of the enterprise (1962a, p. 33). This sets the stage for Feyerabend's two-prong argumentative strategy, which aimed to show that the orthodox accounts of reduction and explanation neither accurately describe actual scientific practice, nor are desirable as methodological prescriptions according to the logical empiricists own methodological views. To support the first descriptive prong, Feyerabend argued that (A) and (B) have not always been enforced in the history of science in episodes of scientific advance, and that they are incorrect as descriptions of the relationship between successive scientific theories. For example, Feyerabend argued that the Copenhagen school tries to enforce both conditions (1962a, p. 88; rp 1981a, p. 88). According to Feyerabend, the Copenhagen interpretation of Quantum Mechanics is

> the first theory after the downfall of the Aristotelian physics that has been quite explicitly constructed with an eye both on the consistency condition and the condition of (empirical) meaning invariance (1963a, p. 12; rp 1999a), p. 85.

In order to support the second prong of his argument, Feyerabend claimed that scientific progress required violating these conditions. The normative arguments turns on showing that if (A) and (B) were to be enforced as methodological prescriptions, then we would have a succession of theories all of which retain *unchallenged* fundamental structural assumptions, and this is unacceptable to contemporary empiricists. Feyerabend summarized his argument as follows:

> Our argument against invariance is simple and clear. It proceeds from the fact that usually some of the principles involved in the determinations of the meanings of the older theories or points of view are inconsistent [qualitatively not quantitatively] with the new, and better, theories. It points out that it is natural to resolve this contradiction by eliminating the troublesome and unsatisfactory older principles and to replace them by principles, or theorems, of the new and better theory. And it concludes by showing that such a procedure will also lead to the

193 Feyerabend cited from Nagel (1961), p. 345 and refers to p. 352.

elimination of the old meanings and thereby to the violation of meaning invariance (1962a, p. 82; rp 1981a, pp. 82-83).

The problem with the principle of meaning invariance, according to Feyerabend, is that it cannot take into account examples of incommensurable concepts:[194]

> Now, it will turn out, in the course of this essay, that any form of meaning invariance is bound to lead to difficulties when the task arises either to give a proper account of the growth of knowledge, and to discoveries contributing to this growth, or to establish correlations between entities which are described with the help of what we shall later call *incommensurable concepts* (1962a, p. 31; rp 1981a, pp. 46-47, italics inserted).

Exactly how *incommensurable concepts* undermines (A) and (B) in support of Feyerabend's argument against them is best illustrated by examining Feyerabend's examples.

The first concept that Feyerabend developed to illustrate incommensurability was impetus as it is used in impetus theory. According to Feyerabend, impetus theory retained the general background of the Aristotelian theory of motion as the actualization of potentiality. In particular, impetus theory retained the idea that "(7) motion is a process arising from the continuous action of a source of motion, or a 'motor', and a 'thing moved'" (1962a, p. 53; rp 1981a, p. 62). However, impetus theory introduced a new idea about that causes that are responsible for the motion of projectiles. According to impetus theory,[195] the motor (for example a hand) transfers an inner 'moving' force (impetus) to a moving body (for example a spear), which is responsible for its continuous motion "which is continually decreased by the resisting air and by the gravity of the projectile" (1962a, p. 53; rp 1981a, p. 62). According to the dynamical characterization[196] of motion of impetus theory, "the motion of a stone thrown would have to be characterized by its inherent impetus, which pushes it along until it is exhausted by the opposing forces of friction and gravity" (1962a, p. 54; rp 1981a, p. 63). The dynamical characterization of motion applied to (7) (Aristotle's theory of motion as the actualization of potenti-

---

194 Compare: "the explanation and reduction of predicates is understandable only if we discuss the conditions according to which theories in a certain domain are replaced by better theories in the same domain. For example, the replacement of such qualitative predicates as pitch, loudness, timbre by quantitative predicates (frequency, amplitude, intensity of Fourier components) can be justified by pointing out that new predicates play an important role within theories which, judged by the standards of explanation, are more powerful, more satisfactory" (1961h, pp. 246ff.).

195 According to (1962a), p. 53; rp (1981a), pp. 62-62, who cites Clagett (1959), p. 425.

196 Feyerabend claimed that there were actually two different competing characterizations of motion at the time. One characterized motion in terms of space traversed. The other characterized motion in terms of the forces that bring motion about in accordance with (7) (see above). See (1962a), pp. 52-53; rp (1981a), pp. 62-62.

ality) results in "(8) The impetus of a body in empty space which is not under the influence of any outer force remains constant" (1962a, p. 54; rp 1981a, p. 63). According to Feyerabend (8) is a "very basic law" of impetus theory that "is not in any quantitative disagreement with anything asserted by Newton's theory" (1962a, p. 54; rp 1981a, p. 63). Thus, particular principles of two theories delineated with incommensurable concepts can be in quantitative agreement.

However, Feyerabend claimed that this concept of impetus is incommensurable with the conceptual apparatus of Newton's Mechanics.[197] Feyerabend explained that by calling a dynamical characterization of the inertial law according to impetus theory[198] *incommensurable* with Newtonian Physics, he meant incommensurable "in the sense that the main concept of the former, viz., the concept of impetus, can neither be defined on the basis of the primitive descriptive terms of the latter, nor related to them via a correct empirical statement" (1962a, p. 74; rp 1981a, p. 76, slightly modified). This is how 'incommensurability' is 'defined' in (1962a).[199]

The reason why the concept of impetus cannot defined on the basis of the primitive descriptive terms of Newtonian Physics, nor related to them via a correct empirical statement, is that, although the dynamical characterization of the inertial law according to impetus theory is in *quantitative* agreement with both experience and with Newton's theory, the "rules of usage" [Nagel's phrase] which are required to explain the inertial law according to impetus theory contain (a) the principle that motion is a process arising from the continuous action of a source of motion, or a 'motor', and (b) the law that constant forces bring about constant velocities. Both of these are *qualitatively* inconsistent with the laws of Newtonian Physics (1962a, p. 74; rp 1981a, p. 76). According to Newtonian Physics, (a) and (b) are simply incorrect. From the perspective of Newtonian theory, any concept of a force which depends on (a) and (b) will have no

197 This example is developed in section 5, "*Second Example: The Problem of Motion*" (1962a, pp. 52ff.; rp 1981a, pp. 62ff.). While this is the second example of a case in which Feyerabend rejected Nagel's claims that there was a successful reduction, the first example of a failed purported reduction, developed in section 4, "*The First Example*" (1962a, pp. 46ff.; rp 1981a, pp. 57ff.), between Galilean science and Newton's Physics, failed for *quantitative* reasons, not because of *qualitative* changes in the nature of the concepts used. In other words, Feyerabend's first example of a failed purported reduction is explicitly not a case involving incommensurable concepts.

198 That is "(8) The impetus of a body in empty space which is not under the influence of any outer force remains constant" (1962a, p. 54; rp 1981a, p. 64).

199 Feyerabend referred to this as a 'definition' of 'incommensurability' (1962a, p. 90; rp 1981a, p. 90). Compare, "Two theories will be called incommensurable when the meanings of their main descriptive terms depend on mutually inconsistent principles" (1965b, p. 227, fn. 19). And compare, (1975a), p. 225, where Feyerabend claimed that incommensurability cannot be defined, and is not susceptible to explication, but can only be shown.

extension. There is no force like impetus according to Newton's theory. The concept of impetus, from the Newtonian conceptual perspective, does not express any actually existing situations (1962a, p. 74; rp 1981a, p. 76).

It follows that, although both theories produce the same numerical predictions concerning the movement of a body, the concepts of the two theories are mutually exclusive given the *qualitative* differences in the theoretical principles that govern their use. That is why they are incommensurable, and this is what Feyerabend meant by calling them 'incommensurable': *They are constructed with mutually exclusive sets of concepts.* Either there are forces like impetus in accordance with (a) and (b), or there are forces in accordance with the principles of Newton's theory.

The main consequence of the realization that scientific advance involves incommensurable concepts is that older concepts are not refined by newer theories, they are *replaced* by newer concepts:[200]

> as soon as this demonstration has been carried out [that they are incommensurable], in the very same moment, the idiom of T′ must be given up and must be *replaced* by the idiom of T. . . Of course, one need not go through the laborious and very uninteresting task of analyzing the context of which T′ is part. All that is needed is the adoption of the terminology and the 'grammar' of the most detailed and most successful theories throughout the domain of its application. This automatically takes care of whatever *incommensurabilities* may arise (1962a, p. 75, italics inserted; rp 1981a, p. 77 slightly modified).

The general interpretation of scientific advance that results is not a refinement and absorption of the concepts of the older theory into the newer theory as it is in Nagel's interpretation of scientific advance (or in some scientific realist interpretations of scientific advance, according to which scientific advance is a series of ever better approximations of the truth). The main consequence of Feyerabend's insight that consecutive scientific theories make use of incommensurable concepts is that the concepts of the older theory are no longer needed, and should not be retained. They have been shown to be inadequate, and thus they should be abandoned. The resulting interpretation of scientific development is a succession of theories that have been *replaced* by their successors. The new theories supply new "rules of usage" for the new concepts:

> What happens *here* [theoretical transitions involving comprehensive theories such as the Aristotelian theory of motion, the impetus theory, Newton's celestial mechanics, Maxwell's electrodynamics, the theory of relativity, and the quantum theory] when transition is made from a theory T′ to a wider theory T . . . is something much more radical than incorporation of *unchanged* theory T′ (unchanged,

200 As we have seen in chapter 3, Feyerabend very often made his account of scientific advance as one of replacement explicitly clear, see footnote 119 above.

> that is, with respect to the meanings of its main descriptive terms as well as to the meanings of the terms of its observation language) into the context of T. What does happen is, rather, a *complete replacement* of the ontology . . . of T′ by the ontology . . . of T . . . The replacement affects not only the theoretical terms of T′ but also at least some of the observational terms which occurred in the test statements (1962a, pp. 28-29, italics in original; rp 1981a, pp. 44-45, cf. Kuhn [1962] 1970a, p. 102).

Feyerabend began to generalize his claim on the basis of just this single example:

> We may generalize this result in the following fashion: consider two theories T′ and T, which are both empirically adequate inside D′ [the domain in which T′ makes correct predictions, see (Feyerabend 1962a), p. 43, (Feyerabend 1981a), p. 55], but which differ widely outside D′. In this case the demand may arise to explain T′ on the basis of T, i.e., to derive T′ from T and suitable initial conditions (for D′). Assuming T and T′ to be in quantitative agreement inside D′, such derivations will still be impossible if T′ is part of a theoretical context whose "rules of usage" [Nagel's phrase] involve laws inconsistent [qualitatively] with T. It is my contention that the conditions just enumerated apply to many pairs of theories which have been used as instances of explanation and reduction. *Many (if not all) of such pairs on closer inspection turn out to consist of elements which are incommensurable and therefore incapable of mutual reduction and explanation* (1962a, p. 75; rp 1981a, p. 77, italics inserted).

The *cause* of incommensurability should by now be clear. Incommensurability results between two different conceptual perspectives, because the concepts that make those perspectives possible are based on incompatible theoretical principles. According to Feyerabend, theoretical principles determine the concepts used to state the theories, and the meaning of the concepts is determined by the theories to which they belong. Thus, meaning change is the cause of incommensurability in (1962a). There can be no doubt that this semantic principle is the basis of Feyerabend's conception of incommensurable concepts:

> To bring about the situation described above (sets of mutually incommensurable concepts), it is sufficient that they [the principles of the context of which T′ is a part] govern the use of the main terms of T′ (1962a, p. 75; rp 1981a, p. 77).[201]

On the following page, Feyerabend suggested a second source of examples of incommensurability: the relation between the concepts of everyday language and scientific theories,[202] but he delayed his discussion of this

201 Note, here, Feyerabend used the term 'use', where to be precise and consistent, he should have used the term 'interpretation' as he distinguished between the two to make his point, and while the theoretical principles determine both the use and the meaning, is the interpretation that causes the incommensurability, not the use.

202 Interestingly, this example has been largely ignored in the secondary literature. For example, see and note the conspicuous absences in the following two sources, which are supposed to be comprehensive analyses of Feyerabend's notion of incommensurability:

example. In the meantime, he developed several other specific examples of incommensurable concepts drawn from two more episodes of theory change in the course of scientific advance. We will follow suit, returning later to this second source of examples of incommensurable concepts: the incommensurability of everyday concepts and the conceptual apparatus of scientific theories.

The second and third examples of incommensurable concepts in scientific theories in (1962a) are drawn from the relation between Phenomenological Thermodynamics and Kinetic Theory and concern the concepts temperature and entropy as they are used in the two theories respectively (1962a, pp. 76ff.; rp 1981a, pp. 78ff.). According to Feyerabend:

> The relation between the thermodynamic concept of temperature and what can be defined in the kinetic theory, therefore, can be seen to conform to the pattern described at the beginning of the present section [the generalized account of incommensurability given above]: we are dealing again with two incommensurable concepts. The same applies to the relation between the purely thermodynamic entropy and its statistical counterpart (1962a, p. 78; rp 1981a p. 79).

The (fourth) example of incommensurable concepts that Feyerabend set out is the classical and relativistic concepts of mass (Feyerabend also mentioned that the classical and relativistic concepts of length and time duration are incommensurable, in the same way as mass, but does not discuss or develop the suggestion further than that) (1962a, pp. 80ff.; rp 1981a, p. 81ff.). According to Feyerabend, in pre-relativistic, Classical Physics, the concept of mass is "*absolute* in the sense that the mass of a system was not influenced (except perhaps causally) by its motion in the coordinate system chosen" (1962a, p. 80; rp 1981a, p. 81). With the advent of relativity theory, however, mass becomes a relational concept: Its specification is incomplete without indication of the coordinate system to which the spatio-temporal descriptions are to be referred.[203] In short, if the concepts of mass in Classical and Relativistic Physics are determined by the theories to which they belong, then they are two different concepts.

Feyerabend argued that although the numerical results of measurements made on classical mass and on relativistic mass will be the same *within the domain of classical physics*, it does not follow that what is measured is the same in both cases:

---

chapters 2 and 3 of Couvalis (1989), and chapter 6 of Preston (1997). Also see Shand (1986).

203 Feyerabend was not the only philosopher of science to emphasize the conceptual differences between these two theories. For example, "the mixture of the theories of Newton and Einstein is intellectually repellent, since the two theories are based on such different fundamental concepts" (Synge, 1964, pp. 296ff. cited in 1970a, p. 84, fn. 49).

> what is measured in the classical case is an *intrinsic* property of the system under consideration; what is measured in the case of relativity is the *relation* between the system and certain characteristic of D′ (1962a, p. 80; rp 1981a, p. 81).

According to Feyerabend, not only are different properties measured:

> It is also impossible to define the exact classical concept in relativistic terms or to relate them with the help of empirical generalizations (1962a, pp. 80-81; rp 1981a, pp. 81-82).

The reason that it is impossible to define the exact classical concept in relativistic terms or to relate the classical and relativistic concepts of mass with the help of an empirical generalization is that any such procedure would imply that the velocity of light is infinitely large, which is false (1962a, p. 81; rp 1981a, p. 82). Feyerabend then generalized on the basis of these three examples of theory change:[204]

> Many more examples can be added to those discussed in the present paper (viz., the impetus theory, phenomenological thermodynamics, and the classical conception of mass). All of these examples show that the postulate of meaning invariance is incompatible with actual scientific practice. That is, it has been shown that in most cases it is impossible to relate successive scientific theories in such a manner that the key terms they provide for the description of a domain D′, where they overlap and are empirically adequate, either possess the same meanings or can at least be connected by empirical generalizations (1962a, p. 81; rp 1981a, p. 82, slightly altered).

To summarize, in (1962a), Feyerabend explicitly developed four examples of sets of incommensurable concepts drawn from three episodes of theory change in the course of scientific advance: the dynamical characterization of impetus in impetus theory of motion and the concepts of force in the conceptual apparatus of Newtonian Mechanics, the concepts of temperature and entropy in the transition from Phenomenological Thermodynamics to Kinetic Theory, and the concepts of classical mass and relativistic mass in the transition from Newtonian to Einsteinian Physics. The reason why these sets of concepts are incommensurable is that they belong to mutually exclusive theoretical perspectives. In subsequent publications, Feyerabend gave one more example of what he takes to be incommensurable theories: geometrical optics and wave optics. For example, Feyerabend explained:

> Two theories will be called incommensurable when the meanings of their main descriptive terms depend on mutually inconsistent principles. Examples are the impetus theory and Newton's celestial mechanics, Newton's celestial mechanics and special relativity, and geometrical optics and wave optics (1965b, p. 227, fn. 19).

204 Or rather 'over-generalized' in that from these three examples, he claims to have shown that his point holds "in most cases" of successive scientific theories.

The main consequence of the insight that successive scientific theories contain mutually exclusive concepts is a revision in the interpretation of scientific advance. Older theories, and the concepts used to state them, *are not corrected and absorbed, and thereby legitimized. They are rejected and replaced, having been falsified.* In this way, the conceptual conservative (or retentive) accounts of scientific advance and explanation of the logical empiricists is undermined by the existence of incommensurable concepts. Feyerabend's own account of scientific advance is one in which newer theories; based on new kinds of incommensurable concepts, replace the older unsatisfactory theories. Scientific advance, on Feyerabend's view, involves a succession of conceptual replacements.

Earlier, we postponed (following Feyerabend's lead) a discussion of a second class of examples of incommensurable concepts that he had mentioned (1962a, p. 76; rp 1981a, p. 78), but only began to develop several pages later (1962a, pp. 82ff.; rp 1981a, pp. 83ff.): The relation of the concepts of everyday language and the concepts of scientific theories. As this source of incommensurable concepts has been very often overlooked,[205] and is less straightforward, it is worth taking some time to establish that Feyerabend did in fact consider the relationship between everyday concepts and the conceptual apparatus of scientific theories to be a source of incommensurability, and to determine what was at stake for Feyerabend when he made this claim. Directly after generalizing his notion of incommensurability, he applied the notion to the relationship between everyday language and scientific theories.

> What has just been said *applies most emphatically* to the relation between (theories formulated in) some commonly understood language and more abstract theories. That is, I assert that languages such as the "everyday language", this notorious abstraction of contemporary linguistic philosophy, frequently contain (not explicitly formulated, that is, but implicit in the way in which its terms are used) principles which are inconsistent with newly introduced theories, and that they must therefore be either abandoned and replaced by the language of the new and better theories even in the most common situations, or they must be completely separated form these theories (which would lead to a situation where it is possible to believe in various kinds of "truth") (1962a, p. 75ff.; rp 1981a, p. 77, italics inserted).

205 In the vast secondary literature discussing Feyerabend's views on incommensurability, I have only seen it discussed or even mentioned just twice, in Smart (1965) and Putnam (1965); rp Putnam (1975a), pp. 117-131. While Feyerabend himself responded to both Putnam and Smart in (1965c); rp (1981a), pp. 104-131, Putnam's main criticism of Feyerabend's views hinges on confusions about the meaning of the term 'meaning', which is somewhat ironic. See Putnam (1975a), pp. 122ff.

And anticipating the objection that the concepts of everyday languages are too vague, and do not contain theoretical assumptions or imply some particular ontology, Feyerabend insisted:

> it is far from correct to assume that the everyday languages are so widely conceived, so tolerant, indefinite, and vague that they will be compatible with any scientific theory, that science can at most fill in details and that a scientific theory will never run against the principles implicitly contained in them. *The very opposite is the case.* As will be shown later, even everyday languages, like languages of highly theoretical systems, have been introduced in order to give expression to some theory or point of view, and they therefore contain a well-developed and sometimes very abstract ontology (1962a, p. 75ff.; rp 1981a, p. 77, italics in original. Also see 1962a, p. 84; rp 1981a, p. 85).

Feyerabend began his discussion of some examples of the relation between the concepts of everyday language and the conceptual apparatus of contemporary scientific theories in the context of responding to a set of possible objections to the argument against meaning invariance. The *first objection* is based on a transition to instrumentalism. According to instrumentalism (as Feyerabend sets it out), new theories are not interpreted as a series of statements, but rather as "a predictive machine whose elements are tools rather than statements and therefore cannot be incompatible with any principle already in existence" (1962a, p. 82; rp 1981a, p. 83). It is the need to defend himself from this objection that led Feyerabend to insist that the concepts of theories are only incommensurable under a certain interpretation, i.e. a realist interpretation. A transition to an instrumentalist interpretation of theories "admittedly cuts the ground from beneath our argument and makes it inapplicable" (1962a, p. 82; rp 1981a, p. 83). The reason that an instrumentalist interpretation of scientific theories undercuts Feyerabend's views is simple and straightforward. If newer scientific theories are interpreted strictly as tools for making predictions (*and not as hypothetical factual statements*), then there is no reason why older concepts need to be replaced. The older concepts can be used to make statements, while the newer concepts are restricted to the role of helping to make predictions (*and not hypothetical factual statements*); and consequently, no conflict between the two sets of concepts can arise.

Feyerabend rebutted this objection on two related grounds. First of all:

> It has never been explained why a new and satisfactory theory should be interpreted as an instrument, where as the principles behind the established usage, which can easily be shown to be empirically inadequate, are not so interpreted. After all, the only advantage of the latter is that they are *familiar*—an advantage which is a psychological and historical accident and which should therefore not have any influence upon questions of interpretation and reality (1962a, p. 82; rp 1981a, p. 83).

Thus, Feyerabend saw no reason to apply an instrumentalist interpretation to contemporary scientific theories, while maintaining a normatively justifies realist interpretation of the concepts that have been superseded in the course of scientific advance.[206] Second, Feyerabend claimed that the attempt to apply an instrumentalist interpretation consistently, i.e., to everyday language as well as scientific theories, and thereby avoid this rebuttal, fails because it amounts to the acceptance of a sense-data account of knowledge, which, as he has argued elsewhere, is impossible.[207]

The second objection Feyerabend considered to his argument against meaning invariance received scathing criticism: "Indeed, I am hesitant to apply the word 'argument' to these expressions of confused thinking, their wide acceptance and asserted self-evidence notwithstanding" (1962a, pp. 82-83; rp 1981a, p. 83). The objection amounts to the suggestion that the layman uses and understands everyday concepts without knowing anything about contemporary scientific theories, and therefore, Feyerabend's suggestion that our concepts are determined by our best theories is obviously incorrect. It is precisely this kind of conservative attitude in linguistic philosophy to which Feyerabend repeatedly objected.

This brings us to the first of two specific examples of everyday concepts that Feyerabend used to illustrate the incommensurability between the concepts used in everyday language and the conceptual apparatus of our best contemporary scientific theories. The example of incommensurable concepts Feyerabend used to illustrate this point is temperature as, in a layman's thought that 'The milk has a temperature of about 10° Celsius', and the contemporary scientific concept of temperature. The objection that Feyerabend considered turns on the point that 'temperature' in a layman's thought about milk has nothing to do with a thought about the mean kinetic energy of molecular motion. This is exactly the kind of objection to Feyerabend that Preston launches much later: "*The use* of a these terms makes it amply clear that their meaning is not fixed by anything *rightly* called theory" (Preston 1997, p. 29, italics inserted. Also see pp. 146ff.). However, Preston does not even mention, let alone respond to, Feyerabend's rebuttal of such objections. [208] In any case, Feyerabend's response to this kind of objection was that, while it is correct that unin-

206 Here Feyerabend is assuming the normatively based principle of realism, not scientific realism.

207 In his (1962a), Feyerabend refer the reader to "On the Interpretation of Microphysical Theories" a title that was never published. In the reprint (1981a), Feyerabend referred his readers to (1960f), for an English version, see also (1999d). The same arguments are also made in (1958a); rp (1981a), pp. 17-36 and (1964d); rp (1981a), pp. 203-206.

208 For an influential critique of Feyerabend's collapse of the theory/language distinction that recommends collapsing the scheme/reality dualism instead, see Davidson (1984). For a discussion supporting Davidson against Feyerabend, see also Preyer (1998).

structed layman do indeed use terms without recourse to scientific theories, this fact is entirely irrelevant to his argument against meaning invariance:

> It is not at all denied by our argument that the "uninstructed layman" may poses a concept of temperature that is very different from the one connected with the molecular theory (after all, some "uninstructed layman," intelligent clergymen included, still believe in ghosts and in the devil). What is denied is that anybody can consistently continue using this more primitive concept and at the same time believe in the molecular theory (1962a, pp. 83; rp 1981a, p. 84).

Moreover, with typical flair, Feyerabend notes:

> By the way, who is this uninstructed laymen? From the purpose for which he is being employed in many arguments, it would seem to emerge that he is not supposed to know much science, or much politics, or much religion, or much of anything. This means that in these time of mass communication and mass education he must be very careful not to read the wrong parts of his newspaper, he must be careful not to leave his television set on for too long a time, and he must also not allow himself to converse too much with his friends, his children, etc. That is, he must be either a savage or an idiot. I really wonder what are the motives which lead to a philosophy where the most interesting language is the language of savages or of idiots (1962a, p. 83, fn. 99; rp 1981a, p. 84, fn. 80).

Whether or not layman actually use everyday concepts in some particular way, is not at issue, and has no bearing on Feyerabend's *normative* claim that one should replace one's everyday concepts with better informed scientific ones. But Feyerabend does not even need to insist on the normative claim that everyday concepts should be entirely replaced. This is because:

> Again, this does not mean that a person may not, on different occasions, use concepts which belong to different incommensurable frameworks. The only thing that is forbidden for him is the use of both kinds of concepts in the same *argument*; for example, he may not use one kind of concept in his observation language and the other kind in his theoretical language. Any such combination—and this is the gist of our considerations in the present section—would introduce principles which are mutually inconsistent and thereby destroy the argument in which it is supposed to occur (1962a, pp. 83-84; rp 1981a, p. 84).

So the incommensurable nature of the conceptual apparatus of everyday language and our best scientific theories does not even necessitate, in Feyerabend's view, the replacement of the one by the other – although this is often Feyerabend's recommendation. All that is required is that the concepts are not joined in a single argument. The reason behind the need strictly to separate the use of the two incommensurable concepts when making one argument is simply that in order for an argument to be valid, the meanings of the terms must be the same. One could support this point by arguing that it is only the relevant community as a collective body that has the resources to determine the meaning of a concept (Putnam

1975b, pp. 227-229). In other words, one could undercut such objections with Putnam's 'socio-linguistic hypothesis', according to which there is a '*division of linguistic labor*'. The argument would insist that the layman does not need to know much about a concept in order to use it adequately in everyday situations. Moreover, the layman need not understand the theory that determines the meaning of the concept, nor must the extension of his term be fixed. The crucial factor is that there are experts within the community who do understand the theories that determine the content of the concept, and which fix its extension, when they are used in science.

Although the objection that layman use everyday concepts without knowledge of scientific theory is arguably irrelevant to Feyerabend's views, Feyerabend spent some time trying to expose the detrimental consequences that such a conservative attitude can have, and he launched an extended argument against the idea that a successful language is to be regarded as adequate and irreplaceable on the grounds that it is successful:

> The objection reflects an attitude that must appear quite incredible to anybody who possesses even the slightest acquaintance with the history of knowledge . . . It insinuates that the existence of an idiom allows us to infer the truth of the principles which underlie this idiom. Or, to be more specific, it insinuates that *what is being used is*, on that account alone, already exhibited as adequate, useful, and perhaps irreplaceable. After all, the reference to the layman's understanding of the word 'temperature' is not made without purpose. It is made with the purpose of preserving the common meaning of this word since, it is alleged, this common meaning can be understood and is not in need of replacement (1962a, p. 84; rp 1981a, pp. 84-85).

Feyerabend briefly interrupted his criticism of this kind of conceptual conservativism to give his second example of incommensurable concepts from everyday language.

The second example from everyday language that Feyerabend used to illustrate incommensurability is chosen in order to illustrate his claim that everyday languages are not vague, but contain implicit principles that conflict with scientific theory, and to show that retaining outdated concepts will lead to the conservation of false ideas (1962a, p. 84; rp 1981a, p. 85, also see 1961b, p. 83). The example is the word pair 'up-down'. Feyerabend claims that Thales and Xenophanes used these terms in an absolute fashion, i.e., without reference to a specified center such as the center of the Earth. Their usage contains two implicit assumptions: That every material object is under the influence of a force, and that this forces acts in a privileged direction in space. The second assumption, about the anisotropic character of space, is Feyerabend's example of a cosmological assumption implicit in a common idiom. Feyerabend claimed that it is to this privileged direction to which the word pair 'up-down' refers, and that this example refutes the thesis that "everyday languages" are fairly free from hypothetical elements, by showing that even the most harmless part of a common idiom may implicitly contain far-reaching theoretical principles.[209] The transition to the Newtonian theory involves replacing the anisotropic notion of space with an isotropic one:

> Any person accepting Newton's physics and the conception of space it contains must, therefore, give a new meaning even to such familiar pair of terms as is the pair 'up-down', and he must now interpret it as a relation between the direction of motion and a center that has been fixed in advance. And as Newton's theory is preferable, on empirical grounds, to the older and "absolute" cosmology, it follows that the relational usage of the pair 'up-down' will be preferable too. Conversely, the attempt to retain the old usage amounts to retaining the old cosmology, and this despite the discoveries which have shown it to be obsolete (1962a, pp. 85-86; rp 1981a, p. 86).

After presenting this example, Feyerabend returned to his attack on the idea that just because some concepts have been successful in the past, they should be retained even in the face of scientific advance. Briefly, as we have seen, he argued against Bohr's suggestion that the concept of Classical Physics are irreplaceable on the grounds that there have been transitions to new, incommensurable conceptual frameworks in the past. There is nothing which rules out the possibility of that happening in this case. In short, a new, incommensurable conceptual framework could be invented that could replace the concepts of Classical Physics; thereby undermining Bohr's principle of complementarity.

In short, in his (1962a), Feyerabend developed two examples of incommensurable concepts from the relation of the concepts of everyday language and the concepts of scientific theories. They were the concepts temperature and the pair up-down. These examples have been largely ignored, as examples of incommensurability, in the secondary literature on the subject. Feyerabend used these examples to expose and criticize a form of conceptual conservativism and to argue for his recommendation that older concepts should be replaced by newer, improved ones.

The third class of examples of incommensurable concepts that Feyerabend discussed in (1962a) is the concepts used in some philosophical problems:[210]

> It will also turn out that these are exactly the difficulties [difficulties for orthodox accounts of reduction and explanation due to the existence of *incommensurable con-*

209 Feyerabend suggested seeing Frank (1950) for his position on this issue.

210 Kuhn also later hinted that there might be incommensurability in philosophical discussions. See Kuhn (1970b) and Kuhn (1970c). The notion of incommensurability in philosophy has been developed as "meta-incommensurability" in Hoyningen-Huene, Oberheim et al. (1996) and Oberheim and Hoyningen-Huene (1997). For a brief discussion of Kuhn's hints about incommensurability in philosophical discussions, see Holcomb (1987).

> *cepts*] we encounter when trying to solve such age old problems as *the mind-body problem*, *the problem of the reality of the external world*, and *the problem of other minds* (1962a, p. 31; rp 1981a, p. 47, italics inserted).

According to Feyerabend, the reason why such age-old philosophical problems have never been solved is because philosophers accept (if only implicitly) the principle of meaning invariance:

> it will usually turn out that a solution of these problems is deemed satisfactory only if it leaves unchanged the meanings of certain key terms and that it is exactly *this* condition, i.e. the condition of meaning invariance, which makes them insoluble (1962a, p. 31; rp 1981a, p. 47, emphasis removed).

Feyerabend argued that the difficulty with the mind-body problem is that the condition of meaning invariance is regarded as necessary condition to its solution. The meanings of some of the key terms such as 'mind' and 'matter', or 'pain' and 'sensation', are:

> *incommensurable in exactly the sense in which this term has been defined at the beginning of the present section* (1962a, p. 90; rp 1981a, p. 90, slightly altered, italics inserted).

That is, they are incommensurable in the sense that the main concept of the former can neither be defined on the basis of the primitive descriptive terms of the latter, nor related to them via a correct empirical statement. It follows that:

> it is of course completely impossible either to reduce them to each other, or to relate them together with the help of an empirical hypothesis, *or to find entities which belong to the extension of both kinds of terms* (1962a, p. 90; rp 1981a, p. 90, italics inserted).

Here, especially the last italicized clause, and elsewhere, e.g. "if a new theory entail that all the concepts of the preceding theory have zero extension or *if it introduces rules . . . which change the system of classes itself*" (1965a, p. 268; rp 1981a, p. 98, italics inserted), Feyerabend can be interpreted as anticipating Kuhn's later notion of taxonomic incommensurability.[211]

The problem cannot be solved in its traditional form is because of a form of conceptual conservativism. A solution "would require relating what is incommensurable, without allowing for a modification of meanings which would eliminate this incommensurability" (1962a, p. 90; rp 1981a, p. 90). According to Feyerabend, we need to be prepared to admit that the "mental connotations of mental terms" may be erroneous and be need of "replacement by physical connotations," according to which mental states such as pains, awareness, and thoughts are complex physical states (1962a, p. 91; rp 1981a, p. 90). This is exactly what Feyerabend rec-

211 See e.g., Kuhn (1983), p. 683, Kuhn (1991), pp. 4-5, Kuhn (1993), pp. 318-319 and pp. 325-326. For discussion, see e.g. Carrier (2001), pp. 69ff., Hoyningen-Huene (1993), pp. 217-218, Irzik and Grünberg (1995), p. 299, Irzik and Grünberg (1998), pp. 211ff. Carrier also comes to this conclusion; see Carrier (2001), fn. 3, p. 88.

ommended with his "Eliminative materialism" (1963b, 1963d, and 1965b). It is also his main example of the incommensurability involved in a philosophical problems.

Feyerabend believed that the mind-body problem could finally be resolved in this way:

> This transition from a point of view which demands that certain 'basic' terms retain their meaning, come what may, to a more liberal point of view which allows for changes anywhere in the system employed is bound to influence profoundly our attitude with respect to many philosophical problems and will also facilitate their solution: Let me take the mind-body problem as an example (1962a, p. 90; rp 1981a, p. 89. Also see 1967a).

It could be resolved by abandoning the conceptually conservative requirement that the meanings of the terms remain unchanged. Feyerabend speculated that once empiricism has become freed of the principle of meaning invariance, these age-old philosophical problems could finally be resolved. They could finally be overcome through the introduction of new concepts, which do not share the older meanings:

> Taking all this into account, we may hope that once contemporary empiricism has been freed from the elements which it still shares with its more dogmatic opponents, it will be able to make swift progress in the solution of the above problems [the mind-body problem, the problem of the reality of the external world, and the problem of other minds] (1962a, p. 31; rp 1981a, p. 47).

In summary, Feyerabend illustrated his notion of incommensurable concepts with three types of examples: Sets of mutually exclusive scientific concepts, concepts form everyday language in contrast to scientific concepts, and concepts used in the construction of some age-old philosophical problems. By calling two concepts incommensurable, he meant that one couldn't be defined in terms of the other, nor related by a correct empirical statement. From one conceptual perspective, the concepts of the other perspective have no extension, and that by using the concepts of one theoretical perspective; the content of the other cannot be properly expressed. It follows that the two concepts are *mutually exclusive* for the purpose of attempting to describe the world. Concepts are incommensurable, because their contents are determined by the theoretical principles that govern their use, and these principles may be *qualitatively* inconsistent, even when two theories, delineated with incommensurable concepts, are in quantitative agreement within the domain of the older theory. According to Feyerabend, the rejection of the existence of all three types of examples of incommensurable concepts is undesirable on the grounds that it supports a kind of conceptual conservativism that hinders conceptual advance.

Feyerabend argued against the logical empiricist accounts of explanation and reduction with a two-pronged argumentative strategy aimed at establishing that these orthodox accounts neither accurately *describe* actual scientific practice, nor are *desirable* as methodological prescriptions. His argument is *ad hominem* in the sense that he assumes empiricist principles such as the principles of empiricism, testability, and realism (normatively), as well as a semantic principle accepted by Nagel, namely that the content of a concept is determined by the "rules of usage" as determined by the theories of which it is apart, and attempted to show that the logical empiricists accounts of reduction and explanation would, if enforced, led to unacceptable consequence *according to their own methodological views.* They would lead to the conservation of untested assumptions through theoretical transitions; and eventually to a situation in which basic assumptions can no longer be tested at all.

Feyerabend used the notion of incommensurability to characterize the relation between two successive 'fundamental', 'universal', or 'comprehensive' scientific theories. By this he meant non-instantial theories that talk about everything that there is.[212] This notion of a fundamental, universal or comprehensive theory may have been drawn from Einstein. One sentence after Einstein used the term 'incommensurable' in a discussion of the difficulties of using empirical evidence to test theories, Einstein wrote:

> Of the 'realm' of theories I need not speak here, in as much as we are confining ourselves to such theories whose object is the *totality* of all physical appearances (Einstein 1949a, p. 23 italics in original).

It is very likely that Feyerabend was well-acquainted with this remark (and Einstein's use of the term 'incommensurable', as in his 1950s work on the Copenhagen Interpretation of Quantum Mechanics, he often cited from the book in which they appear (Schilpp 1949). He even cited from this book in 'Explanation, Reduction and Empiricism' (1962a). Kuhn's notion of incommensurability, by contrast, has a much wider scope of application than Feyerabend's. Kuhn sees incommensurability even in relatively minor conceptual transitions. By restricting his notion to transitions involving such comprehensive theories, Feyerabend intended to stipulate those that have ontological implications; i.e., those that have implications about the very nature of reality. Feyerabend's claim is that theories whose object is

---

212 See (1954c), (1958b), pp. 78-79, and pp. 90-91, (1961i), pp. 7ff.; rp (1999a), pp. 52ff., (1962a), pp. 62-68, 79-82, (1967a), p. 40, and (1970e), p. 233 and p. 246, (1975a) pp. 224-225, 271, 276-277, 279, (1987), p. 272. See Hoyningen-Huene (2005) for a discussion of Feyerabend's restriction of the notion of incommensurability to such 'universal' or 'comprehensive' theories, and a comparison to Kuhn's much wider application of the term 'incommensurability'.

the *totality* of all physical appearances, such as Newtonian Mechanics or Relativity Theory, deeply affect our beliefs about the nature of reality.

By calling two such theories incommensurable, Feyerabend meant that they were conceptually incompatible: The main concepts of one could "neither be defined on the basis of the primitive descriptive terms of the other, nor related to them via a correct empirical statement" (1962a, pp. 74, 90). This is because they give incompatible accounts of the nature of reality. Because the meanings of their terms are provided by the theories to which they belong, when there is theoretical change, there are meaning changes (1962a, pp. 68, 94). Here, Feyerabend can, and did, simply assume the semantic principle that the meanings of their terms are provided by the theories to which they belong, as it was part of his immanent criticism of Nagel's views. He can and did assume it because he target of his critique also accepts it (Nagel [1961] 1966, pp. 352, 357). These changes in meaning not only affect our theoretical and observational terms. They even affect our ontology (i.e. our conception of the nature of reality). When this occurs, there is incommensurability; or as Feyerabend later put it, a "theory is incommensurable with another if its ontological consequences are incompatible with the ontological consequences of the latter" (1981a, p. xi).

Let us consider the *function* of Feyerabend's notion of incommensurability in (1962a). Feyerabend used the notion to attempt to expose a *dogmatic element* that contemporary empiricists share with school philosophies such as Platonism and Cartesianism (from whom they had tried to distance themselves by insisting on an empirical foundation for scientific knowledge).[213] The dogmatic element is due to the assumption that the meanings of empirical terms remain stable through theoretical transitions; or what Feyerabend called "the principle of meaning invariance" (1962a, p. 30). Feyerabend argued that this principle is inconsistent with the existence of incommensurable concepts.

Feyerabend drew two main consequences from the insight that some pairs of successive scientific theories are incommensurable. First, successive scientific theories that are incommensurable have *no logical relations*:

> The order introduced into our experiences by Newton's theory is retained and improved by relativity. This means that the concepts of relativity theory are sufficiently rich to allow us to state all the facts which were stated before with the help of Newtonian physics. Yet these two sets of categories are completely different *and bear no logical relation to each other* (1962a, pp. 88-89).

213 For a discussion, see Hoyningen-Huene (2000b).

Because their concepts have different meanings, they cannot be brought directly into formal relations.[214] This is the basis of Feyerabend's argument against formal accounts of reduction and explanation. Second, a revision in the logical empiricist conception of scientific advance is required. Older theories, and the concepts used to state them, are not *corrected* and *absorbed*, and thereby *legitimized*. Rather, they are *rejected* and replaced, having been *falsified*. In this way, the logical empiricists' *conceptually conservative* (or retentive) accounts of reduction and explanation are undermined by the existence of the development of incommensurable concepts in the course of scientific advance.

The last issue we will address in this section concerns the main influences on Feyerabend's 1962 exposition of incommensurability. As we shall see, Feyerabend's development of the notion of incommensurability was heavily influenced by a number of individuals who were discussing a wide range of different topics. As we have seen chapter 3, in 1962, Feyerabend argued that formal accounts of reduction and explanation are impossible for general theories for *two* reasons. The first reason that formal accounts fail is because *some* pairs of successive scientific theories that are supposed to be formally reducible are actually *logically inconsistent*. (This is not a case of incommensurability, which is the second "qualitative reason" for the failure of formal accounts just discussed.)[215] Feyerabend illustrated the *first* reason by challenging one of Nagel's favorite examples of a purported formal reduction: The reduction of Galilean to Newtonian Physics. A basic assumption of Galilean Physics is that the acceleration of a free-falling body is constant over any finite interval. According to Newtonian Physics, the acceleration of a free-falling body is *never constant*. It always increases during its fall, due to the fact that the body approaches the center of attraction. If the height of the fall is negligible compared to the radius of the earth, then the effect will also be negligible. However, if the height is significant as compared to the radius of the earth, then this effect will be considerable, and Galileo's theory will produce false results. Galilean Physics can indeed be logically derived from Newton's theory, *but only by introducing a false assumption* (that the height of the fall divided by the radius of the earth is zero). According to Feyerabend, "[i]t is therefore impossible, *for quantitative reasons*, to establish a deductive relationship between these two theories" (1962a, p. 47, italics in original).

---

214 See similarly Feyerabend (1958b), p. 83 and Feyerabend (1961c), p. 388.

215 As we have seen in chapter 3, Feyerabend clearly distinguished "quantitative reasons" for the failure of formal accounts (logical inconsistency), discussed in sections 3 and 4, from the "qualitative reason" for the failure of formal accounts (i.e. incommensurability), discussed in sections 5 through 7. (Sections 1 and 2 introduce formal accounts, and the last section, 8, is the conclusion).

Let us briefly focus on the historical origins of this argument. This is instructive for two reasons. First, it will help to clarify a misunderstanding: In his analysis of incommensurability, Preston conflates Feyerabend's first *quantitative* reason for the failure of formal accounts (logical inconsistency) with his second *qualitative* reason (incommensurability), and subsequently unjustly criticizes Feyerabend for being inconsistent about inconsistency and incommensurability (Preston 1997, pp. 105-106). The mistaken view that between the 1960s and the 1970s Feyerabend *changed* his notion of incommensurability from implying logical contradiction to denying the possibility of logical contradiction is the result of conflating Feyerabend's two distinct reasons for the failure of formal accounts (quantitative and qualitative). Second, the historical origins of this argument will help illuminate the historical origins of its compliment, the second qualitative reason for the failure of formal accounts (i.e. incommensurability).

The point that some pairs of successive scientific theories are logically inconsistent had been made by Popper (1957). Moreover, Popper had used exactly the same example as Feyerabend to make the point that, "logic, whether inductive or deductive, cannot possibly make the step from [Galileo's theory] to Newton's" ([1957] 1972, pp. 199-202). Popper also illustrated this point with another example, arguing "Kepler's third law contradicts Newton's theory in precisely the same way as Galileo's" ([1957] 1972, p. 200). In order to establish deductive relations between Kepler's third law (the planet's orbits are perfect ovals) and Newton's theory, one has to make either one of two false assumptions: Either that the mass of the sun is equal to the mass of the planet, or that the mass of the planet is zero. Otherwise, the two theories make quantitatively different claims about the orbits of planets around the sun.[216] As we have seen in chapter 3, Feyerabend acknowledged Popper as the source of this argument, and as the source of these examples as the "starting point" of his investigations in the original 1962 version of his paper.

Popper had acknowledged Duhem as the source of this point and the example of the relation between Kepler's laws and Newton's theory ([1957] 1972, p. 200, fn. 10).[217] However, the point and the example are even older than that. They had been used against Hegel's claim that "[t]he Newtonian formula may be *derived from* the Keplerian law" ([1827] 1989, p. 250, Part C, § 269, italics inserted) by Whewell in 1860.[218] Feyerabend (1962a), following Popper (1957), following Duhem ([1906] 1954), had repeated

216 In 1954, Feyerabend had used exactly this example to make the same point (1954c, pp. 470-471).

217 See Duhem ([1906] 1954), pp. 190-195.

218 See Whewell (1860), appendix H. To my knowledge, it is unclear whether Duhem, Popper or Feyerabend were aware of Whewell's argument.

Whewell's point. Thus, Feyerabend's innovation concerning the first reason for the failure of formal accounts was merely to apply Whewell's point to Nagel instead of to Hegel. The fact that point had been made by Duhem will become especially relevant later in this chapter.

As we have seen, Fleck and Polanyi had used the term 'incommensurable' prior to 1962, and they very likely influenced Kuhn. However, there is no evidence that Feyerabend had read either Fleck or Polanyi before 1962. There is, however, another source of the notion of incommensurability that is common to both Kuhn and Feyerabend. In 1920, Köhler had written a book on the relationship between the 'mental' concepts of psychology and the 'material' concepts of physics (Köhler 1920). A *condensed* version appeared in English in 1938. In the opening sentences, Köhler wrote:

> In order to orient itself in the company of natural sciences, psychology must discover connections wherever it can between its own phenomena and those of other disciplines. If this search fails, then psychology must recognize that its categories and those of natural science are *incommensurable* (Köhler [1920] 1938, p. 17, italics inserted).[219]

Köhler proceeded to suggest that while *Gestalten* are common in psychology, their existence in physics would "violate the fundamentals of exact science" (Köhler [1920] 1938, p. 17), and he emphasized that the "thought and language of physics were established in accordance with other points of view than [psychology, specifically psychological Gestalten]" (Köhler [1902] 1938, p. 17). Köhler's strategy for attempting to overcome this problem with the relation of psychological 'wholes' to the reductionist concepts of physics was to attempt to find such 'organic units' in physics. This is the central topic of his essay. But what connections are there between Köhler's ideas and Feyerabend's 1962 notion of incommensurability?

First of all, one of the central topics of Feyerabend's early philosophical work was Wittgenstein's *Investigations*. This book discusses Gestalt figures at some length (see XI, 193^{e}ff.). In fact, Wittgenstein's later philosophy was heavily influenced by Gestalt psychology, especially by Köhler,[220] who is one of the very few named sources in the entire book (Wittgenstein 1958, XI, 203^{e}).

---

219 The term 'inkommensurable' does *not* appear in the original 1920 German version. Thus, Feyerabend's use of the term 'incommensurable' appears to be the result of W. Ellis' *translation*.

220 See Wittgenstein on the difference between 'seeing' and 'seeing as', as well as the general problems with perceptual psychology that led to the development of Gestalt psychology, Wittgenstein (1958), XI, 193^{e}-214^{e} and Köhler (1947), pp. 81-122.

Second, Kuhn had originally used the notion of a Gestalt switch to illustrate his notion of scientific revolutions and the incommensurability involved (Kuhn [1962] 1970. p. 204). In his 1961 comments on a draft of (Kuhn [1962] 1970a), Feyerabend had even recommended that Kuhn "must read" Wittgenstein on the distinction between 'seeing' and 'seeing as', i.e. exactly those passages of Wittgenstein's views on perceptions in which Wittgenstein cites Köhler ([1960/61] 1995, p. 384). This, by itself, is a possible indication that Feyerabend may have had the term 'incommensurable' from Köhler.

Third, Köhler was not only a founding father of Gestalt psychology; he was also active in philosophy, especially the philosophy of mind.[221] Like Feyerabend, he was very interested in the consequences that the psychology of perception have for philosophical accounts of science,[222] as well as the relationship between folk psychological concepts and physical concepts. This is the topic of (1963b) and (1963c), where Feyerabend argued against the need to reduce the former to the latter in favor of *eliminative materialism.* According to Feyerabend, the meanings of the 'mental' psychological terms should be eliminated, and replaced by 'materialist' physical concepts. In arguing this point, Feyerabend went so far as to claim that "every interesting discussion, that is every discussion which leads to an advance of knowledge, terminates in a situation in which some decisive change of meaning has occurred" (1963b, p. 58). Moreover, Feyerabend used the example of the mind-body problem to illustrate his notion of incommensurable concepts in (1962a), where he first used the term 'incommensurable'. Thus, Köhler had applied the term 'incommensurable' to an example quite similar to the one Feyerabend used to illustrate incommensurability in 1962, and many times thereafter. He applied it to the difficulties in relating the holistic 'mental' concepts of psychology and the reductionist 'material' concepts of physics.

Fourth, as we have seen, Feyerabend first used the term 'incommensurable' to attack Nagel's views on the formal conditions of reduction. They appear in chapter 11, "The reduction of theories", of (Nagel [1961] 1966). In the same chapter, Nagel also considered the issue of "individual wholes" or "organic unities" (such units are supposedly more than just aggregates of independent parts or members), noting that it has been

221 There is even a contribution by Köhler in the first book of which Feyerabend was an editor, see Feyerabend and Maxwell (1966).

222 By the mid-1960s at the very latest, Feyerabend was clearly aware of the basic principles of Gestalt psychology. See e.g. Feyerabend (1965b), p. 220, fn. 8 where he sets them out. But as early as his 1951 doctoral thesis, he was citing works on the psychology of perception, and his close scrutiny of Wittgenstein's *Investigations* in 1952, all suggest that Feyerabend was familiar with, and excited by, the implications that Gestalt psychology could have for philosophical accounts of the nature of scientific knowledge and its advance.

claimed that such "wholes" exist in physics, biology, psychology, as well as the social sciences) (Nagel [1961] 1966, pp. 380ff.). In fact, Nagel's discussion of the issue takes up some 17 pages of the chapter (Nagel [1961] 1966, pp. 380-397). Who did Nagel cite on this issue? Nagel cited Köhler. What did Nagel cite from Köhler? Nagel cited (Köhler [1920] 1938),[223] exactly where Köhler used the term "incommensurable" to describe the relationship between some concepts in psychology and in physics – an example very similar to one of Feyerabend's oft repeated examples of incommensurability. It seems very unlikely that this is a mere coincidence. It seems more likely that when Feyerabend was preparing his criticism of Nagel's account of reduction, he had a look at the main sources Nagel cited on the difficulties with reduction, and there he found the term 'incommensurable'.

Thus, the *term* 'incommensurable' may have found its way into Feyerabend's philosophy of science via the influence of holistic ideas in Gestalt psychology and the potential problems they could cause for formal accounts of reduction. In this sense, the term appears in a 1938 publication by Wolfgang Köhler, and there are some reasons that suggest that Feyerabend had it from that source. This is not, however, the only possible source from which Feyerabend was drawing in his 1962 exposition of incommensurability. As we have seen, there is another prominent discussion of difficulties in judging the merits of universal theories in which the term 'incommensurable' appears with which Feyerabend was acquainted that may well have influenced Feyerabend.

In his 'Autobiographical Notes' (1949a), Einstein discussed two difficulties with assessing the merits of universal theories:

> something of a broadly general nature will first have to be said concerning the points of view according to which it is possible to criticize physical theories at all. The first point of view is obvious: the theory must not contradict empirical facts. However evident this demand may in the first place appear, its application turns out to be quite delicate. For it is often, perhaps even always, possible to adhere to a general theoretical foundation by securing the adaptation of the theory to the facts by means of artificial additional assumptions. In any case, however, this first point of view is concerned with the confirmation of the theoretical foundation by the available empirical facts (Einstein 1949a, p. 22-23).

This is a concise statement of what is well-known in the philosophy of science as the Duhem-Quine thesis. Both the incommensurability thesis and the Duhem-Quine thesis concern difficulties that arise when trying to use experience to judge the merits of theories. Directly after delineating the Duhem-Quine thesis, Einstein used the term 'incommensurable' while struggling to explain the idea that assessing the relative merits of universal

223 See Nagel ([1961] 1966), pp. 392, fn. 22 and fn. 24, and p. 393, fn. 26.

theories involves making difficult judgments about their 'naturalness'. According to Einstein, this requires making judgments based on the reciprocal weighting of 'incommensurable qualities':

> The second point of view is not concerned with the relation to the material observation but with the premises of the theory itself, with what may briefly but vaguely be characterized as the "naturalness" or "logical simplicity" of the premises (of the basic concepts and of the relations between these which are taken as a basis). This point of view, an exact formulation of which meets with great difficulties, has played an important role in the selection and evaluation of theories since time immemorial. The problem here is not simply one of a kind of enumeration of the logically independent premises (if anything like this were at all unequivocally possible), but that of a kind of reciprocal weighing of incommensurable qualities (Einstein 1949a, p. 23).

This use of the term incommensurability does not match Feyerabend's nearly as well as Köhler's. This is because the point Einstein's is making with the term 'incommensurable' does not concern difficulties in comparing two comprehensive theories. Einstein's point has to do with judging the 'naturalness' or 'logical simplicity' of a single comprehensive theory.

Even so, Feyerabend's views in 'Explanation, Reduction and Empiricism' (1962a), where he first used the term 'incommensurable' are strikingly similar in several significant respects to Einstein's as they are set out in (Einstein 1949a). First of all, as we have seen, they both use a distinction between universal theories and those that do not apply to the totality of physical appearances to qualify their respect discussions of the difficulties of using empirical evidence to judge the merits of theories. Second, Einstein's basic metaphysical perspective is a form of neo-Kantianism, much like Feyerabend and Kuhn's (see chapter 6). Einstein's attitude:

> is distinct from that of Kant only by the fact that we do not conceive of the "categories" as unalterable (conditioned by the nature of the understanding) but as (in the logical sense) free conventions. They appear to be *a priori* only insofar as thinking without the positing of categories and of concepts in general would be as impossible as is breathing in a vacuum (Einstein 1949b, p. 374).

This is very similar to both Kuhn and Feyerabend's views.[224] They all see scientific advance as Kant-on-wheels, to use Peter Lipton's expression. In other words, Einstein, Feyerabend and Kuhn all *explicitly* adhere to the Kantian idea that experience is partly determined by the categories we use to organize sense-impressions, but at the same time adamantly reject Kant's belief that these categories are necessary, and thus historically invariant. At the beginning of 'Explanation, Reduction and Empiricism' (1962a), Feyerabend explained that he was working from such a neo-Kantian metaphysical perspective, and investigating the question: If uni-

224 For a detailed discussion of Kuhn's neo-Kantian views, see Hoyninge-Huene (1993).

versal theories determine how we experience the world, how can experience be used to test them? That was Feyerabend's main concern in 'Explanation, Reduction and Empiricism'. Moreover, the main example Feyerabend used in to illustrate how incommensurable theories can be compared on the basis of a crucial experiment was Einstein's quantitative prediction of Brownian Motion as a crucial experiment between Classical and Statistical Thermodynamics (see chapter 8). Einstein discussed this episode at length in his 'Autobiographical notes', where he used the term 'incommensurable'. In other words, one of the main topics of Feyerabend's (1962a) is how Einstein's quantitative prediction of Brownian Motion served as a crucial experiment between incommensurable theories, and that is also a main topic of Einstein's autobiographical notes. Thus, there are some very strong reasons to suspect that Feyerabend was drawing on Einstein when he introduced the term incommensurable, even if used the term in a significantly different sense than Einstein did.

5.3. Niels Bohr and the principle of complementarity (1958)

In his autobiography, Feyerabend acknowledged Niels Bohr's influence on the development of his notion of incommensurability in the 1950s. Feyerabend recalled a conversation in which Bohr had talked about the discovery that the square root of two cannot be an integer or a fraction. According to Feyerabend, Bohr presented the event as having led to the extension of a concept of number that retained some properties of integers and fractions, but changed others; and claimed that the transition from Classical to Quantum Mechanics was carried out in accordance with precisely this principle. Bohr compared this to the transition from classical to Quantum Mechanics:

> He [Bohr] talked about the discovery that the square root of two cannot be an integer or a fraction. To him this seemed to be an important event, and he kept returning to it. As he saw it, the event led to *the extension of a concept of number that retained some properties of integers and fractions but changed others*. Hankel, whom Bohr mentioned, had called the idea behind such an extension the principle of the permanence of rules of calculation. *The transition from classical mechanics to quantum mechanics*, said Bohr, *was carried out in accordance with precisely this principle*. That much I understood. The rest was beyond me (1995a, p. 78, italics inserted).

As Feyerabend used his notion of incommensurability to try to capture the idea of conceptual changes that retain some properties of a concept, while changing others, these autobiographical recollections are evidence of

Bohr's direct influence on Feyerabend's development of the notion of incommensurable concepts in the 1950s.

However, with typical flair Feyerabend immediately turned his notion of incommensurability into an attack on Bohr. In a 1958 paper on Bohr's interpretation of Quantum Mechanics entitled "Complementarity", Feyerabend argued that Bohr's interpretation of Quantum Mechanics employs philosophical assumptions that are an example of an unjustified *conceptual conservativism* in the sciences (1958b, pp. 81-83).[225] He presented Bohr's defense of the principle of complementarity based on the conviction that every experience must *necessarily* make its appearance within the frame of our customary points of view, which is currently that of classical physics (e.g. 1958a pp. 22). Apparently, Feyerabend's attack is directed at the 'correspondence principle', although, Feyerabend does not use that expression.[226] Feyerabend's challenge runs as follows:

> Now does it follow, as is asserted by Bohr, that we can never go beyond the classical framework and that therefore all our future microscopic theories will have to use the notion of complementarity as a fundamental notion? (1962a, p. 88, 1981a, p. 88).

And his answer is:

> It is quite obvious that the actual use of classical concepts for the description of experiments within contemporary physics can never justify such an assumption, even if these concepts happen to have been very successful in the past (Hume's problem). For a theory may be found whose conceptual apparatus, when applied to the domain of validity of classical physics, would be just as comprehensive and useful as the classical apparatus, without coinciding with it. Such a situation is by no means uncommon. The behavior of the planets, of the sun, and of the satellites can be described both by Newtonian concepts and by the concepts of general relativity. The order introduced into our experiences by Newton's theory is retained and improved by relativity. This means that the concepts of relativity

225 Feyerabend often made this argument. See (1954e), (1958a) pp. 22; rp (1981), pp. 148, (1958d), especially p. 50, p. 59, fn. 6; rp (1981a), p. 240, fn. 6, (1960e), see especially p. 321, pp. 323-324, and p. 326; rp (1981a), p. 219, pp. 220-221, and p. 224, ([1960/1961] 1995), pp. 378ff., (1961e), p. 371, and especially section 4: "The Dogmatic Elements in Bohr's Approach" pp. 384-390, (1961f), p. 399, and (1962a), p 43 and 88; rp (1981), p 88 and p. 54, (1963a), (1965b), and (1970c). But around 1968, in a typical Feyerabendian reversal, he abandoned this criticism and completely inverted his estimation of Bohr. In his (1968c), Bohr is no longer a dogmatist tied to conceptually conservative assumptions, but open-minded with respect to future developments. In (1968c), Feyerabend no longer claims that Bohr assumed the necessity of classical concepts once and for all. Rather, Bohr was forced to this conclusion on the basis of physics available, and Bohr only concluded that *at the present*, classical concepts are necessary. See and compare (1961c), pp. 384ff. to (1970a) and (1993), pp. 15, fn. 1, pp. 31-32. Or better compare (1961e) to (1968c) and (1969d), or the heavily revised compilation of the latter two in (1981a), especially section 3 "Measurement: Classical limit", pp. 252ff.

226 I would like to thank Gonzalo Munévar for pointing this out to me.

> theory are sufficiently rich to allow us to state all the facts which were stated before with the help of Newtonian physics. Yet these two sets of categories are completely different *and bear no logical relation to each other* (1958b, p. 83, see similarly 1961e, p. 388).[227]

Here, Feyerabend used the notion of incommensurability in his attack on a dogmatic element in Bohr's principle of complementarity. Feyerabend's general point is that:

> We have to realize that no element of our knowledge, physical or otherwise, can be held to be absolutely certain and that in our search for satisfactory explanations we are at liberty to change any part of our existing knowledge, however "fundamental" it may seem to be to those whose are either unable to imagine or to comprehend alternatives" (1961e, p. 384).

Feyerabend's argument against the necessity of maintaining classical descriptions is that new incommensurable concepts can be invented, which could replace those of classical descriptions. Just because classical concepts have been successful in the past, and because at the moment it may be difficult, or even impossible, for us to imagine how to replace them, it does not follow that the classical framework could not one day be superseded. Consequently, it does not follow that all our future microscopic theories will have to use the notion of complementarity as a fundamental notion.

Instead, according to Feyerabend, a theory may be found whose conceptual apparatus, when applied to the domain of validity of Classical Physics, would be just as comprehensive and useful as the classical apparatus, without coinciding with it. He claimed that such a situation is by no means uncommon, and he used the transition from Newtonian to Einsteinian Physics to bolster his point. According to Feyerabend, while the concepts of relativity theory are sufficiently rich to state all of the facts captured by Newtonian Physics, the two sets of categories are "completely different" and bear "*no logical relations*" to each other.[228] This is clearly a version of the incommensurability thesis. On Feyerabend's fallibilist view of empirical knowledge, no element of our knowledge can be held to be necessary or absolutely certain. In our search for satisfactory explanations, we are at liberty to change any parts of our existing knowledge, however 'fundamental' they may seem, including the concepts of Classical Physics (e.g. 1960e, pp. 323-324, 326).

In summary, Feyerabend's 1958 argument against Bohr's contention of the *necessity* of maintaining classical concepts for the description of

227 An almost identical paragraph appears in (1961e), p. 388, as well as in (1962b); also see the partial reprint in (1981a), p. 323. Cf. (1962a), p. 89. Note that the actual example of incommensurability is drawn from the transition to Einstein's theory from Newton's, not from classical mechanics to quantum mechanics.

228 See (1958b), p. 83; (1961c), p. 388, and compare (1962a), pp. 88-89.

Quantum phenomena was based on the idea that new incommensurable concepts could be invented, and one day could even be used to replace classical descriptions. According to Preston, Feyerabend's notion of incommensurability, though not the term itself, was first published in this 1958 paper on Bohr's thesis of complementarity.[229] As we have seen, one of several chief influences on Feyerabend's development of the notion of incommensurability in the 1950s was unquestionably Niels Bohr. Even so, Preston's claim is incorrect. Feyerabend had already published the idea in German in 1954.

5.4. Physics and ontology (1954)

The first use of the notion of incommensurability in Feyerabend's *published* work appeared in the paper "Physik und Ontologie" (1954c), which discusses the nature of major theoretical advances in physics, and their semantic and ontological implications. Feyerabend argued that major theoretical advances in physics are *not* part of a continuous process of enriching already established ideas, and that scientific progress is *not* a process in which existing ideas are improved or the number of known facts is simply increased. Instead, Feyerabend argued that new theories force a revision in the ontological status of older ideas. After major theoretical transitions, older ideas can no longer be accepted as purported true descriptions of reality. This is because they are *conceptually incompatible* with better ideas that *are* supposed to describe reality. According to Feyerabend (1954c), conceptual replacements that occur in theoretical advance can have ontological implications: When new fundamental theories replace older ideas, a correspondingly new conception of reality is formed. In the course of theoretical advance in physics, *what we take to be real itself* is not simply improved or enriched: "It is completely reordered" (1954c, p. 470, my translation).

Feyerabend argued for this view of scientific advance on the basis of the claim that the meanings of scientific terms are determined by the theories in which they occur. Consequently, given theory change, there is meaning change (1954c, p. 467). This is the first instance of the notion of incommensurability in Feyerabend's published work. According to Feyerabend, ontology and semantics are thus normative, not descriptive: They

229 Preston (1997a), pp. 102-103, cites (1958b) and (1961c).

vary historically in accordance with theoretical advance, and when new theories replace older ones, both our conception of reality and the meanings of our terms describing it change.

Feyerabend used the notion of incommensurability to criticize an implicit *conceptual conservativism* in attempts to separate philosophical accounts of ontology from physics. Purely analytic attempts to explicate what we take to be real presuppose a fixed conception of reality. However, if our conception of the nature of reality is contingent upon our developing theoretical ideas, then ontology cannot be studied independently of the empirical sciences.

5.5. Werner Heisenberg and cumulative progress (1951)

Feyerabend first discussed the relationship between the General Theory of Relativity and Newton's theory in his unpublished doctoral thesis (1951, pp. 87ff.). There, he claimed that Newton's theory was *replaced by* the General theory of relativity, and he emphasized that the latter talks about *completely different relationships* than the former (1951, p. 89ff.). Feyerabend also claimed that these two conceptually incompatible theories may even predict the same experimental results within their shared domains. However, as the theories used to interpret the sentences expressing the experimental result use different concepts, there will be two different interpretations of it:

> The correctness of the General Theory of Relativity can be tested by experiment in such Galilean domains . . .But with that it is not claimed that the results of the General Theory of Relativity statement imply these statements. *Rather, the content of these statements is determined by the new Law of Gravity. The concepts change their meanings according to the new theory* (1951, p. 90, my translation, italics inserted).

According to Feyerabend, the meanings of the concepts of General relativity and Newton's theory are determined by the theories to which they belong. The concepts changed there meaning when one theory *replaced* the other (1951, p. 89).

Feyerabend made exactly these points concerning the same theoretical transition in his more famous 1962 paper, where he added the examples of transition from Aristotelian dynamics to Newton's theory and the transition from classical to Statistical Thermodynamics (see e.g. 1962a, p. 57).[230] The only difference appears to be that in his 1962 paper, he used

230 Here in 1951, as elsewhere (including 1962a; rp 1981a, pp. 44-96), Feyerabend talked about "universal theories" that are supposed to apply to everything. For discussion and examples,

the term 'incommensurable' to describe the situation, where as in his 1951 doctoral thesis he did not. Even so, Feyerabend had clearly employed the *notion* of incommensurability as early as 1951.

Moreover, in his doctoral thesis Feyerabend *used* this basic idea of incommensurability in much the same way in which he used it in his 1954, 1958, and 1962. He used it to attack what he considered to be an unjustified form of *conceptual conservatism*. In this instance, he attacked the conceptual conservativism implicit in the methodological principle that established results must be presupposed by all further research, as set out by Heisenberg (1951, p. 90). Feyerabend interpreted Heisenberg as claiming that once an experimental result has been conclusively established, it sets conditions for further theoretical developments. This conservative attitude is wide-spread. It continues to be defended, even in an introductory textbook to the philosophy of science. For example:

> Certain types of experimental findings serve as benchmarks, permanent facts about phenomena which any future theory must accommodate (Hacking 1983, p. 254).

Feyerabend understood Heisenberg as placing unnecessarily restrictive *a priori* conditions on research. His reaction to this conceptual conservativism was to argue that imposing conditions on further research on the basis of established results could hinder progress. It overlooks that new concepts with different meanings could replace those older ideas. It is not that established experimental results set conditions on further theoretical development. Rather, older experimental results may have to be reinterpreted in light of new theoretical insights. This reaction mirrors exactly his reaction to the conceptual conservativism he found in attempts to separate philosophical and scientific accounts of ontology, Bohr's complementarity thesis, logical positivist theories of meaning, and formal theories of reduction and explanation. It was also based on exactly the same idea: The possibility of developing new theories whose concepts are conceptually incompatible with those of existing theories (i.e. incommensurability).

It follows that Jacobs' assertions concerning Feyerabend's unpaid debt to Polanyi, as the source of his notion of incommensurability, cannot be correct.[231] Polanyi's notion of incommensurability as applied to scientific

---

see e.g. (1954c), (1958b), pp. 78-79, and pp. 90-91, (1961i), pp. 7ff.; rp (1999a), pp. 52ff., (1967a), p. 40, and (1970e), p. 233 and p. 246. Feyerabend restricted his notion of incommensurability to such universal theories. For a discussion, see Hoyningen-Huene (2000a).

231 See Jacobs (2002), p. 106 and Jacobs (2003), pp. 58, 69, 70. Neither Jacobs (2002), (2003) nor Preston (1997b) provide any instance of Feyerabend having mentioned Polanyi before the mid-1970s. They base their claims about the connection between Feyerabend and Polanyi on Feyerabend's 1991 remark that "*somewhere along the way* I read an interesting paper by Michael Polanyi" (1991, 492, 501, italics inserted). It is, however, clear that

theories first appeared in his (1958).[232] Feyerabend had developed his notion of incommensurability *by 1951*, and he had already used it in publications that had appeared in 1954 and 1958. Feyerabend did not draw on Polanyi concerning his notion of incommensurability, and Polanyi had not "presaged . . . Feyerabend's motif of incommensurability" (Jacobs 2002, p. 106).[233] Feyerabend's notion of incommensurability actually *pre-dates* Polanyi's use of the idea as applied to scientific theories, and as we shall see, it comes from very different sources.

Moreover, while there are indeed some significant similarities between Polanyi's notion of incommensurability and Feyerabend's post-1970 ideas, and more so between Polanyi and Kuhn's ideas,[234] there are also some striking differences between Polanyi's notion of incommensurability and Feyerabend's pre-1970 ideas. For example, according to Jacobs, Polanyi used his notion of incommensurability to argue *against* the possibility of rationally comparing incommensurable conceptual systems (Jacobs 2002, pp. 114-115). Between 1951 and 1970, Feyerabend repeatedly insisted that the existence of incommensurability does not by any means preclude rational comparison (e.g. 1965c, pp. 227ff.).

5.6. Logical Positivism and protocol sentences (1949-1958)

Feyerabend discussed his dissertation, and thus the notion of incommensurability, with the Kraft Circle from 1949-1951. The Kraft circle was a student group named after Viktor Kraft (Feyerabend's Ph.D. supervisor) and modeled after the Vienna Circle (of which Kraft was a member). Guests such as Elizabeth Anscombe and Ludwig Wittgenstein attended meetings (1978, pp. 115-116, 1993a, pp. 259-260 and 1995, p. 92). Meeting Wittgenstein and attempting to rewrite the *Investigations* as a treatise had a

---

Feyerabend and Polanyi were acquainted at the latest by 1957. They both participated in the Ninth Symposium of the Colston Research Society held in Bristol in April of 1957, where they discussed each other's contributions. Feyerabend typed up these discussions for publication. There is even a photo of the participants in which Feyerabend is standing directly behind Polanyi. See the title page of Körner (1957).

232 See and compare Jacobs (2002), p. 108 with pp. 111-116, and see Jacobs (2003), p. 61. Polanyi had developed a notion of incommensurability between a scientific and a religious worldview in 1946, but in his ([1946] 1964), he explicitly rejected the idea of incommensurability between scientific theories, see Polanyi ([1946] 1964), p. 66, and Jacobs (2003), pp. 61 and 71, fn. 5 for a discussion.

233 This does not rule out the possibility that Feyerabend's 1975 notion of incommensurability my have owed something to Polanyi.

234 See Jacobs (2003), p. 67 and Jacobs (2002), p. 113.

profound and long-lasting influence on Feyerabend's philosophical development (1995a, p. 94).[235] As we have seen, Feyerabend had even planned to study under Wittgenstein at Cambridge, and had received a fellowship to do so. But Wittgenstein died, so Feyerabend went to London to work with Popper (1995a, p. 86, 1987, p. 312). However, as we have seen in chapter 2, Feyerabend did not develop the idea of incommensurability primarily out of Wittgenstein's *Investigations*. This is clear as in developing the notion of incommensurability, Feyerabend explicitly rejected the Wittgensteinian idea that the meaning of a term is determined by its use, grounding the idea of incommensurability on a contextual theory of meaning. Moreover, Feyerabend emphasized that he first read the *Investigations in detail* in London in 1952, i.e. *after he had submitted his dissertation*, (1978, pp. 115-116). Elsewhere, Feyerabend also claimed to have read a manuscript of the *Investigations* in 1950 (1987, p. 294).

Feyerabend sometimes traced the development of his notion of incommensurability back to discussions he had with Anscombe sometime between 1949 and 1952 (1978, pp. 145ff).[236] The discussions with Anscombe concerned the nature of perception and the foundational role of observation in positivist philosophy.

According to Feyerabend's critique of logical positivism, every positivistic observation language is based upon a metaphysical (i.e. untestable) ontology (1958a, pp. 148-149). The problem lies in the tension between the positivists' maxim of testability and their foundational use of observation claims. On the one hand, knowledge claims should be verifiable. They should be subject(able) to empirical tests. On the other hand, empirical tests are supposedly made on the stable basis of inter-subjectively verifiable observations. These observations are crucial to positivism because they provide the stable, empirical foundation for such tests. In other words, the reason that knowledge claims have a special epistemic status (according to logical positivism) is that they have been (or can be) tested against experience. However, in order for the observation sentences to play this foundational role for testing knowledge claims, they themselves must be stable and inter-subjectively verifiable. Consequently, they cannot themselves be open to tests. Otherwise their foundational role would be undermined. Hence, a metaphysical element has seeped in to science by

235 Anscombe gave Feyerabend an unpublished manuscript copy of the *Philosophical Investigations*. Feyerabend attempted to rewrite it as an essay, see (1955a).

236 Feyerabend claimed that he tried to explain this early notion of incommensurability in Popper's seminar (in 1952), and also to a small group of people at Anscombe's home in London (in 1952); see (1978), p. 67, fn. 114 and 115, (1993a), p. 260 and (1995), p. 92.

way of the foundational role of observation sentences in testing knowledge claims.[237]

In this respect, logical positivism contains a dogmatic or *conceptually conservative* element: The foundational role of observation sentences in theory-testing. Feyerabend's critique of this conservative element was based on the idea that the meanings of the terms of observation sentences are determined by the theories to which they belong. He tried to extend and enforce the maxim of testability across the whole system – not just to theoretical knowledge claims, but to the observation sentences supporting them as well. Not only should the theoretical knowledge claims be testable, but so should the observation sentences which serve as the basis for their test. In this way, Feyerabend used the notion of incommensurability to attack the foundational role of protocol sentences as an unjustifiable form of conceptual conservativism in logical positivism.

In the third English edition of *Against Method* (but *not* in the two English editions that preceded it), Feyerabend emphasized that his notion of incommensurability was originally the result of such considerations about the nature of perception that he had had while studying the literature on the positivists notion of 'protocol' statements (1993a, p. 211). He then proceeded to give the most detailed account of his early development of his notion of incommensurability.

> Following this argument I introduced the assumption that the meaning of observation statements depends on the nature of the objects described and, as this nature depends on the most advanced theories, on the content of these theories. Or as I formulated it in my first English paper on the topic [(1958a)]: the interpretation of an observation language is determined by the theories which we use to explain what we observe, and it changes as soon as these theories change (1993a, p. 211, italics inserted).

Thus, Feyerabend had discussed his notion of incommensurability with Elizabeth Anscombe at the latest by 1950, and he used it to attack a form of conceptual conservativism inherent in the logical positivists' use of protocol statements in theory testing in 1951. Feyerabend also emphasized that it was discussions of protocol sentences, *and not discussions of Wittgenstein's theory of meaning*, that influenced his developing notion of incommensurable concepts (1981a, pp. 49, 125).

237 For an excellent discussion of Feyerabend's critique of foundationalism, see Couvalis (1989).

5.7. Pierre Duhem and theoretical transitions (1906)

According to Giedymin (1982), the origins of the idea of incommensurability in the philosophy of science date back to LeRoy and Ajdukiewicz in Poincaré's conventionalist tradition.[238] While something similar to the notion of incommensurability can indeed be found in Ajdukiewicz's work, there is no *direct* evidence connecting Ajdukiewicz to either Kuhn or Feyerabend.[239]

Moreover, the similarities to Ajdukiewicz's discussion of the 'untranslatability' of scientific theories are only superficially similar to Kuhn and Feyerabend's notions of incommensurability. Ajdukiewicz's concerns were with situations in which two empirically equivalent theories were nonetheless *untranslatable*. Here, the problem is that while they differ semantically, they do not differ empirically. For Kuhn and Feyerabend, incommensurable theories are *not* empirically equivalent, and Kuhn only began to use a special notion of untranslatability to help explicate incommensurability in 1969 (well after he initially introduced his notion of incommensurability). Feyerabend explicitly rejected the idea that incommensurable theories are untranslatable, *insisting that one can very easily translate between two incommensurable theories* (1970d, p. 233). Ajdukiewicz's discussions about the untranslatability of empirically equivalent theories appear primarily to concern the problem of the *under-determination of theory by fact*, whereas Kuhn and Feyerabend's notion of incommensurability has much more to do with *the theory-ladeness of observation.*

However, there was another direct influence on Feyerabend's development of the notion of incommensurability that has been overlooked: The influence of Pierre Duhem. In his Ph.D. thesis, Feyerabend cited several times from *The Aim and Structure of Physical Theory* (Duhem [1906] 1908).[240] Although Feyerabend cited Duhem on a different point than what he later called *thesis I*, one of the locations in which Feyerabend cited Duhem is quite revealing. Feyerabend cited Duhem directly in the middle

238 See Giedymin (1970), p. 257, fn. 1, and Giedymin (1982). See also Preston (1997a), p. 217, fn. 4. The point has also been made by Wisdom (1974), p. 299, who cites Skolimowski.

239 See Ajdukiewicz (1974), pp. 56-57, 67, 75-81, 86-87, (passages originally published 1934). Ajdukiewicz is *indirectly* connected to Kuhn, as Fleck and Ajdukiewicz appear to have been acquainted. There is a reprint of a paper by Ajdukiewicz's in the University Library in Warsaw with a dedication by Ajdukiewicz to Fleck dated 1935, see Cohen and Schnelle (1986), p. 16.

240 See especially (1951), pp. 135, fn. 75 and 142, fn. 177. I will cite from the 1954 English translation, which contains all of the relevant passages to the 1908 German translation from which I will draw.

of explaining his early conception of incommensurability - one sentence before criticizing Heisenberg's conservativism, and three sentences after writing, "rather, the content of their statements is determined by the new law of gravitation. The concepts change their meaning corresponding to this new theory" (1951, p. 90, my translation). Thus, Feyerabend cited Duhem while explaining the notion of incommensurability for the very first time. But what has Duhem got to do with incommensurability?

Duhem had claimed that what a physicist states as the result of an experiment is not simply the recital of some observed facts. Rather, it is the *interpretation* of these facts *on the basis of the theories the scientist regards as true* ([1906] 1954, p. 159). It follows, according to Duhem, that in order to understand the meanings that scientists ascribe to their own statements, it is necessary to understand the theories that they use in order to interpret what they observe. In other words, Duhem had stated what Feyerabend called *thesis I*, which, according to Feyerabend himself, is an early version of his notion of incommensurability. Moreover, exactly this passage is doubly underlined in Feyerabend's personal copy of *The Aim and Structure of Physical Theory* (Duhem [1906] 1908, p. 209).[241] Duhem was one of the most knowledgeable and competent historians of science of the twentieth century. He wrote many volume of history of science, so it is no wonder that incommensurability was obvious to him.[242] After all, it was Kuhn's historical investigations into the natural sciences that contributed to the popularization of the notion in the 1960s.

Preston has presented Feyerabend's development of *thesis I* as based on a poor reading of Wittgenstein's theory of meaning (Preston 1997a, pp. 23ff.). This is incorrect.[243] Feyerabend developed thesis I from a *correct* reading of Duhem. T*hesis I* is an early version of the incommensurability thesis. It follows that Feyerabend developed the notion of incommensurability from ideas he found in (Duhem [1906] 1908). That Duhem was indeed the un-cited source of the second qualitative reason for the failure of formal accounts (i.e. incommensurability) in (Feyerabend 1962) is further substantiated given that Duhem, as we have seen, was belatedly admitted to be the source of Feyerabend's first quantitative reason for the failure of formal accounts of the relation between some pairs of succes-

241 Feyerabend's personal copy of Duhem ([1906] 1908) can be found in the Philosophical Archive at the University of Konstanz, Germany. It is stamped with Feyerabend's name and 1940s Vienna address, and a handwritten date suggests that he acquired it on or before 04.05.1949. The margins are heavily noted, and the fact that Feyerabend sometimes used his German shorthand further suggests that these notes were made in the late 1940s.

242 For an excellent discussion, see Jaki (1969) that introduces Duhem ([1908] 1968).

243 Similar ideas may be found in Wittgenstein, and Wittgenstein and Feyerabend were both heavily influenced by a common factor (advances in the theory of perception that pertain to incommensurability, i.e. Gestalt psychology).

sive scientific theories. That Duhem was also the source of Feyerabend's second reason is not surprising given that Duhem discussed the second reason (incommensurability) forty pages before he discussed the first reason (logical inconsistency) in (Duhem [1906] 1954).

After explaining that the meaning of a term depends on the theory to which it belongs, and that a consequence of theoretical advance is meaning change, Duhem continued:

> If the theories admitted by this physicist are those we accept, and if we agree to follow the same rules in the interpretation of the same phenomena, *we speak the same language and can understand each other. But that is not always the case.* It is not so when we discuss the experiments of a physicist who does not belong to our school; and *it is especially not so when we discuss the experiments of a physicist separated from us by fifty years, a century, or two centuries* (Duhem 1954, p. 159, italics inserted).

To those familiar with Feyerabend and Kuhn's work, this too will sound remarkably familiar. It represents a basic point that both Feyerabend and Kuhn made with their notions of incommensurability. In Feyerabend's copy of (Duhem [1906] 1954, p. 210), he marked this passage, and placed a small star in the margin.

Furthermore, Duhem continued:

> . . . How many scientific discussions there are in which each of the contenders claims to have crushed his adversary under the overwhelming testimony of the facts! . . . How many propositions are regarded as monstrous errors in the writings of those who have preceded us! *We should perhaps commemorate them as great truths if we really wished to enquire into the theories which gave their propositions their true meaning* . . . (Duhem 1954, pp. 160-161, italics inserted).

Once again, this sounds extremely similar to a number of claims both Feyerabend and Kuhn made about incommensurability, and once again, this passage is underlined in Feyerabend's copy of (Duhem [1906] 1954, p. 212).

The passage expresses a basic idea behind one major consequences of Feyerabend's notion of incommensurability. Because the older ideas are misunderstood, as a result of taking them out of the theoretical context that provides them with their correct content, disputants misunderstand each other, both claiming to have the "facts" on their side. As is well known, both Kuhn and Feyerabend repeatedly argued this point, and claimed that in such a situation, even empirical arguments can become circular.[244]

One could counter that Duhem's ideas do not really match those of either Kuhn or Feyerabend, because Duhem suggested that the vocabu-

244 See, for example, Feyerabend (1965b), p. 152 and Kuhn ([1962] 1970a), p. 94. Duhem also recognized the limitations of logic for determining the outcome of theoretical disputes in the course of scientific advance (1954), p. 218, cf. Kuhn ([1962] 1970a), p. 94.

lary of older theories can be translated into that of newer theories (Duhem [1906] 1954, p. 160). However, such an argument fails to consider that Feyerabend rejected the idea that incommensurability implies untranslatability as late as 1970 (1970d, p. 233), and Kuhn only began to use a *special* notion of untranslatability to explicate incommensurability in 1969.[245] It follows that Duhem's remarks concerning translatability do not undermine the similarity to his views concerning incommensurability set out in 1962 by Kuhn and in the 1950s and 1960s by Feyerabend. Rather they bolster the point that Ajdukiewicz's problem of *untranslatability* is significantly different than the Duhem/Feyerabend notion of incommensurability.

Feyerabend developed *thesis I*, and consequently the incommensurability thesis, from Duhem. Duhem had already set out the ideas: (a) that the meanings of observational terms are dependent on the theories to which they belong; (b) that meaning change, even of observational terms, is a consequence of theoretical advance; and (c) that scientists have difficulties understanding members of different schools as well as those separated from them by 50, 100 or even 200 years *because of meaning change due to theoretical advance*. These ideas clearly anticipate the debate often co-credited to Kuhn and Feyerabend in 1962.

There are a number of other significant similarities in Duhem and Feyerabend's ideas. One significant similarity concerns the issue of *realism*. Both Duhem and Feyerabend understood *realism* as a *normative ideal*; i.e., as the idea that our theories should be interpreted as attempts to interpret experience realistically, and *not as a factual claim about the status of scientific knowledge*. Moreover, both Duhem and Feyerabend used the same *methodological* argument in favor of such a normative realism – that by encouraging competition, it better promotes progress than instrumentalism. A further main similarity between Feyerabend, Duhem and Kuhn, one which most significantly divides these individuals from the logical empiricists, is that they agree that empirical adequacy is just one of many *values* that scientists use to judge their theories. Others include simplicity, universality, internal coherence, coherence with other views, fertility, beauty, etc. As opposed to the cut and dry empirical model of the logical empiricists, all three of these philosophers recognized the need to leave room for rational disagreement among scientists.[246]

---

245 For excellent discussions, see Hoyningen-Huene (1993), pp. 213ff., Sankey (1997), and Sankey (1998).

246 For an excellent discussion of Feyerabend's value based rationality, see Farrell (2003). For Kuhn's value based rationality, see Kuhn ([1962] 1970a) and Hoyningen-Huene (1989). For Duhem, see Duhem ([1906] 1954). Also see Einstein (1949).

There are two main differences between Feyerabend and Duhem's notions of incommensurability. First, Feyerabend emphasized that incommensurability only occurs when *fundamental* theories change, as only they affect our ontology; i.e., only when theoretical transitions result in a new conception of reality. This restriction fits with Einstein's concerns about the difficulties in theory comparison, but marks a significant difference to Kuhn's notion of incommensurability, which is not so restricted Second, unlike Duhem, Feyerabend's conception of incommensurability is best understood in the context of his repeated attacks on various forms of conceptual conservativism. Feyerabend was not simply describing incommensurability as a historian. He was trying to use it to promote progress. If the notion of incommensurability in Feyerabend's philosophy is based on the idea that scientific terms receive their meaning from the theories to which they belong, and these meanings change in the course of theoretical advance (as I have argued that it should), then the origins of Feyerabend's idea of incommensurability should be traced directly back *at least* to Pierre Duhem ([1906] 1954).[247]

5.8. Conclusions

Feyerabend used his notion of incommensurability to mean that two fundamental theories are *conceptually incompatible.* The basic idea is that because the meanings, even of observational terms, are determined by the theories to which they belong, when there is theory change, there are meaning changes that can result in a new conception of reality. As a consequence, logical relations cannot correctly characterize the relationship between certain pairs of successive scientific theories.

Feyerabend used the notion of incommensurable concepts to attack several prominent forms of conceptual conservativism, such as Heisenberg's methodological principle (1951), attempts to distinguish philosophical accounts of ontology from physics (1954c), the foundational role of the stability thesis in logical positivism (1951, 1958a), Bohr's principle of complementarity (1958a, 1961c, 1962a, 1965b, 1970c), and the logical empiricists' models of explanation and reduction (1962a, 1963a). He also used the idea of incommensurability to attack forms of conceptual conservativism implicit in the monistic test-models promoted by classical empiricists, the logical positivists, the logical empiricists, Popper, and

247 This also supports Giedymin's claim concerning LeRoy's contribution to the notion of incommensurability, as at the outset of the chapter in which Duhem develops the notion of incommensurability, he acknowledged LeRoy as the source of his ideas, see Duhem ([1906] 1954), p. 144, fn. 1.

Kuhn (1958a, 1962a, 1963a, 1965b, 1970c); Kantian transcendental necessities (1962a, 1965b, 1965c); traditional philosophical conundrums like the problem of the existence of the external world, the problem of the existence of other minds, and especially the mind-body problem (1962a, 1965b, 1967, 1969c, 1970c); certain aspects of Wittgenstein's later philosophy as well as the linguistic philosophical movement that takes the utility of everyday concepts as given and merely attempts to analyze them, without changing their meanings (or 'arguments from synonymy') (1962a, 1963a, 1965b, 1969e, 1970c); Popper's conception of verisimilitude (1970b); and scientific realist characterizations of scientific knowledge and scientific advance (1970b). The notion of incommensurability is the central notion of Feyerabend's philosophy. It is the basis for his repeated attacks on conceptual conservativism.

Kuhn and Feyerabend's developments of their notions of incommensurability differed significantly. Kuhn developed his notion of incommensurability out of studies in the history of science, and he was directly influenced by Fleck, and Polanyi, and perhaps indirectly (through Fleck) by Ajdukiewicz. Feyerabend, on the other hand, developed his notion of incommensurability in the late 1940s and early 1950s through his studies of the bearing of the psychology of perception on the logical positivists' foundational use of protocol statements. Köhler, and the idea from Gestalt psychology that observation is theory-laden, heavily influenced both Kuhn and Feyerabend's developments of their notions of incommensurability. But the notion of incommensurability is by no means original to Kuhn and Feyerabend, as it is usually taken to be. In the first half of the last century, the idea of incommensurability, and even the term 'incommensurable' can be found in many areas of the philosophy of science. Various different versions of the basic insight that *logical relations are inadequate for capturing the relations between certain pairs of successive scientific theories because meanings change in the course of theoretical advance* had been proposed by Duhem, Ajdukiewicz, Fleck, and Polanyi – before the more famous introductions of the term 'incommensurable' by Kuhn and Feyerabend in 1962. Feyerabend's 1962 use of the term 'incommensurable' appears to owe much to Köhler. Köhler (1938) used it to describe the relationship between some concepts in psychology and in physics – an example very similar to one of Feyerabend's oft-repeated examples of incommensurability. All of these individuals contributed in their own ways to the development of the contemporary notions of incommensurability in the philosophy of science.

Chapter 6
Incommensurability and scientific realism

Feyerabend was very interested in the question of realism from early on in his career:

> I was interested in the realist position, I had tried to read every book on realism I could lay hands on [Feyerabend is referring to the late 1940s and early 1950s] (1978a, p. 112).

It proved to be a main topic of his philosophical deliberations from his doctoral thesis right through to his last unfinished book *Conquest of Abundance*. Kuhn consistently maintained that incommensurability challenges some elements of scientific realism. It is well know that Kuhn argued that the fact that incommensurable concepts have indeed actually been used by scientists in the course of scientific advance challenges some particular central components of scientific realism, especially the notion of progress as an advance on truth:

> There is, of course, no way, even in an enriched Newtonian vocabulary, to convey the Aristotelian propositions regularly misconstrued as asserting the proportionality of force and motion or the impossibility of a void. Using our conceptual lexicon, these Aristotelian propositions cannot be expressed—they are simply ineffable . . . *It follows that no shared metric is available to compare our assertions about force and motion with Aristotle's and thus to provide a basis for a claim that our (or, for that matter, his) are closer to the truth* (Kuhn 1970a, p. 206, italics inserted).[248]

Feyerabend's conception of the relation of incommensurability to the issues of *realism* and *scientific realism* is much less straightforward, and has not yet been sufficiently clarified. The main aims of this chapter are to clarify Feyerabend's views on realism and how they developed, and to explicate how they bear on his notion of incommensurability.

248 Also see Kuhn (1979b), p. 418, and Kuhn (1993), p. 330. For discussions of how incommensurability is supposed to challenge some elements of scientific realism according to Kuhn, see e.g. Brown (2001), Caneva (2000), pp. 113ff., Hoyningen-Huene (1993) as well as Sankey (1994), Sankey (1997a), Sankey (1997b), and Sankey (1998).

6.1. Scientific realism vs. normative realism

There is a widespread misunderstanding of Feyerabend according to which he was initially a scientific realist, who then retreated from scientific realism. For example, Preston has titled a prominent article "Feyerabend's Retreat from Realism" (Preston 1997a). It begins:

> It sometimes comes as a surprise to those who encounter him [Feyerabend] through *Against Method*, or one of his later productions, to hear that *Feyerabend had previously been an advocate of scientific realism* (Preston 1997a, p. 421, italics inserted).[249]

The reason that this may sometimes be surprising is that Feyerabend often attacked scientific realism in these later publications. Moreover, Kuhn often argued that incommensurability undermines the scientific realist conception of scientific progress as an advance on truth.[250] The perceived threat that incommensurability poses to the scientific realist conception of scientific progress as advance on truth is based on radical ontological change evidenced in the history of science. For example, Kuhn views the transitions from Aristotle, to Newton, and then to Einstein, as having "no coherent direction of ontological development" (Kuhn 1970a, p. 206). From a historical perspective, Kuhn sees "no historical evidence for a process of zeroing in [on nature's real joints]" (Kuhn 1979b, p. 418). Kuhn developed incommensurability as an attack on the scientific realist notion of advance on truth. In the 1970s, Feyerabend followed suit, arguing that incommensurability undermines a conception of scientific progress as an increasingly better approximation to the truth (1970c, pp. 219-220; rp 1981b, pp. 152-153).

Preston correctly points out that Feyerabend argued that scientific theories should be interpreted as *attempts* to describe the world (Preston 1997a, p. 421), and Preston's reconstruction of Feyerabend as a *conjectural* realist provides an insightful presentation of some of Feyerabend's views. However, Preston goes much too far with the claim that:

> Feyerabend even goes further, insisting that theories tell us about what things *are*, their very *nature*, in a world which exists independently of measurement and observation (Preston 1997a, p. 421, italics in original).

249 Preston is not the only one to confuse Feyerabend's realism with scientific realism, also see e.g. Llyod (1996), p. 252.

250 For a discussion, see Sankey (1998), pp. 11ff.

Exactly this claim also appears in 'Feyerabend's Retreat from Realism" (Preston 1997, p. 61). In neither case is there a citation or reference to any statement in any of Feyerabend's publications.[251]

As the notion of scientific realism is notoriously slippery, as there have been many different characterizations of it. For the moment we can simply adopt Preston's characterization of scientific realism as set about above; that is, according to scientific realism, theories tell us about what things are, their very nature, in a world that exists independently of measurement and observation. Was Feyerabend ever such a scientific realist?

Although Feyerabend often argued for *realism* in the 1950s and 1960s, he was not arguing for *scientific realism* in those publications.[252] For example in his (1962a), Feyerabend adopted a non-realist metaphysical position that was explicitly neo-Kantian.[253] Although Feyerabend often argued that science *should* be conceived of *as if* it were 'an attempt at a realistic interpretation of experience' (the title of the 1958 version of his dissertation),[254] Feyerabend's realism is not a form of *scientific realism*. Preston, however, insists that Feyerabend was indeed a *scientific realist*, and that Feyerabend developed an unorthodox view of *scientific realism* (Preston 1997a,

251 Also see Oberheim and Hoyningen-Huene (1999), p. 227, and compare Preston (1999), p. 236. As we pointed out, another example of Preston failing to support his claim about Feyerabend is Preston's claim that Feyerabend's failed attempts to satisfy his critics, who had doubted the intelligibility of incommensurability, led to a "*tirade* which rests on the *very worst part* of *AM* [*Against Method*], an *indefensible* relativism with respect to logic" (Preston 1997, p. 186, italics inserted). Once again, Preston provides no justification, explanation or even citation to back up this strongly worded dismissal of Feyerabend's ideas. Preston responds to our review: "H-H & O [Hoyningen-Huene and Oberheim] joke that my book lacks sufficient citations, when I give up to eight citations for almost everything I quote," and he then provides a couple of the missing citations that we had used to make the point that he *often* makes specific claims about what Feyerabend wrote without providing any citation (Preston 1999, p. 236). In so doing, he conceded that, in fact, he did not provide sufficient citations. We were not joking. To make matters worse, he continues, "the relativism with respect to logic which *AM* contains isn't all that hard to find" (Preston 1999, p. 236). Then why not provide the citation? The point is not that we doubt that Feyerabend made relativistic claims with respect to logic. After all, as early as 1956, in the context of a discussion about Kant, Feyerabend had already argued that "the relation between logic proper and scientific theories is not clearly apprehended: For the growth of, say, physical theories might necessitate a revision of logic" (1956c, p. 263). Rather, the point is that meeting basic standards of scholarship is especially important in Feyerabend's case, because Feyerabend published and republished many different variations of ideas that have been widely misunderstood. There are four significantly different published versions of *Against Method* in English alone. Cf. (1970d), (1975a), (1988) and (1993a). So where exactly did Feyerabend adopt "relativism with respect to logic" in *Against Method*?

252 See Oberheim and Hoyningen-Huene (1999), especially pp. 229ff.

253 For an excellent discussion of such Neo-Kantianism and why it is not a scientific realist position, see especially Boyd (1992). For a recent critical examination of Neo-Kantian metaphysical elements in Kuhn's position, see Bird (2002), pp. 459ff.

254 See (1958a); rp (1981a), pp. 17-36.

p. 61, and p. 421). According to Preston, Feyerabend endorsed "conjectural realism", but "he never really embraced the important Popperian claim that our theories exhibit increasing correspondence with reality ('verisimilitude', or convergence to the truth)" (Preston 1997a, p. 61). But this is just how Preston appears to understand progress: "to say that we have made progress . . . [is to say] we move closer to the truth" (Preston 1997a, p.153). Preston also contradicts himself claiming that "By 'progress' *Feyerabend* usually . . . means progress towards the truth" (Preston 1997a, p. 153, italics inserted). With such a conception of 'progress', it is almost impossible to make sense of Feyerabend's arguments.

Confusions about Feyerabend's views on realism cause Preston difficulties in interpreting Feyerabend's other ideas. For example, Preston wrote:

> for Feyerabend, the nature of observation, the nature of meaning, and the relation between theory and observation are all determined by our decision to adopt particular theories about those matters. *It is hard to see how this can be squared with the realism* that Feyerabend simultaneously professes, for it subverts the idea that these 'theories' of observation and of meaning are about *mind-independent phenomena*, phenomena whose natures do not depend on our decision (Preston 1997a, p. 49, italics inserted).

Instead of trying to tease out what Feyerabend was getting at by arguing that science *should be understood as a realistic interpretation of experience*, while at the same time maintaining that the nature of observation, the nature of meaning, and the relation between theory and observation are all determined by our *decisions* to adopt particular theories about those matters, Preston quickly turned the apparent tension in Feyerabend's ideas into an unwarranted criticism. Preston recognizes that:

> the exact nature of [Feyerabend's] realism is an issue about which there can certainly be different interpretations (Preston 1999, p. 238).

While this is indeed the case, not all of the different interpretations of Feyerabend's notion of realism are equally well supported by Feyerabend's texts.

As Hoyningen-Huene and I pointed out in a review of Preston's book, some of things Feyerabend wrote are very difficult to square with scientific realism. For example:

> The realism [in my early work][255] invites us to *reject common sense* and to announce the discovery: *objective reality has been found to be a metaphysical mistake* (1981a, p. xii, italics inserted).

255 Feyerabend referred to "the realism of Vol. 1 ch. 2.6, ch. 4.6, chs. 9 and 11", which are the collection of his papers that deal with the topic: (1958a), (1962a), (1969e), and (1964d) respectively.

This sounded like an odd thing for a scientific realist to say. However, the fact that Feyerabend himself had claimed that the realism he developed in all of the major papers specifically on the topic in the years between 1958 and 1969 invite us to *reject* common sense in favor of the view that *objective reality has been found to be a metaphysical mistake* is simply waved aside by Preston, even though it is a clear and concise rejection of a central claim of scientific realism.

Confronted by such a criticism of his portrayal of Feyerabend's views on realism (Oberheim and Hoyningen-Huene 1999), Preston admits that Feyerabend *may* have adopted an anti-realist stance in the mid-1960s based on his interpretation of Quantum Theory (ignoring the fact that Feyerabend referred to papers which range from 1951/1958 to 1969). However, Preston continues to insist that he is:

> still not convinced that this is what his realism amounted to . . . I am still persuaded that he believed in a mind-independent reality. I think of his *liberal use of the concept of truth*, for example (Preston 1999, p. 239, italics inserted).

In other words, Preston continues to insist that Feyerabend, before the mid-1960s, was a scientific realist who believed that science describes a mind-independent world ('its very nature'). In his response to out review, Preston now provides six sources to try to back up his claim about Feyerabend's scientific realism. All of the sources are supposed to support Preston's interpretation of Feyerabend as a scientific realism because in each of them Feyerabend allegedly makes liberal use of the concept of truth.[256]

Instead of taking Feyerabend at his word (which admittedly can often be quite misleading), and having a look at the major articles Feyerabend wrote on the issue of realism between 1951 and 1969, Preston disregards Feyerabend's own statements *and* Feyerabend's major articles on the topic, and continues to insist that he is still right in his interpretation of Feyerabend's early ideas as *scientific realist.* He now basis his interpretation of Feyerabend's views on scientific realism, *especially his belief in a mind-independent reality*, on a remark made in a 1957 "Discussion note" that is less than two pages long, his interpretation of parts of two papers that Feyerabend specifically identified as an invitation to reject the idea of an objective, mind-independent reality (see 1958a; rp 1981a, pp. 17-36 and 1958b), a 10 page review on Reichenbach's interpretation of Quantum Mechanics, and a review of a book by Bohm. In sum, Preston based his claim on four short discussions of other people's views, and two papers that explicitly reject the view that Preston tries to pin on Feyerabend.

256 They are (1957d), p. 182, sections 3 and 7 of (1958a), (1958b), section 6 of (1958d), (1960e), and "perhaps even the difficult last section of" (1960f). See Preston (1999), p. 239.

Feyerabend did assume that there is some kind of external reality. This is obvious as according to his pragmatic or causal theory of observation, observations are caused by an external world. The pragmatic theory of observation is supposed to contrast with Berkeley's idealism, and with logical positivist accounts of sense-data as the objects of perception. Feyerabend based his pragmatic theory of perception on the causal theories of Descartes and Locke (1965b, p. 147), as well as on Russell (see Russell 1948, p. 203ff). Feyerabend also drew from Popper, Carnap and Neurath in the 1930s to defend and develop his causal theory of perception (although Carnap supposedly dropped it when he adopted the view that the objects of perception are sense-data). But the claim that there is an external world that causes our us to have perceptions is a fry cry from the claim that Feyerabend believed that we have access to a mind-independent reality, or that Feyerabend insisted "that theories tell us about what things *are*, their very *nature*, in a world which exists independently of measurement and observation" (Preston 1997a, p. 421, italics in original). After all, a belief in the existence of a mind-independent world is compatible with, or even a part of, many prominent forms of anti-realism,[257] such as Kuhn's neo-Kantianism, Putnam's internal realism,[258] and Van Fraassen's constructive empiricism; all of which reject scientific realism.[259] The question relevant to the issue of scientific realism concerns the kind of epistemic access we have to it.

There are indeed some passages in Feyerabend's work where he appears to be using a realist notion of truth (e.g. 1957d). As Feyerabend was well aware, scientific realist assumptions are very difficult to avoid as they are practically built into our language and our common sense view of the world. However, according to Feyerabend, this is not a reason to suspect that they are correct. Quite to the contrary, it makes them very difficult to test, or even argue for:

257 The term 'non-realism' is more appropriate than 'anti-realism', because the positions that are usually labeled 'anti-realist' are not diametrically opposed to realism as has often been assumed. Moreover, they do not result *because of* a denial of realism. Rather, properly understood, they differ from realism because they are developed with the intention of capturing different features of science and its dynamic. For example, the fact that for a neo-Kantian metaphysical interpretation of science, the question of entirely extra-linguistic reference cannot be sensefully answered is not a weakness of that position. Instead, it is a feature of a neo-Kantian metaphysical interpretation of science that such a question cannot be sensefully posed. See Oberheim and Hoyningen-Huene (1997), p. 462, fn. 10.

258 For a reconstruction of Kuhn's ideas as a form of neo-Kantianism, see Hoyningen-Huene (1993). For a more recent discussion, see e.g. Bird (2002), pp. 458ff. For internal realism, see e.g. Putnam (1981).

259 See e.g. Feyerabend (1957d), Van Fraassen (1980) and Van Fraassen (1984). For critical discussions, see Churchland and Hooker (1985), especially Musgrave (1985).

> I found that the arguments for realism worked only when the realist assumption had already been introduced (Feyerabend 1978a, p. 112).[260]

Moreover, as we shall see, Feyerabend practiced pluralism in philosophy, temporarily adopting different sides of the same issue, so it would come as no surprise to find Feyerabend adopting scientific realism in order to launch immanent criticisms of it, or even just to try out some different ideas, as on his own view, the merits of non-realist positions are best appreciated in contrast to the strengths and weaknesses of scientific realism. However, in the vast majority of his writings, and certainly not in the 1950s and 1960s, Feyerabend certainly did not adopt any kind of the position that a *scientific realists* would promote: That we have *access* to a mind-independent world, or that scientific knowledge discovers the true, objective structure of the mind-independent world.

Let us begin by examining some of the sources Preston belatedly used to back up his scientific realist interpretation of Feyerabend (belatedly, because he provided them only after he was criticized for not having supported his claim that Feyerabend was a scientific realist). The first two sources are (1958a) and (1958b). Both argue against Bohr's contention that classical concepts will always be needed. They argue that new kinds of concepts (which have no logical relations to the old ones; i.e. new kinds of incommensurable concepts) that do not fit within the classical framework could be invented in order to reinterpret Quantum Mechanics. The new kinds of concepts would jettison the idea of a mind-independent reality. As Feyerabend himself pointed out that his realism in these two essays invites us to conclude that the idea of a mind-independent reality was a metaphysical mistake, let us concentrate on some of the other pieces Preston uses to try to support his claim that Feyerabend was a scientific realist, and then try to reconstruct Feyerabend's notion of realism and compare it to scientific realism.

In section 6 of (1958d), which Preston cited as an instance of Feyerabend's liberal use of the concept of truth in support of his claim that Feyerabend believed in a mind-independent reality, Feyerabend rejected three arguments Reichenbach had used to support the claim that three-valued logic, and it alone, can provide an adequate interpretation of Quantum Physics (1958d, p. 55; rp 1981a, p. 243). In this two-page section, there is only a single instance of the term 'true' and none of 'truth', and there is no discussion of whether or not there is a 'mind-independent reality'. Here is the single instance of the term 'true':

> Reichenbach's third argument consists in pointing out that in his interpretation all the laws of quantum mechanics are statements which are either *true* or false,

260 For an elaboration of this point, see Oberheim and Hoyningen-Huene (1997).

> but never indeterminate (whereas within the Copenhagen interpretation they may be meaningless). We have shown above this statement is incorrect. We may summarize this section by saying that none of Reichenbach's arguments in favor of his own position and against the Copenhagen interpretation are tenable (1958d, pp. 56-57; rp 1981a, p. 245, italics inserted).

None of these statements, nor any other in section 6 of (1958d; rp 1981a, pp. 243-245), support Preston's interpretation of Feyerabend as a scientific realist, who believed in a mind-independent reality on the basis of his liberal use of the concept of truth.

In (1960e; rp 1981a, pp. 219-235), which according to Preston definitively demonstrates Feyerabend's realism (Preston 1997, p. 236), Feyerabend did argue for a from of 'realism'. However, according to this realism

> the facts of experience . . . are not regarded as unalterable building stones of knowledge; they are regarded as capable of analysis, of improvement . . . (1960e, p. 327; rp 1981a, p. 224)

Feyerabend cited (Popper 1957b) as the source of this conception of realism, where Popper argued for realism as a normative methodological ideal, not a factual description of the nature of scientific knowledge and its advance. Scientific realism does not regard the facts as capable of improvement. According to scientific realism, the facts are the facts, independently of our views.

However, sometimes it does indeed sound as if Feyerabend had been a scientific realist:

> This tradition [realism, see above] proceeds from the very reasonable assumption that our ideas *as well as* our experiences may be erroneous and that the latter give us at most an approximate account of what is going on in reality. Bohm's own point of view is closely connected with this tradition (1960e, p. 327; rp 1981a, p. 225).

Here Feyerabend may seem to assume a mind-independent reality. However, it is sometimes difficult to decipher exactly which statements represent the views of Bohm that Feyerabend is presenting and discussing, and which statements represent Feyerabend's own views. Take for example the following claim:

> *Nature itself forces man to speak the truth*, and it also forces him to speak in such a way that his theories have objective reference (1960e, p. 335; rp 1981a, p. 232).

This certainly sound like a contemporary *scientific realist*; and thus *appears* to support Preston's interpretation of Feyerabend as a scientific realist. Perhaps this is the basis of Preston's claim that Feyerabend was a scientific realist. But as is so often the case with Feyerabend, looks can be deceiving. Feyerabend continued (directly):

> This, then, is the epistemology behind *Bohm's belief* that every theory, however absurd it may seem at first sight, has some kind of truth in it and correctly mirrors

> what exists in the universe: the ill success of a theory which is outright wrong and does not describe anything whatever is corrective which after a very short time forces us to abandon it . . . *Knowledge is a natural process which leads to a mirroring, in the head of man, of the properties of the universe.* The mirror-image may be distorted at the edges . . . [but] this distortion does not reach into the center of the mirror which perfectly represents the situation at a certain level (1960e, p. 335; rp 1981a, p. 233, italics inserted).

Thus, it is *Bohm's* conception of realism here at discussion, not Feyerabend's own views. What does Feyerabend think of Bohm's realist views? He continued (directly):

> *I do not believe that this account of knowledge is a correct one.* . . clearly Hegel's notorious 'Alles Vernünftige ist wirklich' [Everything rational is real – my translation] is here lurking in the background . . . (1960e, p. 337; rp 1981a, p. 233, italics inserted. Cf 1995b, p. 355).

This book review does not support Preston's claim that Feyerabend was a scientific realist, whose liberal use of the concept of truth exposes a belief in a mind-independent reality. In fact, this book review contains a clear rejection of the scientific realist view that our knowledge mirrors the real structure of the objective world. But why does Preston think that Feyerabend, in his early work, was a scientific realist in the first place, and what were Feyerabend's views on realism?

6.2. Feyerabend's normative realism

The impression that Feyerabend was a scientific realist arises from the fact that throughout his major publications he endorsed "realism" over "instrumentalism".[261] However, in these publications, the *realism* for which Feyerabend argued was a *normative* claim about how best to set the aim of science. Feyerabend argued that scientists should interpret their theories as *attempts* to interpret experience realistically: Hence the title of the English version of his dissertation, "An Attempt at a Realistic Interpretation of Experience". The title is a concise statement of Feyerabend's *normative* conception of science. According to Feyerabend, *science should be understood as an attempt at a realistic interpretation of experience.* As we have seen, from this perspective, he developed "*thesis I*: the interpretation of an observation

261 For example, see (1958a); rp (1981a), pp. 17-36, (1958b), (1960a); rp (1981a), pp. 37-43, (1963a); rp (1999a), pp. 50-77, and especially (1964d); rp (1981a), pp. 161-175. It has been argued that because Feyerabend failed to distinguish between different kinds of instrumentalism, he has been incorrectly perceived to have argued against Dewey's ideas, see Moore (1972). For a short discussion of Feyerabend's position on the realism/ instrumentalism debate that aims to sort out the relevant issues, see Jakovljevic (1996).

language is determined by the theories which we use to explain what we observe, and it changes as soon as those theories change."[262]

Thus, Feyerabend's early realism was a prescriptive claim about sound scientific method, *not a factual claim about the nature of scientific knowledge and its advance*. Feyerabend often made this clear:[263]

> The issue between positivism and realism is not a factual issue which can be decided by pointing to certain actually existing things, procedures, forms of language, etc., it is an issue between ideals of knowledge (1958a, p. 155; rp 1981a, p. 33-34).[264]

Scientific realism is supposed to be a factual issue about the nature of scientific knowledge and its advance. Feyerabend's conception of realism, on the other hand, is not such a factual issue, it is an issue about what we should set as an ideal for science.

Feyerabend claimed that sound scientific method requires interpreting scientific theories as *attempts* to describe our experiences realistically, and not merely as instruments for the systemization of experience.[265] The reason that such a *realism* is better suited to science, according to Feyerabend, is that interpreting theories realistically both increases the severity of tests of theories, and it promotes the development of new points of view.[266] According to Feyerabend, realism (as a normative ideal) promotes scientific progress better than instrumentalism.

Tests of theories are more severe on a realistic interpretation than an instrumentalist interpretation, because on a realistic interpretation theories are forced to compete. One the assumption that nature has but one nature, only one of two realistically interpreted theories can be correct. On an instrumentalist interpretation, two conflicting theories are acceptable as they are only tools, not attempted descriptions. Similarly, conceiving of theories as a realistic interpretation of experience better promotes the development of new points of view, because one cannot simply accept inconsistencies between theories with the excuse that one or the other (or

262 See (1958a), p. 163; rp (1981a), p. 32.

263 See additionally (1951), (1958a); rp (1981a), pp. 17-36, (1958b), and especially (1961i); rp (1999a), pp. 50-77, where Feyerabend makes this explicitly clear. See also especially (1964d); rp (1981a), pp. 161-175.

264 Here again, for his views on realism, Feyerabend originally pointed to both Popper and Kraft, see (1958a), p. 155, fn. 25. In the reprint, the reference to Popper has been removed; compare (1981a), pp. 33-34, fn. 24.

265 For a discussion, see Preston (1997), pp. 61ff. Also see e.g. (1954c), pp. 473ff., (1961b), p. 83, (1962a); rp (1981a), pp. 44-96, (1964d); rp (1981a), pp. 161-175, (1965a); rp (1981a), pp. 97-103, and especially (1970d), pp. 82ff.

266 Feyerabend often made this clear, see e.g. (1958a), p. 143; rp (1981a), p. 17), (1958b), especially pp. 101ff., (1958d), pp. 50ff.; rp (1981a), pp. 237ff., (1962a), pp. 82-82; rp (1981a), p. 83, and (1962b).

both) are just instruments for the systemization of experience. Confronted with difficulties, theories, interpreted realistically, must be improved or exchanged. The difficulties cannot merely be accepted as limitations of the conceptual tool (1958a, pp. 143ff.; rp 1981a, pp. 17ff.).

This exposes the bases for Feyerabend's often repeated claim that theories are only incommensurable on a realist interpretation.[267] His point is simply that if one of two theories (or both) are interpreted instrumentally, i.e. merely as tools and not as attempted descriptions, then they cannot conflict as attempts to describe nature. The result is that there is no competition between them, and consequently no need to improve them or to find a better theory. Conceptually incompatible tools can be used for different purposes, and there is nothing stopping us from using such tools at different times. However, conceptually incompatible attempted descriptions can and do conflict. On a realistic interpretation, theories expressed with incommensurable concepts are mutually exclusive. This promotes competition; and consequently, progress.

Whether or not Feyerabend's arguments against positivist or instrumentalist interpretations of scientific knowledge on the basis of promoting progress succeed or not is beside the point. The point is that for *scientific realists* the claims that there is anthe external world and that our theories correspond to the external world are supposed to be factual claims. They are also supposed to be justified on the basis of science itself. However, according to the kind of *normative realism* that Feyerabend adopted:

> The step beyond experimental results can be made only with the help for which little or no initial justification is given. A point of view that is used for providing knowledge about reality is such a system of hypotheses . . . [these hypotheses] are unjustified conjectures for which no foundation can be given (1963e, pp. 321-322).

According to Feyerabend, the only support for the unjustified conjecture that there is an external world behind the phenomena of experience is methodological considerations, not *proof* or *induction* (1963e, p. 322. Also see 1958b, p. 103). It is a matter of convention. Thus, *normative* realism differs significantly from *scientific* realism, both in the type of issues at stake (normative versus factual), as well as the justification for the claims made (methodological versus inductive).[268]

267 See e.g. (1962a), pp. 40ff.; rp (1981a), pp. 52ff., or even as late as (1977b), p. 365, n. 1; also see (1978a), p. 68.

268 Scientific realists tend to base the justification on their position on a form of induction called 'abduction' or inference to the best explanation. For some background on abduction, see e.g. Lipton (1991). Note, however, that Lipton does not condone the use of abduction to support scientific realism, see Lipton (1991), pp. 158ff. It has been argued

Feyerabend did not originally adopted this *normative* realism solely from Popper, whom he sometimes cited in support of his views on realism (e.g. 1960e, p. 327; rp 1981a, p. 224). He had it from his Ph.D. thesis supervisor, Viktor Kraft, who had argued that "realism cannot be *proved* by philosophical reasoning" (1966a, p. 4, italics in original). Feyerabend adapted Kraft's normative form of conjectural realism to Popper's views.[269] But this normative form of conjectural realism does not imply the main claims of contemporary *scientific* realism; i.e., that scientific theories actually succeed in capturing the true structure of the world. In fact, according to the normative perspective of Feyerabend's early realism (i.e. not the descriptive perspective of scientific realism), the nature of scientific knowledge is not even a factual issue. It is the result of our epistemic decisions. Thus, whereas Preston characterized Feyerabend as having initially been a 'scientific realist', and then having 'retreated from realism', Feyerabend was actually never a scientific realist.

To make things even worse for Preston's account, very early on, i.e. before Feyerabend adopted a version of *normative* realism, he was initially very skeptical about even such a limited form of realism. Recall that in the 1940s and 1950s logical positivism would have rejected scientific realism as baseless metaphysics. In fact, in the late 1940 and early 1950s, Feyerabend used logical positivist arguments to reject any such realist ontological speculations as baseless metaphysics. In his autobiography, he recounts how at university, he would attend seminars in the social sciences, and then scoff at their discussions as meaningless gibberish, as they could provided verification conditions for their claims. As for his attitude toward realism in the late 1940s and early 1950s:

> I had heard realistic arguments before; as far as I was concerned they were all circular: you made an assumption that contained a realistic core and then proved realism by revealing the core. *It took Walter Hollitscher two years to convince me that the circularity is practical, not vicious, and that it was an asset, not a disadvantage.* Walter started by pointing out that scientific research was conducted in realistic terms. I countered that scientists, unfortunately, had not yet shed their metaphysical egg shells. Metaphysics or no, said Walter, scientists achieve results that are accepted by everybody, including positivists, where as if they had adopted an antiseptic language and strict logic they would never have gotten anywhere. That shut me up for a while—but a residue of doubt remained (1995a, p. 72, italics inserted).

These 1995 recollections of the late 1940s and early 1950s are supported by the account of his early views on realism that he had given in 1966:

---

that the use of such adductive arguments to support scientific realism are circular in Fine (1984a), Laudan (1984a), and Oberheim and Hoyningen-Huene (1997). For a defense of abduction in support of scientific realism, see Psillos (1999).

269 See Popper (1963), especially, "Three views concerning knowledge", pp. 97ff. See also Preston 1997, for an excellent discussion.

> We were not satisfied by his [Hollitscher's] arguments [for realism], for we thought, in accordance with Kraft's ideas, that it was possible to interpret science in a positivistic manner and that such an interpretation did not require an external world . . . (1966a, p. 4).

According to Feyerabend, it was a combination of Feigl's "Existential Hypotheses" and the ideas Popper had explained in Alpbach in 1948 and 1949 that greatly diminished his (and the rest of the Kraft circle's) doubts about realism (1966a, p. 4). Thus, in the late 1940s, while at university, Feyerabend defended positivism *against normative* realism (not *scientific* realism), but eventually sometime after 1950, Feyerabend adopted a realistic point of view (normatively). This is a far cry from Preston's claim that Feyerabend was originally a scientific realist, who then retreated from that position.

In fact, Feyerabend did not retreat from scientific realism. He was never a scientific realist. He started as a positivist who rejected the hypothesis of the external world as baseless metaphysics. Then, he became a normative realist who insisted that although the hypothesis of the existence of the external world is factually baseless, it is methodologically fruitful. Feyerabend's early normative realism was not a factual issue to be supported on the basis of the inductive sciences or by an abductive inference. As we shall see, actually Feyerabend defended and developed a form of anti-realism from the 1950s to the 1990s.

6.3. Feyerabend's anti-realism

According to Howard Sankey:

> it would undoubtedly be an important discovery, if it were shown that Kuhn and Feyerabend meant all along for incommensurability to be understood in the context of an anti-realist position of the kind described by Hoyningen-Huene (Sankey 1997b, p. 430).

The kind of anti-realist (or non-realist) position that Hoyningen-Huene reconstructed on the basis of Kuhn's ideas developed the incommensurability thesis in the context of a neo-Kantian metaphysical position (Hoyningen-Huene 1993). According to such a position, scientists do not discover objective truths about a mind-independent reality. They invent theories to contribute to the creation of how they experience the world. Such 'phenomenal worlds', in which scientists live and work, are 'co-constituted' by 'subject-sided' and 'object-sided moments' (Hoyningen-Huene 1993). Such a non-realist, neo-Kantian position is also the kind of metaphysical perspective from which Feyerabend developed his notion of incommensurability in 'Explanation, Reduction and Empiricism' (1962a).

Although he wasted few words developing his basic metaphysical position, Feyerabend made this explicit in a concise statement in the introductory section:

> The position may be said to consist of two ideas. The first idea is that the influence, upon our thinking, of a comprehensive scientific theory, or of some other general point of view goes much deeper than is admitted by those who would regard it as a convenient scheme for ordering facts only. According to this first idea scientific theories are ways of looking at the world; and their adoption affects our general beliefs and expectations, and thereby our experiences and our conception of reality. *We may even say that what is regarded as "nature" at a particular time is our own product in the sense that all the features ascribed to it have first been invented by us and then used for bringing order into our surroundings.* As is well-known, it was Kant who most forcefully stated and investigated this all persuasive character of theoretical assumptions (1962a, p. 29; rp 1981a, p. 45, italics inserted).

The second idea of Feyerabend's position is to temper this Kantian perspective with a critical attitude bent on innovative theoretical advancement. The "categories" that shape our conception of reality are not fixed, but vary historically (cf. Feyerabend 1962a, Einstein 1949b, and Hoyningen-Huene's 1993 reconstruction of Kuhn). The paradoxical situation with which Feyerabend was struggling in (1962a) is that although the theoretical assumptions of our universal theories shape our conception of reality and thereby all of our experiences of it, our theories must still be open to test on the basis of experimental evidence. The requirement that theories must be testable is needed to maintain the possibility of making progress:

> However, Kant also thought that the very generality of such assumptions and their omnipresence would forever prevent them from being refuted. As opposed to this, the second idea implicit in the position to be defended here demands that our theories be testable . . . It is this second idea which makes science proceed to better and better theories . . . (1962a, pp. 29-30; rp 1981a, p. 45).

Thus, Feyerabend explicitly adopted the Kantian idea that our conception of nature has been invented by us in order to bring order into our surroundings, and he made this into a form of neo-Kantianism by stressing that categories that shape what we experience as real are historically variable. They change in the course of scientific advance.

Feyerabend developed his Kantianism through a combination of his work on Wittgenstein's nominalism, Bohr's complementarity, and perhaps Einstein's (1949b).[270] Einstein stresses that our conception of nature is the

270 For Wittgenstein, see (1954a), (1954b), (1954d), (1955a); rp (1981b), pp. 98-130. For Bohr, see (1954e), (1957b), (1957c); rp (1981a), pp. 207-218, and (1958b). With regards to Einstein, Feyerabend never cited directly from Einstein (1949b) in his (1962a). However, Feyerabend did cite from Bohr's contribution to Schilpp (1949) many times in the 1950s in his work on foundational problems in Quantum Mechanics and directly in his (1962a) as

result is the result of a free play of ideas, and that his views are similar to Kant's but without *necessary*, *unalterable* categories and forms of pure understanding. For example:

> A few more remarks of a general nature concerning concepts and [also] concerning the insinuation that a concept - for example that of the real - is something metaphysical (and therefore to be rejected). A basic conceptual distinction, which is a necessary prerequisite of scientific and pre-scientific thinking, is the distinction between "sense-impressions" (and the recollection of such) on the one hand and mere ideas on the other. . . . we shall make use of this distinction unconcerned with the reproach that, in doing so, we are guilty of the metaphysical "original sin". We regard the distinction as a category which we use in order that we might the better find our way in the world of immediate sensations. The "sense" and the justification of this distinction lies simply in this achievement. But this is only a first step. We represent the sense-impressions as conditioned by an "objective" and by a "subjective" factor. For this conceptual distinction there also is no logical-philosophical justification. But if we reject it, we cannot escape solipsism. It is also the presupposition of every kind of physical thinking. Here too, the only justification lies in its usefulness. We are here concerned with "categories" or schemes of thought, the selection of which is, in principle, entirely open to us and whose qualification can only be judged by the degree to which its use contributes to making the totality of the contents of consciousness "intelligible." . . . *The theoretical attitude here advocated is distinct from that of Kant only by the fact that we do not conceive of the "categories" as unalterable (conditioned by the nature of the understanding) but as (in the logical sense) free conventions. They appear to be* a priori *only insofar as thinking without the positing of categories and of concepts in general would be as impossible as is breathing in a vacuum* (Einstein 1949b, pp 666-667).

The italicized portion express the same kind of position as Feyerabend delineated in the introduction of the essay where he first used the term 'incommensurable' in 1962.[271]

This position contrasts sharply from scientific realism. According to scientific realism, our conception of nature results from the structure of nature itself. Our conception of reality is supposed to mirror the true structure of the world, and it is supposedly *caused* by the real, objective structure of the world. According to Feyerabend's position, on the other

well. This is the same book in which Einstein (1949a) and (1949b) appear. As we have seen in chapter 5, Feyerabend (1962a) bears many remarkable similarities to Einstein's two discussions in Schilpp (1949). Moreover, in a German version of *Against Method*, Feyerabend claimed that his ideas would have seemed trivial to Einstein (1983a, p. 12), and later Feyerabend often cited Einstein's characterization of the scientist as appearing like a scrupulous opportunist to the systematic epistemologist, which is in Einstein (1949b). Given Feyerabend's lack of reliability when it comes to identifying his sources, these are very strong reasons for suspecting that Feyerabend was drawing from Einstein's 1949 publications. Feyerabend was also well-versed in Kant's ideas from other sources (e.g. 1956c).

271 For an excellent, well-referenced, development of Einstein's neo-Kantianism, see Rosenthal-Schneider (1949) also in Schilpp (1949).

hand, our conception of nature is our own product, invented by us, and projected as reality. when Feyerabend defended and developed the term 'incommensurable concepts', he explicitly described his position as a form of neo-Kantianism.

One might object that in his specification of the first Kantian point, Feyerabend only claims that it is our conception of reality, not reality itself, that is shaped by our theoretical assumptions. But if we adopt a Kantian perspective, and that is what Feyerabend is explicitly doing, then we adopt a distinction between the world we experience, i.e. our conception of the world, and the world in-itself. Feyerabend is claiming that the world we experience as real depends on the theories we bring to it, and these experiences change when we change the theories that contribute to its constitution. It is Kant-on-wheels.

We do not, however, need to rely on this single passage to support the case for Feyerabend's anti-realist neo-Kantianism. In order to strengthen the case for the view that Feyerabend adopted a non-realist position in his (1962a), and not a *scientific realist* position, let us consider some of the other remarks that Feyerabend made on the relevant issues in that piece, and compare them to *scientific realism.* In order to do this, a brief explication of scientific realism as it is understood in contemporary philosophy of science will be necessary. Unfortunately, this is somewhat more complicated then it may at first sound, as contemporary scientific realists do not share a uniform position.

The usual starting point for contemporary discussions of scientific realism in the philosophy of science is the Boyd-Putnam formulation according to which the terms in mature scientific theories typically refer, and the theories accepted in a mature science are typically at least approximately true.[272] This formulation was an explicit attempt to capture what most scientific realists would agree upon, and many of the subsequent papers in the realism debate specifically refer back to it. However, ensuing discussions have shown that there is no single agreed upon statement of the scientific realism. In fact, scientific realists continue to disagree about many different issues. These include views on the nature of explanation, views on the nature of the meanings of terms, views on the nature of the reference relation, views on the nature of causation, views on the nature of natural kinds, views on the nature of essences, truth, objectivity, causation, and so on. The wide range of issues over which scientific realists disagree among themselves makes it difficult to delineate a basic position acceptable to all of them. This has led some authors to suggest that attempting to explicate scientific realism as a position is the wrong ap-

272 Putnam credited Boyd with this characterization in Putnam (1975c), p. 73.

proach. According to this view, all scientific realists share a certain general form of reasoning. The success of some aspect of science justifies a belief in the approximate truth of that aspect of science. While realists tend to disagree about where and why to apply such an inference, as well as about the exact nature of the conclusion of the inference in each particular case, they tend to infer truth from some form (explanatory, predictive, technological) of success. As we shall see, Feyerabend rejected exactly this type of abductive reasoning to the conclusion of the truth or approximate truth of scientific theories.

The many attempts to explicate scientific realism, and to delineate arguments both in favor and against scientific realist and non-realist positions has resulted in the realism debate.[273] None of the discussions of scientific realism just cited take up the same issue that Feyerabend addressed with the term 'realism' in his 1950s and early 1960s publications. The contemporary realism debate concerns the correct description of scientific knowledge and its development, not a normative issue about which ideal of knowledge is best to pursue.

Scientific realist positions usually contain both a characterization of scientific knowledge and a characterization of scientific advance. Scientific realist characterizations of knowledge usually have two interrelated components. There is an *epistemic* claim about the nature of knowledge: Our best theories are at least approximately true, and there is a *metaphysical* claim about the nature of existence:[274] the objective world exists mind-

273 By the realism debate, I mean the vast literature on the issue of scientific realism, as it is addressed in the contemporary literature in the philosophy of science. See e.g. Boyd (1981), Boyd (1983), Boyd (1984), Boyd (1985), Boyd (1989), Boyd (1990), Boyd (1992), Bunzl (1994), Carrier (1991), Carrier (1993), Cartwright (1983), Chalmers (1999), Churchland (1979), Churchland and Hooker (1985), Couvalis (1997), chapter 7, Devitt (1984), Devitt (1997), Devitt (2001), Douven and Brakel (1995), Ellis (1985), Fine (1984a), Fine (1984b), Fine (1986a), Fine (1986b), Fine (1991), Ghosh (1998), Glymour (1982), Glymour (1992), Hacking (1983), Hacking (1984), Hacking (1993), Hesse (1977), Kitcher (1993), chapter 5, especially p. 150, Kukla (1994), Kukla (1995), Laudan (1977), Laudan (1984a), Laudan (1984b), Leplin (1984a), Leplin 1(984b), Mackinnon (1979), McMullin (1984), McMullin (1987), McMullin (1991), Musgrave (1982), Musgrave (1988), Musgrave (1989), Musgrave (1999), Newton-Smith (1981), Psillos (1994), Putnam (1975a), Putnam (1981), Putnam (1983), Putnam (1984), Putnam (1987), Putnam (1990a), Putnam (1990b), Putnam (1994a), Putnam (1994b), Reiner and Pierson (1995), Salmon, Earman et al. (1992), Sankey (1994), Sankey (1997b), Sankey (1998), Sankey (2000), Sankey (2001), Sankey (2002), Van Fraassen (1980), Van Fraassen (1984), Van Fraassen (1985), Weber (2004), chapter 9, and Worrall (1982), among many others.

274 Because the term 'metaphysical' is understood in different ways by the various participants in the realism debate, I will briefly elaborate. Scientific realists make a metaphysical claim because they purport to tell us what exists, and how it exists. It is the world and its properties which exist, and they exist mind-independently; that is, they exist independently of the acts and ideas of people. *This is a claim about the nature of existence.* This makes it a metaphysical claim. Some participants continue to understand 'metaphysical' as meaning

independently. These two components are inter-related in that theories can only be true if the entities that they posit exist; and to justify believing in the existence of an entity, at least some approximately true description of its properties is needed. In other words, scientific realism requires that it is the fact that the entities posited by our best theories exist in a mind-independent word that makes those theories true. For this reason, most scientific realists subscribe to both the epistemic and metaphysical claims, as well as a correspondence theory of truth to connect the two.

This does not hold for Cartwright and Hacking, who develop a position called "entity realism", according to which we are justified in believing that the objects exist, but not that the theories are true. [275] In other words, they accept (b) but reject (a). However, entity realism has been justly criticized exactly for trying to separate these components.[276] To see that without the correspondence theory of truth, the epistemic and metaphysical claims do not necessarily add up to a scientific realist position, consider the following example. The epistemic claim can be developed with some version of a coherence theory of truth, according to which a statement is true in virtue of its relations to other statements, not in virtue to a correspondence to the world. On such theories, what makes statements true does not depend on its relations to reality.[277] One could even also accept that the existence of a mind-independent world. However, the resulting position is not a scientific realism, because it is not the objective properties of the world which makes scientific claims at least approximately true. Here is another example that illustrates why scientific realism requires both the metaphysical claim and the correspondence theory of truth in addition to the epistemic claim. One could accept the metaphysical claim of scientific realism (that there is an objective, mind-independent world), but then deny the epistemic claim by insisting that we have no access to it. For example, an 'internal realist' position (which is a from of non-realism) can accept the claim that scientific theories are at least approximately true, but according to internal realism, truth is only relative to one particular conceptual perspective (Putnam 1981). Consequently, an internal realist is not really a scientific realistic position at all. Similarly, the neo-Kantian metaphysical position that Hoyningen-Huene reconstructed from Kuhn's ideas can also accept the claim that mature scientific theories

non-empirical (as in the sense of logical positivism), while at the same time the move toward naturalism has made it fashionable to discuss an empirical metaphysics; see e.g. Devitt (1984), and Devitt (1997), as well as Devitt (1996), pp. 48ff.

275 See e.g. Cartwright (1983), Hacking (1984), Hacking (1984).

276 See e.g. Reiner and Pierson (1995) and Weber (2004), chapter 9.

277 See e.g. Lehrer (1990), especially pp. 132ff., Blanshard (1941), pp. 260-308, and Kirkham (1992).

are at least approximately true, but they are true of a particular phenomenal world, not a mind-independent world (Hoyningen-Huene 1993). This is what makes it a non-realist, and not a scientific realist position. In order for a position to be scientific realist, it must accept the metaphysical claim as well as the correspondence theory of truth: Scientific knowledge is at least approximately *true of the mind-independent world, and it is the objective properties of the world which makes scientific knowledge at least approximately true.* As for scientific advance, according to scientific realist characterizations of it, scientific advance is an ever better approximation to the truth. It is implicit within this characterization of scientific advance that progress is leading toward the one true description of the objective world. The question is: Was Feyerabend using or arguing for a scientific realist position in his (1962a)?

In his (1962a), Feyerabend proposed his pluralistic test-model and argued its superiority to monistic test-models on the grounds that it better promotes progress. He claimed:

> Persistence of a single point of view will lead to the gradual establishment of well-circumscribed methods of observation and measurement; it will lead to codification of the ways in which those results are interpreted; it will lead to a standardized terminology and to other developments . . . the gradual acceptance of a theory by an ever-increasing number of people must finally bring about a transformation even of the most common idiom that is taught in very early youth. In the end, all the key terms will be fixed in an unambiguous manner, . . . and the idea that they are copies of unchanging entities and that change of meaning, if it should happen is due to human mistakes—this will now be plausible. Such plausibility reinforces all the maneuvers which may be used for the preservation of the theory (elimination of opponents included) (1962a, p. 69).

This is how Feyerabend presented the conception of scientific knowledge and scientific advance on the view that our theories are ever better approximations to the truth that refer to unchanging entities. He did not defend such a view. He attacked it on the grounds that it does not promote progress, but inhibits it:

> it will seem that at last an absolute and irrevocable truth has been arrived at. Disagreement with facts may of course occur, but, being now convinced of the truth of the existing point of view, its proponents will try to save them with the help of *ad hoc* hypotheses. . . The result will be absolute truth, but at the same time, it will decrease in empirical content to such an extent that all that remains will be no more than a verbal machinery which enables us to accompany any kind of event with noises (or written symbols) which are considered to be true statements by a theory (1962a, pp. 69-70; rp 1981a, pp. 75. Cf. 1963a; rp 1989a, pp. 114-115; rp 1999a, p. 95).

Feyerabend added a footnote to this negative assessment of the realist idea that scientific advance is a process that leads to the one correct description of the world. Part of the footnote makes it very clear that he was not a scientific realist:

> *Even the stability of a testable hypothesis cannot be regarded as a sign of its truth* because it may be due to the fact that, owing to some particular astigmatism on our part, we have as yet overlooked some very decisive test. *There is no sign by which factual truth may be recognized* (1962a, p. 70, fn. 86, italics inserted; rp 1981a p. 75, removed entirely).

Feyerabend is claiming that even the fact that some hypothesis may have passed all of the tests that have been suggested so far, i.e. even a very successful stable hypothesis, should not be regarded as true, because it may be that we have simply overlooked some decisive test that would refute it. Success is a function of popularity, where as 'truth' purports to describe a unique match between theory and experience. According to Feyerabend, *the success of our theories does not warrant a belief in their truth.* Some participants in the realism debate claim that it is acceptance of exactly this principle that distinguishes scientific realists from anti-realists (Fine 1991, p. 82, and Ellis 1985). Feyerabend concluded that:

> whereas unanimity of opinion may be fitting for a church, or for the willing followers of a tyrant, or some other kind of 'great man', variety of opinion is a methodological necessity for the sciences and a fortiori, for philosophy (1981a, p. 71, italics in original).

And on the basis of his pluralistic conception of knowledge, Feyerabend argued against the scientific realist idea that scientific advance converges on the one true theory:

> At all times, the existence, within certain traditions, of a variety of opinions (or a variety of theories) has been regarded as proof of the unsoundness of the method adopted by the members of this tradition. *It was assumed, as being nearly self-evident, that the proper method must lead to truth, that the truth is one,* and that the proper method must therefore result in the establishment of a single theory and the perennial elimination of all alternatives. Conversely, the existence of various points of view ... was always regarded as a sign of confusion (1962a, p. 87, italics inserted).

Feyerabend proceeded to argue against the idea that the we should be confident that the distinctions in our language capture real distinctions in the world, because it has been successful. We should not assume that our language captures the real structure of the world. He argued *against* the idea that:

> The existence of *certain distinctions in a language may be taken as an indication of similar distinctions in the nature of things,* situations and the like. And the reason for this is the fact that people who are in constant contact with things and situations will soon develop the correct linguistic means for describing their properties. In short: human beings are good inductive machines (1962a, p. 87, italics inserted).

In conclusion, Feyerabend explicitly set out a non-realist, neo-Kantian position in his (1962a), where he first set out the term 'incommensurable'. According to the position, what we experience as reality is shaped by our theoretical assumptions. They are products of our own invention. A number of other passages clearly demonstrate that Feyerabend was not a scientific realist in that publication. He rejected the scientific realist ideas that science converges on the one true theory, that a theory's success warrants belief in its truth, and the belief in the view that the distinctions in our language mirror real distinctions in the world on account of their usefulness. Preston's baseless claim that Feyerabend insisted that theories tell us about what things are, their very nature, in a world which exists independently of measurement and observation is simple bogus (Preston 1997a, p. 421).

Although Feyerabend clearly used Kantian considerations in the development of his metaphysical position, this should *by no means* be taken to suggest that Feyerabend was a proponent of Kant's transcendental philosophy, or that he believed in transcendental necessities. He was not, and did not. Throughout the 1960s, Feyerabend explicitly and repeatedly rejects transcendental necessities.[278] Feyerabend's use of Kant's ideas to develop his metaphysical position does not support the idea that there is some set of conceptual presuppositions that "necessarily underpin our descriptions of the worlds." This is exactly the kind of conceptual conservativism that Feyerabend repeatedly attacked. At most, Feyerabend's use of Kant's ideas supports the idea that, necessarily, there are some conceptual presuppositions that underpin our descriptions of the world. The qualification "at most" signifies the further restriction that, according to Feyerabend, even this limited requirement is the result of a decision to interpret our best theories realistically, i.e., as purportedly referring; and this decision is also not forced upon us by some transcendental necessity. But which set of presuppositions we choose to make is up to us. It is not forced upon us by nature. In other words, in order to make descriptions of the world, we must rely on some conceptual presuppositions. Feyerabend used a Kantian idea, that our theoretical assumptions affect our conception of reality, to present his metaphysical position.[279] At the same

278 See (1962a), pp. 69-70; rp (1981a), pp. 74-75. Also see (1965b), fn. 116, p. 247 and accompanying text, and (1965c), p. 244; rp (1981a), p. 126. Eventually Feyerabend added the philosophy of science to his list or worthless endeavors such as ordinary language philosophy and Kantianism: "I think that fields such as the philosophy of science, or elementary particle physics, or ordinary language philosophy or Kantianism should not be reformed, but should be allowed to die a natural death" (1978a p. 12).

279 Traces of Feyerabend's Kantianism can be found throughout (1958b), see e.g. 81, 83, and 84. Also see (1958a), pp. 150-151; rp (1981a), pp. 22-23. For a discussion of Feyerabend' non-realism as a form of Berkeley's idealism, also see Ghosh (1998).

time, he repeatedly and adamantly rejected the Kantian idea of transcendental necessities as an unjustifiable and undesirable form of conceptual conservativism. Feyerabend attempted to collapse the distinction between interpretations of phenomena and the facts very early on:

> In such a situation a distinction between phenomena and interpretations on one side and phenomena and objective facts on the other cannot readily be drawn (1958a, p. 158; rp 1981a, p. 28).

Feyerabend also made the Kantian nature of his own metaphysical picture extremely clear in his review of Kraft's *Erkenntnislehre* that he praised as "a masterpiece of clarity and incisive argument [which] should be read by anyone intending to take a problem of epistemology or of the philosophy of science" (1963e, p. 319). Feyerabend cited Kraft,

> the hypothesis of a world of objects which exists outside the reality of our experiences can only be tested on the basis of experience; . . . the reason is that all other facts already involve the hypothesis to be tested (1963e, p. 320).

This is a concise statement of a fundamental problem with trying to justify realism, i.e. "all other facts already involve the hypothesis to be tested" which makes independent, non-question begging justifications of realism difficult (cf. Oberheim and Hoyningen-Huene 1997). Feyerabend explained Kraft's position as having "great similarity to Kant's point of view" (1963e, p. 322):

> The domain of experience corresponds to Kant's domain of phenomena or to his 'nature' which is just phenomena connected by laws. And the real objects seem to be very similar the *Ding an sich* of Kant: they are not accessible to direct observation, they are postulated in addition to the perceptions and they seem to be altogether useless (1963e, p. 322).

Although Feyerabend did not develop these ideas in much detail, clearly the picture is supposed to be some form of 'Kant on wheels'.[280] There are, however, two main differences between Kant and Kraft, according to Feyerabend. First, Kraft "establishes a connection between the real posited objects and the experience which allows him to control properties of the former by reference to that latter" (p. 322) (this seems somewhat anti-Kantian but is not further developed), and instead of transcendental deduction, a methodological, normative justification is given for the otherwise unjustifiable supposition of the external world. For example:

> The hypothesis of the existence of the material objects is therefore not only an essential part of our thinking, in the sciences and within common sense, a part

280 The expression is due to Peter Lipton. See e.g. Lipton (1993) and (2001). Feyerabend had the basic idea already in (1956c), p. 263. For a critical exposition of Feyerabend as such a Kantian, see Rorty (1979), pp. 269ff.

> without which much apparently valuable knowledge would simply collapse . . . it is also capable of support. (1963e, p. 323).[281]

Keep in mind that by 'hypothesis', Feyerabend explicitly meant unjustified conjecture. Scientific realists do not usually call the postulation of the external world an unjustifiable conjecture, and if they were to agree that this is correct, they would no longer be scientific realists. Moreover:

> The support is not by 'proof' or 'induction' it is by methodological argumentation that is by reference to some of the norms which constitute or epistemology (1963e, p. 323).

In other words, for Feyerabend, the unjustified conjecture of the existence of the external world cannot be supported by facts, but only by norms. The idea that the existence of the external world cannot be supported by facts is unacceptable to a scientific realist.

This neo-Kantianism can be found throughout Feyerabend's philosophy. Feyerabend also presented and discussed a version of such a 'Kant on wheels' metaphysical position in a number of other publications. For example,

> . . . Kant's proposal to derive the structure of the mind from the structure of a theory then widely held. This proposal rested on the features of a *single* theory. The possibility that other theories might bend the forms of perception and the categories to accommodate *their* structure was either not consider, or rejected outright, and without any experimental investigation (1965c, pp. 246-247; rp 1981a, p. 128, italics in original).

As well as in later publications, as for example:

> Now when examining traditions "from the inside" we must adopt the ideas and procedures of the participants, and we must try to reconstruct the world as it looked to them (their 'phenomenal world'). If the participants do not distinguish real object and theoretical object, then we must not draw this distinction either, and 'symptomatic readings' which import external criteria, are out (1978a, p. 171, fn. 38).[282]

In the 1970s, Feyerabend continued to endorse conjectural realism, but insisted that "no method can guarantee an approach to the 'truth'" (1970a, p. 65). And once again, he stated his position as a modified form of neo-Kantianism (1970a, p. 73). the neo-Kantian interpretation of Kuhn that Hoyningen-Huene reconstructed is presented partly in Hegelian terms. Specifically, a key element to Kuhn's metaphysical disagreement with realism is that the phenomenal world is co-constituted by subject-sided and

281 Compare, "objective reality has been found to be a metaphysical mistake" (1981a, p. xii).

282 For some other instances of Feyerabend expressing his position as a dynamic form of neo-Kantianism in the 1960s, also see (1965b), p. 259, fn. 161, p. 216, and p. 221, fn. 8.

object-sided 'moments' *that are inseparably bound.*[283] The points that there is no way to tell exactly which part of our experience is determined by us. And which part by the world-in-itself. For this reason, according to Hoyningen-Huene, Kuhn's anti-realist theory of scientific development shares many similarities with Hegel's *Phenomenology of Spirit* (Hoyningen-Huene 1993, p. 222, ft. 139). Feyerabend also used such Hegelian ideas to urge non-realist interpretations of science.[284] For example, according to Feyerabend:

> We must proceed dialectically, i.e., by an *interaction* of concept and fact (observation, experiment, basic statement, etc.) that affects *both* elements. The lesson for methodology is, however, this: Do not work with stable concepts. Do not eliminate counterinduction. Do not be seduced into thinking that you have at last found the correct description of 'the facts' when all that has happened is that some new categories have been adopted to some older forms of thought, which are so familiar that we take their outlines to be the outlines of the world itself (1970d, p. 36, italics in original).

Feyerabend's development of Hegel's ideas to support his ideas is specifically directed against the idea that concepts and real things can be treated separately (1970d, p. 35). He is quite explicit about his rejection of realism based on its failure to recognize the inseparability of subject and object:

> A 'realist' may admit changes of the observational matter, but he will insist on the separation of the subject and the object and he will try to restore it whenever research seems to have found fault with it. Believing in an 'approach to truth,' he will also have to set limits on the development of concepts. For example, he will have to exclude incommensurable concepts from a series of succeeding theories.

283 See Hoyningen-Huene (1993), pp. 64ff. McDowell couches a very similar idea in Kant's terms, i.e., 'sensibility' and 'receptivity'; as for example, when he claims that "we must not suppose that receptivity makes an even notionally separable contribution to its contribution with spontaneity" if we are to avoid falling into the 'Myth of the Given' McDowell (1994), p. 51. McDowell attempted to make sense of Kant's notions of experience and knowledge by focusing on Kant's slogan: "Thoughts without content are empty, intuitions without concepts are blind." According to Kant, knowledge springs from two sources of the mind. The first is the capacity to receive sensations (receptivity) and the second is the power of knowing an object through these representations (spontaneity): "Without sensibility no object would be given to us, without understanding no object would be thought." Hence, "Thoughts without content are empty, intuitions without concepts are blind" Kant (1781), p. 92-93 A51/B75. One of the main points of McDowell's book is quite similar to the point Hoyningen-Huene expressed in Hegelian terms – that the world we experience is co-constituted by subject-sided and object-sided moments *that are inseparably bound.*

284 For a clear introduction to the Hegelian aspects of Lakatos' works to which Feyerabend was referring, and which Feyerabend himself developed in the original version of *Against Method,* (1970d), pp. 31-36; see Larvor (1998), especially pp. 22-30 and pp. 62-70, and compare to Hacking (1970). For praise of the idea of applying Hegel's insights on the limitations of methodological rules (as well as Mao's and Lenin's) to the philosophy of science, see, e.g., (1970b), ft. 27B, p. 101.

> This is the traditional attitude, up to, *and including* Popper's critical rationalism. As opposed to it, the attitude about to be discussed regards any prolonged stability, either of ideas and impressions which are capable of test or of background knowledge which one is not willing to give up (realism; separation of subject and object, commensurability of concepts), as an indication of *failure* pure and simple. Any such stability indicates *that we have failed to transcend an accidental stage of knowledge, and that we have failed to rise to a higher stage of consciousness and of understanding* (1970d, p. 30).

Feyerabend proceeded to criticize such stability that results in "our thinking los[ing] its spontaneity until we are reduced to the 'bestial and google-eyed contemplation of the world around us'" (citing Hegel's flowery prose) (1970d, p. 31). Feyerabend also praised Lakatos for being a Hegelian, but criticized him for not coming out of the closet about it: "One must praise Lakatos for having made such excellent use of his Hegelian upbringing [in Lakatos (1963-1964)]. On the other hand one must perhaps also criticize him for not revealing the source of his inspiration in a more straightforward manner . . ." Feyerabend goes on to chastise Lakatos for preferring to be seen as indebted to Wittgenstein, "a much less comprehensive and more mechanical school of thought", and insinuates in a provocative tone that Lakatos' "temporary membership in this school made him lose his sense of perspective" (1970d, p. 115, ft. 91). Feyerabend used the notion of incommensurability to attack certain aspects of realism already in 1970.[285] By 1978, Feyerabend is well known as a non-realist: "Incommensurability also shows that a certain form of realism is both too narrow and in conflict with scientific practice" (1978a, p. 69). The realism that incommensurability rules out is that "[science] makes the discoveries without changing the objects, properties, relations discovered" (1978a, p. 69ff.).

6.4. Conclusions

Feyerabend did not retreat from scientific realism. He was never a scientific realist. In Vienna in the late 1940s, he started as a positivist who rejected scientific realist accounts as unjustifiable metaphysics. Then in the early to mid-1950s, he adopted a form of normative realism based on ideas he found in Kraft and Popper, and on the persistent arguments about the possible methodological advantages that Feyerabend had with Hollitscher. In this phase, he continued to reject scientific realism. As Feyerabend moved away from normative epistemology toward the historical school at the end of the 1950s and in the early 1960s, Feyerabend de-

285 See (1970c), pp. 218-219; rp (1981b), pp. 151-152.

fended and developed a non-realist, neo-Kantian metaphysical stance, and he very often challenged many aspects of *scientific* realism. Farrell argues that Feyerabend returned to a form of realism in his 1990s publications (Farrell 2002). However, in *Conquest of Abundance* (1999b), Feyerabend continued to refine his neo-Kantianism. In fact, his 1990s positions is very similar to that of the ontological ideas he discussed in his early publications on Quantum Mechanics. Just as nature reveals its secrets to complementary approaches, 'Being' responds differently to different conceptual approaches.

Feyerabend's initial positivist rejection of scientific realist claims was based on the idea that claims that cannot specify how they could be empirically tested are meaningless. His early normative realism concerns an ideal aim of science, and it is justified on the methodological grounds that it better promotes scientific advance than instrumentalism. It promotes progress better because it encourages competition between theories as the one true account, whereas instrumentalism does not. Feyerabend's early normative realism is not a descriptive claim about the nature of scientific knowledge. The two should not be conflated. According to Feyerabend, while scientists should be realists when doing research because it forces them to test their theories more severely, scientists should also be aware that upon a little philosophical reflection, this belief is completely unjustifiable as a factual claim.

Feyerabend developed his notion of incommensurable concepts in 1962 in the context of a non-realist, neo-Kantian metaphysical position. However, Feyerabend repeatedly and adamantly rejected any form of transcendental necessity. He believed that we improve our theories by forcing them to compete against each other, not by comparing them to an unchanging reality. Or, as he put it:

> These thinkers [Ionesco and Bacon] do not criticize and so improve ideologies by pitting one against another; they criticize or mean to criticize *by comparing ideologies with an unchanging reality*, a reality which no longer contains any element of ideology. . . But who, one may ask, can guarantee that in jettisoning existing ideologies we are not falling victim to an even more exalted ideology (1967b, p. 309, italics in original).

Part III. Feyerabend's Philosophical Pluralism

> the history of the Scientific Revolution is an excellent example of the necessity of pluralism for the purpose of progress (1965b, p. 227, fn. 22).

The primary aims of Part III are to delineate the basic nature of Feyerabend's conception of knowledge, and to show how it complements and supports his conception of philosophy. Instead of trying to reconstruct Feyerabend's diverse ideas into a single philosophical position, Feyerabend's work is best understood as a series of constructive, immanent critiques within the context of his pluralistic conception of philosophy. Feyerabend did not have a single, coherent philosophical position from which he addressed the speculative epistemological issues that he treated. Instead, he used a variety of different philosophical approaches and ideas to discuss specific issues; especially what may be needed to promote and to safeguard the progressive nature of scientific advance. Feyerabend was a pluralist in two closely related senses: He argued for the use of a plurality of competing alternative views in the pursuit of knowledge on the grounds that such pluralism encouraged, advanced, and preserved their continual improvement: He was a theoretical pluralist with respect to knowledge. Moreover, he adopted and developed a plurality of different ideas and points of views within philosophy to promote his theoretical pluralism: He was a methodological pluralist with respect to philosophy. Feyerabend's theoretical pluralism about knowledge is supported and complemented by his methodological pluralism in philosophy. From early on, he was an epistemological opportunist and a philosophical pluralist: He adopted a plurality of philosophical positions to launch constructive immanent criticisms in order to promote progress in the pursuit of knowledge. Throughout his entire career, Feyerabend adapted this strategy to the changing times – without sustaining an allegiance to any one particular point of view.

Chapter 7
Feyerabend's methods

Many commentators have attempted to reconstruct Feyerabend's ideas by coalescing them into a single, new methodological position. This chapter argues that Feyerabend did not have a single, coherent philosophical position. He did not attempt to develop a new methodology. Feyerabend's work is best understood a series of immanent critiques. Consequently, appreciating Feyerabend's ideas requires a piecemeal approach.

7.1. A new methodology?

To many, Feyerabend's philosophy appears largely incoherent because of the blatant inconsistencies it contains. For this reason, a standard response to Feyerabend's philosophy has been simply to reject it as inconsistent. Preston, for example, has repeatedly criticized Feyerabend's philosophical work on the grounds that it is inconsistent.[286] As one reviewer put it:

> Preston's book very clearly shows the role certain underlying assumptions play in Feyerabend's views and convincingly argues that one cannot consistently and plausibly adhere to all of its various strands (Couvalis 1999, p. 215).

Feyerabend's work is indeed full of inconsistent ideas. One cannot consistently adhere to all of its various strands *at one and the same time*. While this is generally acknowledged with regard to Feyerabend's post-1970 publications, it also holds for his pre-1970 publications. As for *Against Method*, it is not an attempt to set out a new methodology. It is a series of criticisms that Feyerabend melded together from his earlier publications. That is why he called it a 'collage'. Lakatos's half of the book was supposed to argue 'For Method'; hence the planned title, "For and Against Method". Preston documents many minor inconsistencies in Feyerabend's ideas, so there is no need to document all of them here (Preston 1997). However, *pace* Preston, the inconsistencies in Feyerabend's diverse publications are

286 See Preston (1997), as well as Oberheim and Hoyningen-Huene (1999) and Oberheim and Hoyningen-Huene (2000).

not an immediate grounds for criticism. This is because *Feyerabend never set out to supply a consistent philosophical position, nor did he need one.*

This section briefly considers and then criticize the standard view of Feyerabend's early philosophy as a failed attempt to develop a new methodological position. Specifically, Preston's attempt to squeeze Feyerabend's philosophical considerations into a single coherent methodological position will be challenged on the basis that it is unsubstantiated. A detailed examination of Feyerabend's ideas undermines the evidence that Preston uses to support his reconstruction. This sets the stage for the an alternative pluralistic account of Feyerabend's philosophy. In other words, after showing *that* Preston's approach misrepresents Feyerabend's ideas, I will show *how* it misrepresents them.

According to Preston, Feyerabend's work is most profitably understood as a contribution to the project of developing *a rational model of science.*[287] Such a model specifies both an aim for science as well as a methodology for attaining that aim (Preston 1997, p. 13). According to Preston, "Feyerabend took knowledge of scientific statements . . . to be the aim of science," and the methodology for achieving that aim is falsificationism (Preston 1997, p. 13 and p. 15). According to Preston, Feyerabend (until 1970) was committed to a single philosophical position: critical rationalism, and he practiced a single specific brand of philosophy: normative epistemology (Preston 1997, pp. 133ff.). Moreover, according to Preston, the diverse ideas in Feyerabend's philosophical publications are best understood as part and parcel of a single model for the acquisition of knowledge.[288] In developing this model, Preston's states his goal as follows:

> to show how Feyerabend's version of scientific realism, his theory of meaning, his argument for theoretical pluralism and his radical materialist philosophy of mind hang together within [his model for the acquisition of knowledge] underpinned by his conception of methodology (Preston 1997, pp. 7-8).

Preston attempts to pull together many of Feyerabend's diverse views on different topics into a single methodological framework. This will be called a 'unificationist approach' to Feyerabend's publications. This label is intended to capture two related points: First, it draws on ideas expressed in publications on different topics from different years to construct a single model for the acquisition of knowledge: *a new methodological position.* And

287 See Preston (1997), pp. 13ff. The first seven and a half chapters, which will serve as my foil, concern Feyerabend's early work (roughly through 1969, but Preston also sometimes includes (1970c), see e.g. pp. 137-138 and the accompanying endnotes; also see (1981b), pp. 203-204.

288 For a more detailed summary, see e.g. Oberheim and Hoyningen-Huene (1999) and Oberheim and Hoyningen-Huene (2000), pp. 363-368.

second, this new position is supposed to *unify* Feyerabend's ideas on various topics, such as realism, meaning and mind.

Preston reconstructed Feyerabend's model for the acquisition of knowledge as a new, normative, methodological position consisting of seven principles (Preston 1997, pp. 137-138): Falsification,[289] revision,[290] empiricism,[291] testability,[292] realism,[293] proliferation,[294] and tenacity.[295]

Preston's Portrayal of Feyerabend's New Pluralistic Methodological Position

The principle of falsification: Take refutations seriously.

The principle of revision: Admit no unrevisable statement into the body of our knowledge. In any conceptual change, regard no statement, whether theoretical or observational, as incorrigible, irrefutable, unalterable, or *a priori*.

The principle of empiricism: Maximize the empirical content of existing scientific theories.

The principle of testability: Use only theories that are testable. Put theories in a form in which they are maximally testable, and test them (attempt to falsify them) relentlessly.

The principle of realism: Develop theories in their strongest possible form, i.e. as attempts at universally quantified descriptions of reality rather than as mere instruments of successful prediction.

The principle of proliferation: Invent, and elaborate, theories that are inconsistent with the accepted point of view, even if the latter should happen to be highly confirmed and generally accepted.

The principle of tenacity: From a number of theories, select the one that has the most attractive features and that promises to lead to the most fruitful results. Stick to it even if it is inconsistent with the evidence, or encounters considerable difficulties.

---

289 See (1956c), p. 262, (1958a), pp. 169-169; rp (1981a), p. 35, (1958d), pp. 49-50; rp (1981a), p. 236, (1960e), p. 337; rp (1981a), p. 233 and p. 235, (1961c), p. 158, and (1968b), p. 253.

290 See (1958a), pp. 168-169; rp (1981a), p. 35, (1960f), p. 64; rp (1999d), p. 42, (1961i), p. 54; rp (1999a), p. 70, (1962a), pp. 39-40; rp (1981a), pp. 51-52, (1965b), p. 184, and p. 196, (1965c), p. 243; rp (1981a), p. 125.

291 See (1958a), pp. 168-169; rp (1981a), p. 35, (1962a), p. 49 and p. 66; rp (1981a), p. 60 and p. 72, (1963a), p. 29; rp (1999a), p. 93, (1964c), pp. 352-353; rp (1981a), p. 206, (1965b), p. 176 and p. 196, (1975a), p. 41, (1988), p. 29, (1993a), p. 29.

292 See (1960e), p. 337; rp (1981a), p. 235, (1961d), p. 279, (1962a), p. 29; rp (1981a), p. 45, (1963e), p. 322, (1964d); rp (1981a), pp. 200-201, (1965a), p. 272; rp (1981a), p. 102, (1965b), p. 196, (1965c), p. 223; rp (1981a), p. 105. Also see (1995b), p. 362.

293 See (1958a), pp. 168-169; rp (1981a), p. 34, (1962a), p. 29; rp (1981a), p. 45, (1964d); rp (1981a), p. 200, (1966c), pp. 247-248; rp (1981b), p. 63.

294 See (1961i); rp (1999a), pp. 50-77, (1963a); rp (1999a), pp. 94ff., (1965c), pp. 223-224, p. 252, and pp. 254-255; rp (1981a), p. 105, (1969a), p. 280; rp (1999a), pp. 107ff., (1970c), p. 205 and p. 210; rp (1981b), p. 139 and pp. 143-144, (1970d), p. 26, (1975a), p. 47, (1988), p. 29, (1993a), p. 29.

295 See (1958b), p. 78, (1961c), p. 158, (1961f), p. 400, (1965c), p. 224; rp (1981a), p. 107, (1968a), p. 131, (1969a), p. 280; rp (1999a), pp. 107ff., (1970c), pp. 203-204; rp (1981b), pp. 137-138 and pp. 143-144.

While Preston admits that "there is no indication that Feyerabend ever put these rules together in his mind as a single pluralistic methodology" (Preston 1997, p. 138), just a page later, he asserted that a "pluralistic test model is intended to be *a single methodology for all scientific inquiry*" (Preston 1997, p. 139). Such a unificationist approach to Feyerabend's ideas must contend with at least two factors. *First*, it must confront Feyerabend's own later testimony about his former self, because Feyerabend later claimed that even in his very early publications, he did not have a single position, but often changed his views.[296] And *second*, it must dramatically distinguish his early views from his later views, because with the advent of his epistemological anarchism in the 1970s, Feyerabend explicitly was *not* trying to develop a single methodology for all of science. In these later works, Feyerabend explicitly rejected all such attempts to do so (1970d, 1975a, 1988, 1993a). If in fact Feyerabend was trying to develop a single methodology for all of science in his pre-*Against Method* publications, then in a dramatic about face, he must have rejected his own former methodological views in 1970 with the advent of epistemological anarchism.

In order to address both of these factors, Preston attempts to qualify and restrict his reconstruction of Feyerabend's views *temporally*:

> he [Feyerabend] glamorized what he thought of as his own tendency to move swiftly from one view to another. Because of this, I have sought within each chapter to evaluate views as a package only if they appear in publications produced at about the same time (Preston 1997, p. 7, italics inserted).

By 'about the same time', Preston appears to mean from the late 1950s to 1970 as he draws from publications ranging from (1958a; rp 1981a, pp. 17-36 to 1970c; rp 1981b, pp. 131-161) in substantiating his interpretation of Feyerabend's purportedly new methodological position.[297] However, as Feyerabend *glamorized* this tendency exactly with respect to these publications, Preston's temporal distinction actually simply ignores Feyerabend's statements.

The standard approach to Feyerabend's philosophy is to divide and separate Feyerabend's views into two distinct phases: pre-1970 and post-1970. For example, according to Preston, in the first phase, Feyerabend, the Popperian, set out to develop a single, new methodology as outlined above (Preston 1997, pp. 9ff.). According to Preston, in the second epistemological anarchist phase, Feyerabend adopted methodological pluralism and thereby rejected his earlier methodological position, arguing that

296 According to 1981 statements introducing his 1950s and 1960s papers, Feyerabend claimed to have often changed his views among them. See (1981a), p. xiv.

297 See e.g. Preston (1997), pp. 124ff., especially pp. 137-138 and accompanying endnotes. As (1958a) is supposed to be a condensed version of (1951), one could argue that Preston's 'about the same time' actually spans about two decades.

there are no such universally valid methodological principles: "This [his alleged new pluralistic methodology] is among the things that he later comes to believe does not exist" (Preston 1997, p. 136. Cf. pp. 138-139, p. 219, fn. 12, and especially p. 218, fn. 11). According to Preston, as a result of this major shift in viewpoints, Feyerabend was forced to abandon the methodological position he had developed in his pre-1970 publications.

The grounds on which Preston bases his claims about Feyerabend's intention to develop a single, new methodological position is not exactly clear, especially given that Preston himself already admitted that there is no direct evidence suggesting that Feyerabend intended to do so (Preston 1997, p. 138). However, Preston does provide a single passage from Feyerabend to support his standard unificatory approach to Feyerabend's ideas (Preston 1997, p. 13):

> (1) *Proliferation.* In writing the papers which are discussed, and criticized by Profs. Smart, Putnam, and Sellars, my aim has been to present an abstract model for the acquisition of knowledge, to develop its consequences, and to compare these consequences with science (1965c, p. 223; rp 1981, p. 104).

This certainly sounds as if Feyerabend had indeed tried to develop a new methodological position, and based on this statement (*and this statement alone*), Preston justified his unificationist approach; presenting Feyerabend as having attempted to set *out a single, new, normative methodology* (Preston 1997, p. 13 and pp. 138-139).

However, this passage does not actually support Preston's claim that Feyerabend tried to set out a new methodological position. To see this, as so often with Feyerabend's texts, first we must put the passage into its proper context, and then carefully check the fine print. The passage is part of a response to criticisms of his papers. It did not appear in the papers in which the ideas were set out. More importantly, the endnote to the passage advises readers that it would be a mistake to interpret him [Feyerabend] as trying to develop some kind of new philosophical position:

> Perhaps unintentionally he (Smart) creates the impression that a new philosophical position is in the making, a kind of neorealism, and that I have contributed to its development. Such an impression would be both incorrect, and unfortunate (1965c, p. 251, fn. 1; rp 1981a, p. 104, fn. 1).

In the very same note, Feyerabend proceeded to ridicule the idea of 'the One True Method'. In other words, attached to the sole passage Preston uses to support his characterization of Feyerabend's philosophy as an attempt to develop a new position setting a new methodology for science is an note advising readers against just such an interpretation, and which also contains a denunciation of the whole idea of the 'One True Method'. This footnote qualifies the same statement that Preston uses to ground his interpretation of Feyerabend's early ideas as an attempt to setout the one true set of universal methodological rules that make up the one true method as distinct from Feyerabend's later views that supposedly jettison this project.

Preston's unificatory approach to Feyerabend's ideas is not only unsubstantiated. It is also leads to incoherence. To see this, let us compare the methodological principles that Preston presents as the core of Feyerabend's new methodological position. As we have seen, Preston would have us believe that these normative principles were intended to be universal methodological rules, which when combined, make up the a single methodology for all of science. However, Feyerabend never suggested that these seven methodological principles were supposed to be *universally valid*, nor part of a single correct methodological position. This is rather obvious as these principles do not really cohere with each other (as Preston does not seem to notice). For example, the *principle of falsification* advises scientists to take refutations seriously. This means that scientists should not use *ad hoc* means to insulate theories from falsifying evidence. Instead, they should give up falsified theories and try to develop alternatives that can cope with the otherwise recalcitrant evidence. The *principle of tenacity*, on the other hand, advises scientists to do just the opposite. According to this principle, scientists should stick to a theory even if it is inconsistent with the evidence. In other words, it advises scientists not to take refuting evidence all too seriously, as it may be possible to adopt the theory to the apparently recalcitrant evidence, or to undermine the evidence itself, or reinterpret the evidence so that it supports the theory. *These two principles make conflicting recommendations.* Moreover, Feyerabend often defends the principle of proliferation. He claims to have developed the principle of proliferation from Popper and the philosophy of possibilism as developed by Naess. The idea, however, in not new. According to Feyerabend, many scientists acted in accordance with it since the Presocratics.[298] The principle of proliferation advises scientists to invent and elaborate *many different theories.* This directly conflicts with the principle of tenacity, which recommends picking and sticking to just *one.*

How are these inconsistencies in Feyerabend's alleged methodological position to be reconciled? Are we really to believe that these conflicting recommendations were supposed to be *universally valid* methodological rules within a *single* coherent methodology applicable to all of science? If this were the case, then Preston's general complaint that Feyerabend's ideas are inconsistent would indeed be justified. But perhaps Feyerabend

298 See (1965c), pp. 252-253, fn. 5; rp (1981a), p. 106, fn.5, removed. For a concise accurate discussion, see Meynell (1978), pp. 243-244. Compare Tibbets (1976), pp. 363-364. For a more recent critical discussion, see Achinstein (2000). For an admirably deep analysis of this principle, see Hoyningen-Huene (1995).

was suggesting that *sometimes*, scientists should retain their theories even in light of apparently falsifying facts, whereas *at other times*, scientists should take falsifying facts seriously, and decide to pursue alternative views? Was Feyerabend suggesting that sometimes some scientists should proliferated theories, while at other times, other scientists should stick to just one? The obvious discrepancies among these principles suggest that *they were not supposed to be universally valid prescriptions in the first place*, but rather context dependent prescriptions. Whether or not a particular scientist should contest or accept the results which appear to falsify a theory, whether to be tenacious, or to proliferate – these issues will depend on the exigencies of the particular situation, not some abstract universal methodological principles.

Moreover, these inconsistencies among the principles also suggest that they were never intended to be part of a *single* new methodological *position* in the first place. The principle of proliferation itself recommends developing alternative views, not the pursuit of a single 'correct' position. It is supposed to hold both in science and philosophy. Presenting a list of methodological principles and claiming that Feyerabend was trying to develop a new methodological position out of them is also an extremely misleading way of representing these ideas, their relations, and the different roles that they play in Feyerabend's ideas and arguments.

To begin with a minor point, in most of the cases cited (except for the 'principles' of proliferation and tenacity, Feyerabend did not usually refer to these as 'principles', nor especially as principles of his own methodological position. Second, Feyerabend did not usually argue for the first four standard methodological principles. Instead, he merely stated them in the course of *ad hominem* arguments. For example, he simply states that according to empiricism, one should take refutations seriously; or that according to the demands of a reasonable methodology, one should not admit unrevisable statements into science.[299] If Feyerabend had really tried to develop and defend a new methodological position based on these methodological principles, we would expect to find Feyerabend developing and defending these principles.[300] This is not often the case with the *first four* principles. Feyerabend usually merely assumed these principles as uncontroversial in the setting in which he was using them. Although the instances documented in the first four principles of Preston's list do indeed represent cases where Feyerabend *used* these methodological demands to make arguments against theses proposed by individuals *who*

299 See the citations attached to the list of the seven methodological principles above.

300 As Nordmann put it (albeit with reference to Feyerabend's later philosophy): "Feyerabend does not present a doctrine that is right or wrong, and which we are asked to accept or reject. He confronts us with a challenge" (Nordmann 1990, p. 325).

accept these four principles, they do not represent Feyerabend's unconditional commitment to these methodological demands more generally. As we shall see, for the most part, they are used in the context of the normative, *ad hominem* half of Feyerabend two-pronged argumentative strategy; and consequently they do not necessarily represent Feyerabend's own views.

However, Feyerabend did argue for the last three methodological principles on Preston's list. For example the principle of proliferation is defended against Kuhn's claim (and others who claim) that alternative points of view should be developed only after an otherwise successful theory has run into empirical difficulties, and the principle of tenacity is defended against empiricists who argue that theories must be given up in the face of recalcitrant evidence.[301] Here, Feyerabend used the possibility of developing new incommensurable concepts that could falsify and otherwise successful theory to attack Kuhn's view of normal science.[302] The last three principles: realism, proliferation, and tenacity, are *not* standard methodological demands, and they were not widely accepted, like the first four principles. Moreover, Feyerabend argued for the last three principles *on the basis of* the first four principles. He was trying to convince his empiricist opponents that as they accept the first four principles, they should also accept the last three. A proper presentation of these methodological principles in Feyerabend's philosophy should not delineate them as part of some new pluralistic methodological position that Feyerabend was trying to develop, but should distinguish between the first four, which Feyerabend assumed and used to argue for the last three.

In summary, Preston's account of Feyerabend's new methodology misrepresents the relationship of the methodological principles that he list. Preston fails to recognize that Feyerabend simply *assumed* the first four principles as part of *ad hominem* arguments (as they were generally acceptable to his many diverse opponents), and then used them to *argue* in favor of the last three principles, which were not generally acceptable to his diverse opponents. Feyerabend very often argued that the realism, proliferation and tenacity follow from the more standard methodological demands. According to Feyerabend, given the first four, it is reasonable to accept the last three:

301 For a discussion, see e.g. Hoyningen-Huene (2000a).

302 See for example (1962a), pp. 39-40; rp (1981a), p. 52, removed.

Methodological Principles in Feyerabend's Philosophy

(I) Four principles Feyerabend assumed in *ad hominem* arguments

The principle of falsification: Take refutations seriously.

The principle of revision: Admit no unrevisable statement into the body of our knowledge. In any conceptual change, regard no statement, whether theoretical or observational, as incorrigible, irrefutable, unalterable, or a priori.

The principle of empiricism: Maximize the empirical content of existing scientific theories.

The principle of testability: Use only theories that are testable. Put theories in a form in which they are maximally testable, and test them (attempt to falsify them) relentlessly.

(II) Three principles for which Feyerabend argued on the basis of (I)

The principle of realism: Develop theories in their strongest possible form, i.e. as attempts at universally quantified descriptions of reality rather than as mere instruments of successful prediction.

The principle of proliferation: Invent, and elaborate, theories that are inconsistent with the accepted point of view, even if the latter should happen to be highly confirmed and generally accepted.

The principle of tenacity: From a number of theories, select the one that has the most attractive features and that promises to lead to the most fruitful results. Stick to it even if it is inconsistent with the evidence, or encounters considerable difficulties.

7.2. Constructive critiques

If Feyerabend was not trying to develop a new methodology, then what was he trying to do? Feyerabend's major publications are mainly constructive criticisms of the view of others, rather than developments of certain positions (or 'position papers'). The stated aims of his major publications makes this clear. A comparison of the opening passages of Feyerabend's major papers (excluding the many reviews, responses, discussions, and notes) exposes a revealing pattern that bears on his critical style and mode of philosophy. As we shall see, in many cases, Feyerabend's papers very often begin by *explicitly* stating the ideas (and/or their proponents) that Feyerabend will be attacking. Because the explicitly stated goals of Feyerabend's main publications were critical, Feyerabend's texts should be approached in a piecemeal fashion.

Let us begin with "An Attempt at a Realistic Interpretation of Experience" (1958a; rp 1981a, pp. 17-36). Here, I will consider more than just the opening passage. My goal is to show that it is not a position paper, but rather a constructive criticism based on an *ad hominem* argumentative strategy. As we have already examined the content of this paper at length, in this section I will give only a brief overview of the structure of Feyerabend's considerations.

Although the title may appear to suggest a position paper, the article begins by stating the ideas under consideration, and then states the general *critical* aim of the essay:

> 'The task of science' writes Niels Bohr, is both to extend the range of our experience and to reduce it to order.' 'Science', echoes a modern philosopher, 'is ultimately intended to systematize the data of our experience.' *In the following paper I shall try to show that these two statements, in spite of their simplicity and their apparently innocuous character, have consequences which are at variance with scientific method and reasonable philosophy* (1958a, p. 143; rp 1981a, p. 17, italics inserted).

The explicit aim of the paper is a criticism of the views of others, not the development of Feyerabend's own position on some issue or a new methodology of his own. This is not to deny that Feyerabend did indeed make some positive suggestions. However, of the seven sections which follow, only the last two, composing less than a third of the whole, are even *partially* dedicated to Feyerabend's positive suggestion, which as we have seen is *thesis I*.

The structure of the paper is as follows: The first section introduces the views under attack, and summarizes Feyerabend's strategy for attacking them. The second section further develops the ideas under attack. The third section criticizes those ideas, explicitly emphasizing the *ad hominem* nature of the criticism (1958a, p. 149; rp 1981a, p. 21). Sections 4 and 5 further criticize different versions of the ideas on the grounds that they are at odds with 'reasonable' methodological principles (not universally valid ones, nor ones for which Feyerabend argues). Feyerabend did not bother arguing for the reasonableness of these methodological principles because these methodological principles are *supposed to be acceptable to the positivists he is criticizing*. Feyerabend does not argue for, or develop, his own methodological position. He merely uses reasonable methodological principles as part of *immanent* criticisms of some aspects of positivism. Section 6 also criticizes the ideas of the positivists, but now the criticism is on the basis of Feyerabend's own positive suggestion. In section 7, Feyerabend considers the logical basis of his critique of positivism and the positive suggestion he draws from it.

The structure of the paper makes its critical nature clear. The paper is almost entirely an *ad hominem* critique of positivism that concludes with a positive suggestion. It is not a position paper that is part of the development of some new methodological position. As we have seen, Preston draws from this paper in support of his reconstruction of Feyerabend as having attempted to develop a single, new model for the acquisition of knowledge, based on seven methodological principles. In so doing, he ignores that in this paper, Feyerabend uses these principles in an *ad hominem* fashion. The original 1951 German version, on which it is based, made the critical nature of Feyerabend's approach even more clear. It is entitled "On the Theory of Basic Statements". It is a *sustained attack* on the notion of basic statements and the role that they play in positivist philosophy. The positive suggestions that Feyerabend makes are couched as results of the failures that he exposes in the positivists account. It is even less of a position paper than the reworked English version, if that is possible.

Just as "An Attempt at a Realistic Interpretation of Experience" (1958a; rp 1981a, pp. 17-36) begins with a direct and concise statement explicitly naming the targets of Feyerabend's attacks (namely Bohr, Hempel and Carnap's views about the role of experience in science), "Explanation, Reduction, and Empiricism" (1962a) also begins with a direct and concise statement of the views he attacked (namely Hempel, Oppenheim and Nagel's views on explanation and reduction). In the process of criticizing these views in an *ad hominem* fashion, Feyerabend introduces some ideas of his own, for example incommensurable concepts, and again ends by considering the logical basis of his arguments. The basic structures, *ad hominem* critical natures, and positive suggestions of the two papers are almost identical. Both paper's attack versions of the stability thesis on similar grounds, i.e. on the basis of the existence of incommensurable concepts. The main difference is whereas (1958a; rp 1981a, pp. 17-36) attacks logical positivist views on the role of experience in testing theories, (1962a; rp 1981a, pp. 44-96) attacks logical empiricist accounts of explanation and reduction.

While these may be the most relevant examples to the ideas developed in this investigation, they are not the only examples of the critical nature of Feyerabend's philosophy as evident by their opening passages. For example, Feyerabend's major early papers on Quantum Mechanics are also a series of constructive criticisms that do not attempt to set out, or to establish, his own methodological position or his own interpretation of Quantum Mechanics. The title of (1958d; rp 1981a, pp. 236-246) ("Reichenbach's Interpretation of Quantum-Mechanics"), makes it clear that it concerns the specific views of others, and that it is not an exposition of his own views. The stated aim of Feyerabend's "On the Quantum Theory of Measurement" (1957c; rp 1981a, pp. 207-218) and (1957b) is "to show the inadequacy of this [Bohr, Heisenberg, and von Neumann's] theory of measurement" (1981a, p. 207). In "Complementarity", Feyerabend explains his aim: "I have set myself the task of making the assumptions upon which this analysis [Bohr's use of philosophical analysis to support his idea of complementarity] rests" (1958b, p. 75). This paper is also a critical analysis of someone else's views, not a paper that sets out Feyerabend's position. Feyerabend began "On the interpretation of scientific theories" with:

> According to positivism the interpretation of scientific theories is a function of either experience, or of some observational language. . . In the present paper I shall discuss two objections . . . Indeed, it seems to me that the difficulties of positivism as revealed by these objections cannot be overcome by inventing a new and ingenious connection between theoretical terms and observational terms, but only by altogether dropping the idea that meaning of theoretical terms depend upon such a connection (1960a; rp 1981a, p. 37).

Again, in this paper, we are not confronted with Feyerabend setting out to develop his own interpretation of scientific theories, or to develop his own set of methodological principles. Rather, Feyerabend set out to criticize other people's views, and to discuss some of the consequences of his criticisms.

What about Feyerabend's "How to be a good empiricist: a plea for tolerance in matters epistemological" (1963a; rp 1999a, pp. 78-103)? This certainly sounds like it might be a position paper that might contain the development of a particular methodological position, such as the one Preston tried to pin on Feyerabend. Yet, the critical nature of this paper (which is also another source from which Preston often drew in support of his unificationist approach to Feyerabend's ideas), is also made clear right from the start:

> It is the purpose of the present paper to show that empiricism in the form in which it is practiced today cannot fulfill this hope ["that only a thoroughly observational procedure can exclude fanciful speculation and empty metaphysics as well as . . . that an empiristic attitude is most liable to prevent stagnation and to further the progress of knowledge"] (1963a; rp 1999a, p. 78).

Once again Feyerabend attacks certain dogmatic versions of empiricism. He does not attempt to develop his own new form of it. He merely exposed its shortcomings, and made some suggestions about how it could potentially free itself from its dogmatic elements.

In fact, Feyerabend's major papers follow a revealing pattern. They begin by stating the target that he will attack. They proceed to challenge someone else's views in an *ad hominem* fashion. And they end with some positive suggestions in light of these criticisms. As such, they support the kind of piecemeal interpretation of Feyerabend's ideas that I recommend. They are also evidence against Preston's attempt to develop a single methodological position out of them.

Feyerabend's "Problems of Empiricism" is also, as the name suggests, largely critical (1965b). This long essay explicitly draws from his earlier work. However, in a rare break from his usual style, Feyerabend *did* seem to suggest that he will set out his own methodological position:

> The attempt will also be made to formulate a methodology that can still claim to be *empirical* but that is no longer beset by the problems of a *radical* empiricism (1965b, p. 149).

But as it turns out, the promised new methodological position is merely another version of Feyerabend's pluralism:

> Put in a nutshell, the methodology requires a theoretical pluralism instead of the theoretical monism that is the ideal of much contemporary empiricism, of philosophy in general, as well as of almost all "primitive" thought (1965b, p. 149).

In fact, Feyerabend's main goal is not to develop some new form of empiricism, or to set out a set of universally valid methodological rules, but rather to fight for *pluralism* and *tolerance* in matters epistemological:

> The fight for tolerance in scientific matters and the fight for scientific progress that was so important part of the lives of the early scientists must still be carried on. *There are still attempts made to arrest progress and to establish a doctrine.* What *has* changed is the domination of the defenders of such doctrines. They were priests, or "school philosophers," a few decades ago. Today they call themselves "philosophers of science," or "logical empiricists." . . . [they] proceed under the flag of progressivism and empiricism and thereby deceive a good many of their followers. Hence although their presence is noticeable enough, they may almost be compared to a fifth column, the aim of which must be exposed in order that its detrimental effect be duly appreciated. It is our purpose to contribute to such an exposure (1965b, p. 149, italics inserted).

Here in 1965 (not 1970 or 1975), Feyerabend made it quite clear that his goal was *not* to establish some new doctrine in the philosophy of science (as Preston would have us believe), but to criticize such attempts to do so. Feyerabend did not attempt to set out the one correct scientific method under the flag of progressivism. He argued that this is the wrong approach to scientific knowledge, because it tends to stifle scientific progress, which requires the kind of opportunistic pluralism Feyerabend preached and practiced.

Before leaving this point, it is opportune to note that between 1952 and 1974, although Feyerabend had over 100 publications, *he did not publish a single book* (Oberheim 1999). In fact, his first properly published book (the 1975 version of *Against Method*) came almost 25 years after he finished his dissertation, and he often called that book "a collage", and for good reasons:

> *AM is not a book*, it is a collage. It contains descriptions, analyses, arguments that I had published, in almost the same words, ten, fifteen, and even twenty years earlier (1995a, p. 139).

This is a correct description of *Against Method*, and as such, it is rather difficult to square with Preston's claim that Feyerabend abandoned his earlier methodological ideas with the advent of epistemological anarchism. As is well known, in *Against Method* Feyerabend argued *against* the standard view of scientific method as a set of universally binding methodological rules. He called his own positive suggestion "epistemological anarchism", according to which the best methodological approach is to avoid developing a single position. *In the 1960s*, he did not set out to delineate his own position. He chose to write a series of papers criticizing the views of others. Attempts to reconstruct Feyerabend's ideas into a stable methodological position misrepresent Feyerabend's ideas. What is worse, they misrepresent how he practiced philosophy.

7.3. *Ad hominem* arguments

When reading Feyerabend's philosophical texts, identifying the specific elements to which Feyerabend is committed can be extremely difficult, because of his extensive use of immanent criticisms, or *ad hominem* arguments.[303] The significance of Feyerabend's extensive use of *ad hominem* arguments derives from the fact that his use of a particular concept or distinction cannot necessarily be taken as a clear indication that he accepted that the concept or distinction. Instead, the concept or distinction is temporarily adopted from the position that he is attacking for the sake of making a cogent constructive criticism of it. For this reason, Feyerabend was not necessarily obliged to defend the concept or distinction that he had used. In order to justify his use of such concepts or distinctions, he must merely show *that* the targets of his criticism are actually using, or are committed to, the concepts or distinctions in question– even if the concepts or distinctions are *inconsistent with the positive suggestions that he make* in light if the criticism.

While Feyerabend's use of such *ad hominem* arguments minimized his own presuppositions, it also facilitates misreading his texts, and misidentifying his own claims for those of others. This leads to misrepresentations of his views. To illustrate with just one more example from Preston's book, let us consider Preston's critical reconstruction of Feyerabend's notion of a theory. Preston provides two citations clarifying the term 'theory'. They are:

> [T]he term 'theory' will be used in a wide sense, including ordinary beliefs (e.g. the belief in the existence of material objects), myths (e.g. the myth of the eternal

303 The points made in this paragraph were also made in Oberheim and Hoyningen-Huene (2000), pp. 369ff.

> recurrence), religious beliefs, etc. In short, any sufficiently general point of view concerning matter of fact will be termed a 'theory' (1965b, p. 219, fn. 3).

And:

> When speaking of theories I shall include myths, political ideas, religious systems, and I shall demand that a point of view so named be applicable to at least some aspects of everything there is. The general theory of relativity is a theory in this sense, "all ravens are black" is not (1965c, p. 252, fn. 5; rp 1981a, p. 105, fn. 5).[304]

Preston then criticizes Feyerabend, "The first is intolerably vague, but the second is worse."[305] Preston continues, "However, Feyerabend does, in one place, give a more useful definition of theory." And he cites from Feyerabend:

> In what follows, the usual distinction will be drawn between empirical generalizations, on the one side and theories on the other. Empirical generalizations are statements, such as 'All A's are B's' (the A's and B' are not necessarily observational entities), which are tested by inspection of instances (the A's). Universal theories, such as Newton's theories of gravitation, are not tested in this manner. Roughly speaking their test consists of two steps: (1) derivation, with the help of suitable boundary conditions, of empirical generalizations; and (2) test, in the manner indicated above, of these generalizations. One should not be misled by the fact that universal theories, too, can be (and usually are) put in the form 'All A's are B's'; for, where as, in the case of generalizations, this form reflects the test procedure in a very direct way, such an immediate relation between the form and the test procedures does not obtain in the case of theories. Many thinkers have been seduced by the similarity of form into thinking that the test procedures will be the same in both cases (1962a p. 28, rp 1981a, p. 44, slightly modified).

Preston then proceeds to present Feyerabend as having espoused this "more sophisticated conception" of theories, and to show how this purportedly illuminates Feyerabend's views about Wittgenstein's contextual theory of meaning, as well as Feyerabend's own views about how the meaning of a term is determined.

However, Preston's general attribution of the distinction between universal theories and general statements of the form all A's are B's to Fey-

---

304 Of course, Preston is correct that Feyerabend often did use a wide or loose notion of theory, see e.g. (1965b), p. 219 fn. 3. However, Preston has left out the rest of the passage, which attempted to clarify Feyerabend's use of 'theory' *and explains why he used such a wide notion of theory*: "There are similarities between my use of 'theory' and Quine's use of 'ontology', Carnap's 'linguistic framework', Wittgenstein's 'language game', Pareto's 'theory', Whorff's 'metaphysics', Kuhn's 'paradigm', etc. I prefer this 'accordion' use of the term because it provides a single name for problems, which in my presentation are intimately related. However the 'philosophical music' . . . I intend to make will not be extracted from the fact that the word is so used, but from the common features of the problems described" (1965b, p. 252, fn. 5).

305 The whole sentences reads: "The first is intolerably vague, but the second is worse, since any empirical generalization of the form 'All A's are B's' is applicable to everything there is', being restatable in the form 'Everything is either an A or is not-B', which is incorrect.

erabend, on the basis of this citation, is unjustified and mistaken in this context. Feyerabend, as we have seen, basis the distinction on the fact that such general statements can be tested empirically directly against specific instances, whereas in order to test theories empirically, one must first derive specific instances from empirical generalizations with the help of suitable boundary conditions. Only then can theories be tested against specific instances. Of course, this is the very same method that Nagel used to distinguish theories from instantial generalizations (Nagel 1961, pp. 81ff.). This is no mere coincidence. It is Nagel's notion of a theory, not Feyerabend's, which Feyerabend set out in this passage. Feyerabend adopted Nagel's notion of a theory, and Nagel's method and justification of distinguishing between theories and instantial generalizations, as part of his immanent criticism of Nagel's views about the relationship between successive scientific theories.[306]

Because it is Nagel's notion of theory that Feyerabend used, it should not be attributed to Feyerabend, nor need Feyerabend defend it in this particular context, nor should it be used to try to reconstruct critically other aspects of Feyerabend's philosophy. All of this should haven been made clear by the rhetorical devices Feyerabend used. He did not write, 'according to my notion of a theory . . .'. He wrote: "*the usual distinction* will be drawn between empirical generalizations, on the one side, and theories, on the other" (1962a, p. 28, fn. 1, original italics removed, italics inserted, rp 1981a, p. 44, fn. 1, slightly altered, italics inserted).

Moreover, Preston has no business taking *this* specific instance of the notion of a theory as representative of Feyerabend's notion of a theory more generally. Again, Feyerabend had explicitly limited the scope of his claim about theories with a rhetorical device: the caveat "in the following" (1962a, p. 28, fn. 1, rp 1981a, p. 44, fn. 1), which apparently Preston has simply ignored. In sum, Preston takes Feyerabend's statement, "In the following, the usual distinction will be drawn between empirical generalizations . . . and theories," to be an expression of Feyerabend's own views on theories, ignoring the rhetorical clues. He then tries to conjure some lessons about Feyerabend's views on meaning and their relation to Wittgenstein from it. As we have seen, the views on meaning Feyerabend

306 Ironically, some fifty pages after Preston ascribes this notion of theory to Feyerabend in order to draw some Wittgensteinian lessons about Feyerabend's conception of meaning from it, Preston notes that, "Nagel's model of science relies on a distinction *closely related* to the one we saw Feyerabend draw between experimental generalizations and theories" (Preston 1997, p. 80). Generally speaking, Feyerabend's conception of a theory is not closely related to Nagel's. It is much, much wider. Moreover, the conception of theory that Feyerabend specifically adopted in (1962a); rp (1981a), pp. 44-96 is not 'closely related' to Nagel's. It is Nagel's.

adopted in this essay were drawn from Duhem, and shared by Nagel, which is why Feyerabend did not bother to defend them.

Before leaving this example, it is worth while briefly to point out that the fact that Feyerabend adopted Nagel's notion of a theory as distinct from instantial generalizations is also inconsistent with Preston's more general project of presenting Feyerabend in the early 1960s as a strict adherent to Popperianism. This is because Popper does *not* make this distinction between universal theories and empirical statements (Popper 1934, chapter 3, and Popper 1959, pp. 426ff.). This, it would seem, is why Feyerabend belabored the point about the distinction in the first place. In other words, he was trying to highlight the fact that he was not simply adopting the Popperian account of theories, and then criticizing Nagel's account of theory reduction and theoretical explanation from Popper's perspective. Rather, Feyerabend was adopting Nagel's account of theories, in order to make a criticism of Nagel's account of theory reduction *that sticks*. To this purpose he temporarily adopted Nagel's account of a theory as part of his *ad hominem* argumentative strategy.

Thus we can see how Feyerabend's use of *ad hominem* arguments has made it difficult for some critics to identify and understand Feyerabend's own views, and to distinguish them from those of others, especially from those of the targets of his criticisms. The reason is that *Feyerabend often temporarily adopted many components of his opponents positions in order to make legitimate criticisms of them.* Expressions of his own views are difficult to identify and isolate without a rigorous analysis of the texts, because of his *ad hominem* argumentative strategy.

In order to better appreciate some of the advantages of an *ad hominem* argumentative strategy, we can contrast it to *non ad hominem* argumentative strategies, which attack a thesis from some particular perspective that is not shared by the target of the attack. The main disadvantage of restricting one's self to such *ad hominem* arguments, is that one cannot use them to set out a new position. Of course, setting out new positions on specific topics is also an important project in the philosophy of science. Often, such strategies proceed to defend the perspective that has been assumed, by providing independent arguments for the unshared assumptions that have been made.

A non *ad hominem* argumentative approach can be illustrated with two very relevant examples. First, let us consider the argumentative strategy employed by Sankey in his book *The Incommensurability Thesis* (Sankey 1994). Sankey analyzes the incommensurability thesis from his own preferred realistic perspective. He uses an approach that presupposes the kind of scientific realist metaphysics that Kuhn and Feyerabend explicitly rejected with their developments of the incommensurability thesis. The scientific realist presuppositions that Sankey brings to the discussion attack, and subsequent re-develop, Kuhn and Feyerabend's notion of incommensurability. These realist assumptions are not shared by Kuhn and Feyerabend. In fact, and to make matters worse, both Kuhn and Feyerabend used their notion of incommensurability to attack the very scientific realist metaphysics that Sankey presupposes in re-developing a realist version of the incommensurability thesis that no longer threatens scientific realism. Having thus shown that an incommensurability thesis can be made compatible with scientific realism, Sankey subsequently rejects Kuhn and Feyerabend's versions of incommensurability and the threat they purportedly pose to scientific realism (Hoyningen-Huene, Oberheim et al. 1996 and Oberheim and Hoyningen-Huene 1997).

Sankey is well aware that the use of this argumentative strategy commits him to defend the scientific realist metaphysics that he had presupposed. In the last three chapters of his book, Sankey brings independent arguments in favor of scientific realism, in order to bolster his claims about incommensurability. He also tries to discredit the kind of non-realist metaphysical views employed by Kuhn and Feyerabend. Again, Sankey needs to argue for scientific realism, because he uses scientific realist presuppositions in his development of a realist version of the incommensurability thesis. If realism should become circumspect, then Sankey's development of incommensurability would suffer accordingly. Moreover, Sankey's arguments concerning incommensurability are only convincing to those who are already convinced, or have been convinced by Sankey's arguments, that scientific realism is correct. By contrast, an *ad hominem* argumentative strategy that adopts the presuppositions of the parties under attack relies on fewer controversial elements, and can convince a broader audience.

Another relevant example of the use of a *non ad hominem* argumentative strategy is Preston's repeated criticisms of Feyerabend (Preston 1997). In his response to a 'Symposia Review' whose title even emphasizes the point that I am now making ("Radical Fallibilism vs. Conceptual Analysis"), Preston admits that he approaches and attacks Feyerabend's ideas from a point of view rather different from Feyerabend's. He has not adopted an *ad hominem* argumentative strategy:

> Couvalis and Munévar have a good idea 'where I'm coming from', and I like the formers final summing-up of our different approaches very much indeed. *I agree completely that mine is not an 'immanent critique', since not only do I depart fundamentally from some of Feyerabend's presuppositions*, but I have (relatively recently) come to think well of entire kinds of philosophy for which he had little time (Preston 1999, p. 235, italics inserted).

Preston even seems to be aware of the limitations that this can have on his ability to understand and present Feyerabend's ideas properly. He continues (directly):

> I am very much alive to Munévar's complaint that my more linguistic approach keeps me from fully grasping Feyerabend's vision (Preston 1999, p. 235, italics inserted).

Here Preston implicitly admits that his linguistic approach caused him to misunderstand Feyerabend's use of the expression 'absolute truth' (see chapter 1). But apparently, Preston does not realize the full consequences that these limitations have for the potency of his criticisms of Feyerabend's ideas:

> But I can't help finding the presuppositions of that vision [Feyerabend's] deeply problematic, so I think it worth explaining how they can be critiqued (Preston 1999, p. 235, italics inserted).

Because Feyerabend's ideas seem problematic from his Wittgensteinian perspective, he criticizes them. But he does not criticize them by adopting Feyerabend's perspective and showing how it runs into difficulties. He criticizes Feyerabend's ideas by assuming the correctness of, and on the basis of, his own analytical philosophical approach. However, unlike in Sankey's case, Preston offers no independent justification for the presuppositions he uses in his criticisms of Feyerabend's ideas. Moreover unlike in Sankey's case, he does not identify what those presuppositions are, and how they differ from Feyerabend's. Yet, the presuppositions Preston brings to his attempt to reconstruct and criticize Feyerabend's ideas are part of the very same tradition that Feyerabend repeatedly and vehemently attacked.

The core of Feyerabend's philosophical ideas is not the contrived methodological position Preston makes it out to be. The real core of Feyerabend's philosophy has to do with a certain kind of philosophical method: Feyerabend launched *ad hominem* attacks against particular views on the grounds that they would inhibit progress. As we have seen, in his pre-1970 publications, Feyerabend often argued in an *ad hominem* fashion. He used *ad hominem* arguments against Carnap's view that the task of philosophy consists in explicating the notions of science, Nagel's theory of reduction, the Hempel-Oppenheim theory of explanation, the classical empiricist, logical positivist, logical empiricist, Popperian and Kuhnian *monistic* models of theory testing, Kuhn's functional argument for normal science, and instrumentalist interpretations of science (to name a few of the more prominent examples discussed in this investigation). According to Feyerabend, none of these methodological proposals accurately capture the nature of progressive developments in the sciences (descriptively speaking), and all of these, if enforced methodologically (normatively speaking) would contribute to inhibiting progress and the advancement of the sciences given some reasonable methodological assumptions about scientific knowledge and its advance accepted by the proponents of these proposals. Later, Feyerabend used *ad hominem* arguments against the proponents of the concepts of 'Reason, Obligation, Duty, Morality, Truth, the Gods' (1975a, especially p. 180), and of course the *universally valid* methodological rules.

Hoyningen-Huene has suggested that the exceptional lack of commitment tied to this kind of argumentative strategy is a reflection of Feyerabend's deep quest for independence, both on the professional and the personal level (Oberheim and Hoyningen-Huene 2000, p. 369). This type of *ad hominem* argument is central to Feyerabend's approach to, and style of, philosophy (Oberheim and Hoyningen-Huene 2000). It is the foundation of his pluralistic philosophical method and he used it to support his pluralistic conception of knowledge. In addition to the specific individual cases of the *ad hominem* argumentative strategies outlined above, there is a more general sense in which Feyerabend can be considered to be *ad hominem* philosopher. Feyerabend did not have his own stable position. His use and development of different sides of the same issue is part of his pluralistic philosophical method. This pluralistic philosophical method may *contain* inconsistent elements, if different arguments are conflated in an attempt to tie Feyerabend's diverse considerations into a single position, but it is *not* inconsistent with Feyerabend's wider pluralistic conception of knowledge, and his related views on the role of the proliferation of alternative views. On Feyerabend's view, the chief task of philosophy is to promote progress by challenging the dogmatic assumptions hidden in authoritative views – not the development of another school philosophy.

Whereas in *Against Method*, he was like an undercover agent playing the game of reason in order to undercut the authority of reason, in the 1960s, he was like an undercover agent playing the game of empiricism, in order to undercut the authority of experience.

Chapter 8
The role of alternatives in promoting progress

The aim of this chapter is to reconstruct Feyerabend's pluralistic conception of knowledge and the arguments for pluralism that he used to support it. The first section begins with an examination of Feyerabend's pluralistic test-model. Section 8.2. investigates incommensurability as an argument for pluralism. Sections 8.3. and 8.4 develop Feyerabend's *general methodological argument for pluralism* and his *in principle argument for pluralism* respectively. The chapter concludes by examining Feyerabend's pluralistic conception of knowledge and its advance, and by contrasting it with more standard views.

8.1. Feyerabend's pluralistic test-model

As we have seen, following Popper, Feyerabend emphasized the *critical* nature of scientific knowledge and its development. He argued both that the potential to criticize and to challenge any aspect of science is science's distinguishing characteristic, and that this critical element of scientific research is responsible for science's progressive nature. Science is thus distinguished from myth, which is dogmatic. Feyerabend often contrasted science as a critical enterprise with myths that are held dogmatically. (e.g. 1961i; rp 1999a, pp. 50-77). However, Feyerabend did not accept any of the predominate views on exactly how theories are critically put to the test. Both the Popperian and logical empiricist models of theory testing are based on the idea that a single theory is tested against the given facts:

> All these investigations [such Popper and logical empiricists' into questions of confirmation and test] use a model in which a *single* theory is compared with a class of facts (or observation statements) which are assumed to be '*given*' somehow (1963a; rp 1999a, p. 91, italics inserted. Also see 1962a, p. 30; rp 1981a, p. 46).

According to Feyerabend:

> This manner of discussion [a monistic test-model] does not allow us to give an adequate account of crucial experiments (1965b, p. 216).

In Feyerabend's view, any such test-model is an oversimplification that errs in two closely related respects: (A) Facts are not simply 'given', and (B) Theories are not just compared with facts. They are also compared against each other. They compete.

Ad (A) Facts are not simply 'given'. They are interpretations. Feyerabend argued that this holds for both observable and unobservable facts.[307] As we have seen, Feyerabend routinely argued that scientific advance involves reinterpreting the accepted facts in light of theoretical advancements. Having earned a Masters in Astronomy, Feyerabend often illustrated this point by discussing the significance of the red shift (the observation that the visible parts of the spectrum of electromagnetic radiation that are emitted by galaxies are shifted to its red extremity, which is usually explained by the Doppler effect; i.e. that the galaxies are moving away from each other). According to Feyerabend, even everyday observables like witnessing a brown barn, and then making the claim that "The barn is brown" involves theoretical interpretations, already at the level of perceptual organization.[308] For example, the brownness appears to be a property of the barn. According to Feyerabend, scientific advance, however, has shown that colors are not primary or even secondary qualities. Rather, brown (to continue with the same example) is a relation that depends upon the relative motions of the object observed and the observer. In other words, according to Feyerabend, barns are not brown. Brownness is not even a property of objects. Objects such as barns only appear to be brown given particular conditions. Brownness is a property of a relation between an object and observer.

Feyerabend's general point about facts as theoretical interpretations is now generally acknowledged with regards to the experimental facts of the natural science (see e.g. Culp 1995). However, from the claim that experimental facts are interpretations of data, it does not follow that facts cannot be 'objective'. Culp claims that although experimental facts are indeed the products of theoretical interpretations (responding directly to Feyerabend among others), scientists have developed techniques (including the use of various independent tests of the same experimental fact) which can be used to gage the reliability, and thus objectivity, of their claims. In any case, it is now widely recognized that the data that results from an experiment are not simply 'given'.[309] The point has been discussed at length in

---

307 Feyerabend argued this at length in (1960f). For an English translation, also see (1999d).

308 Feyerabend was keenly aware of, and heavily influenced by, contemporary developments in the psychology of perception; i.e. Gestalt psychology.

309 See e.g. (1958a), especially pp. 164-165; rp (1981a), pp. 32-33, (1958b), (1960a); rp (1981a), pp. 37-43, especially (1960f), (1962a), pp. 29-31; rp (1981a), pp. 45-46, (1965b), p. 198, and

the literature on the theory-ladeness of observation as well as the literature on the significance of the distinction between observable and theoretical entities.

Feyerabend challenged the notion of a theoretical entity as something that is unobservable to the unaided senses *as an epistemically significant distinction*. As Feyerabend repeatedly argued (beginning already with his 1951 dissertation on the role of basic sentences in empiricist philosophy of science), *there is no justifiable reason to assume that human senses are a good guide to the real nature of things*. There is ample evidence suggesting that the immediate results of human senses often need to be reinterpreted in light of theoretical knowledge. Examples can be drawn from everyday experience. They include simple optical illusions, such as the appearance of the sun going down, when in fact, the earth spins into the way. According to Feyerabend, what we see is always a theoretical interpretation of some raw data. He based this claim on significant advances in the psychology of perception that can be found in Köhler (1947). From there, the point became popular in philosophy via Wittgenstein and Hanson.[310] Feyerabend argued that there is no epistemic significance to the distinction between observable and unobservable, not because theory-ladeness arguments have succeeded in challenging the objectivity of theoretical entities, but because common observable facts have been shown to be just as bad off epistemically as the theoretical entities.

Theories are not only used to interpret the data that experiments produce. They are also used in designing and running the experiment in the first place.[311] Thus, theories are used in at least two stages of producing an experimental fact. Consequently, experimental facts are not simply 'given'. They are just as open to criticism and improvement as the theories, which the facts are supposed to challenge or support. Experimental facts are no better off epistemically than the theories used to construct the experiment and the theories used to interpret the data. The *legitimacy* of the results of some experiment depends on the theories both used to design the experiment and the theories used to interpret the data. Only if those theories are assumed to be correct can the experimental fact serve as a reliable tool for criticism or support of some hypothesis. These are exactly the kinds of

(1969b); rp (1981a), pp. 132-135. Also see (1954c), p. 470, (1961b), p. 82, (1961g), p. 443, and especially (1965b), p. 219-224, fn. 8.

310 See Wittgenstein (1958), XI, pp. 193e-214e who cites Köhler, and also see Hanson who cites Wittgenstein (Hanson 1958). From Hanson and Wittgenstein, the point was picked up and emphasized by Feyerabend and Kuhn. See e.g. (1951).

311 For illustrative discussion of both of these points, as well as dramatically different positions and arguments as to their bearing on the issue of the *objectivity* of experimental facts, see and compare Culp (1995) and Collins (1985).

considerations that Feyerabend used to justify the principle of tenacity. Feyerabend often stated the point forcefully:

> There is no 'neutral experience', with which theories can be confirmed or against which theories can be directly falsified (1960f; rp 1999d, p. 48).

Instead, experience comes organized by the theories we use to interpret it:

> experience arises *together with* theoretical assumptions, *not* before them, and . . . an experience without theories is just as incomprehensible as is (allegedly) a theory without experience (1969b, p. 792; rp 1981a, p. 133).

The "allegedly" is apparently because Feyerabend argued that it is possible to have "Science without experience" (1969b). Note that here by "experience", Feyerabend means subjective sensations. Throughout his publications, he argued that scientific advance corrects beliefs based on misinterpretations of subjective sensations (e.g. 1963b).

As his early ideas matured, Feyerabend expressed these ideas by attacking what he called the "autonomy principle". According to the autonomy principle, "facts *exist, and are available independently of whether or not one considers alternatives to the theory to be tested*" (1963a; rp 1999a, pp. 91ff.).[312] Feyerabend did not believe this to be the case. On his view, what are accepted as the facts depends on the theoretical viewpoint used to interpret the relevant experiences, and even to decide what the relevant experiences are. Feyerabend argued that theories are not confirmed or falsified by statements describing refuting facts by themselves.[313] Rather, *theories are tested against facts as they are interpreted by some theory.*

This insight underwrites much of Feyerabend's ideas on the nature and role of experience in the production of knowledge and in scientific advance. It is the kernel of his conception of knowledge and its dynamic, and it can be traced all the way back to the original German version of his doctoral thesis (1951, p. 89ff., and especially p. 89, p. 104, and p. 105.) (although it is most clearly formulated and developed in his major 1960s publications). It is also the basis of one of his main complaints (if not *the* main complaint) against any interpretation of theory-testing that does not make room for it, from classical empiricism, to logical positivism, to logical empiricism, to Popper.[314] (Many of Feyerabend's publications contain

312 In this particular citation, Feyerabend is assuming Popper's conception of empirical content. According to Popper, the empirical content of a theory is the class of basic statements contradicting that theory, see e.g. Popper (1963), p. 385. Popper attempts to distinguish between a theory's logical content and its empirical content, for a discussion, see Dilworth (1981), pp. 41ff. However, the autonomy principle also hold for other less formal conceptions of empirical content, such as Mach's.

313 For an excellent discussion, see Van Fraassen (1997).

314 See e.g. (1962a), pp. 29- 31, especially p. 31. For Popper, see (1981a), pp. 47ff.

criticisms of central tenets of empiricism based on it.)[315] As we shall see, it also furnishes him with the tools to construct his main arguments for pluralism.

For these reasons, it is worthwhile taking a moment to appreciate the central role it played in Feyerabend's ideas, and then to probe its context and historical origins. This will help to set the stage for developing the image of scientific advance that Feyerabend believed it promotes.

The centrality of the view that theories are confirmed or refuted by facts as interpreted by some theory (not by given facts) in Feyerabend's ideas can best be appreciated by considering that the idea is at the foundation of the pragmatic theory of observation, as well as the contextual theory of meaning. As Feyerabend's views on observation and meaning are the founding pillars with which he launched his attacks on empiricism; arguing instead for a more tolerant pluralism that recognizes incommensurability and the possibility of rational disagreement, it is clear that these idea underwrites much of his ideas.

Feyerabend's basic conception of perception and the relation between theoretical knowledge and observations mirrors his conception of meaning: *scientific terms receive their meaning from the theories whose truth is presupposed when the term is used to describe or explain some feature of the world.* Scientific observations, according to Feyerabend, are *caused* by physical objects, but they receive their *content* from the theories whose truth is presupposed in *making* the observation. Feyerabend's views on the meanings of terms and on the content of observations are both top-down: *Scientists use theories to give content to their language and to their observations.*

Feyerabend's pragmatic theory of observation denies the standard empiricist conception of an asymmetry between theory and observation: Observation statements eliminate theories, but theories do not eliminate observation statements. Feyerabend claimed that the "history of science has given ample evidence for the fact that it is 'sound scientific method' not to take experience on its face value, . . . but to try to explain it as the result of process which are not immediately accessible to observation" (1960e, p. 330, fn. 1; rp 1981a, p. 227, fn. 16). According to Feyerabend, *theories are invented in order to criticize observational results* (1965b, p. 152.)

Much of Feyerabend's philosophical work should be seen in the context of his having developed this idea out of developments in the psychology of perception (under the influence of Köhler, Anscombe, Wittgenstein and Duhem, and having brought it to bear on the central tenets of

315 See e.g. (1958a), especially pp. 164-165; rp (1981a), pp. 32-33, (1958b), (1960a); rp (1981a), pp. 37-43, especially (1960f), (1962a), pp. 29-31; rp (1981a), pp. 45-46, (1965b), p. 198, and (1969b); rp (1981a), pp. 132-135. Also see (1954c), p. 470, (1961b), p. 82, (1961g), p. 443, and especially (1965b), pp. 219-224, fn. 8.

logical positivism, logical empiricism, Bohr's principle of complementarity, and critical rationalism. In the late 1940s and early 1950s in philosophy and in psychology, in the wake of the development of Gestalt psychology and its influence on Wittgenstein's later philosophy, a great deal of emphasis was put on the idea that our mental faculties play a fundamental role in ordering our experience. Placed within a larger and longer tradition in the philosophy of science, the basic idea should be understood as an attempt to invert the standard principle of empiricism.[316] According to empiricism (at least since Bacon), scientific inquiry should begin with the facts as they are witnessed by the senses and then work up to theories about them.[317] Bacon suggested that scientists reason from particular facts to general conclusions:[318]

> [searching into and discovering truth] derives axioms from the senses and particulars, rising by a gradual and unbroken ascent, so that it arrives at the most general axioms last of all. This is the true way, but as yet untried (Bacon 1620, xix).

In other words, Feyerabend argued that instead of a bottom-up model which begins with bare experiences and works up to a theory about them (such as in Bacon's methodology), scientific advance is best understood as top-down:[319] Theories are used to interpret, correct, and reinterpret our experiences (1965b, p. 208):

> Observational findings are not at all final barriers for theories, although they are usually presented as such. Observational findings can be reinterpreted, and can perhaps be made to lend support to a point of view that was originally inconsistent with them (1965b, p. 202. Similarly 1961i; rp 1999a, pp. 50-77).

In order to illustrate his point, Feyerabend developed the example of how observational facts that apparently support a mentalistic point of view

---

316 While Feyerabend should be understood a following Duhem in arguing against Bacon's empiricism, Feyerabend accepted the idea of crucial experiments, but insisted that they are never absolutely final. See Duhem (1954) and Feyerabend (1970e). For a discussion, see Preston (1997), pp. 115ff. For Bacon on crucial experiments, see Book II, section 36 of Bacon (1620).

317 Bacon's place in the history of discussions about the scientific method is controversial as interpretations differ widely. For example, one contemporary encyclopedia of philosophy claims that Bacon's reputation reached its lowest point as the result of Karl Popper's methodology, which rejects induction, while others see Bacon as a forerunner of Popper's falsificationism and the hypothetico-deductive method, as he emphasized negative instances, *anticipating* Popper. See and compare the end of Milton (1999) to Honderich (1995), p. 75.

318 To be fair, Bacon was also aware of the obvious need to correct the unaided senses in light of newly gained theoretical knowledge. See e.g. Bacon (1620), especially the 'Idols'.

319 For a direct criticism of Bacon's version of empiricism, see e.g. (1965b), pp. 156ff.

about psychological states could be reinterpreted to support a materialistic point of view.[320]

The basic idea of the causal or "pragmatic" theory of perception[321] has become so deeply entrenched in our everyday, common-sense view of our relation to the world, that it is only through a contrast to the views of some of the early logical positivists that its significance can be adequately appreciated. Some logical positivists originally subscribed to the view that observation statements are about sense-data; i.e., observations are about *psychological events*. Neurath, Carnap and Popper objected to this view, arguing that observations statements are about external *physical objects, not internal psychological events*. Feyerabend defended this later view, arguing that (a) observations are *caused* by physical objects, and (b) observation statements are *about* physical objects, not *about* psychological events. These two claims make up the pragmatic theory of observation.

However, Feyerabend also distinguished sharply between the *cause* of an observation statement and the *content* of an observation (or its *meaning*). While observations may be *caused* by physical objects, and observation statements are *about* physical objects, according to Feyerabend, the *content* (or *interpretation)* of an observation and the corresponding observation statement is based on theoretical knowledge, not the objects themselves:

> a statement will be regarded as observational because of the *causal context* in which it is being uttered, and *not* because of what it means. According to this theory, 'this is red' is an observation sentence, because a well-conditioned individual who is prompted in the appropriate manner in front of an object that has certain physical properties will respond without hesitation with 'this is red'; and this response will occur independently of the interpretation that he may connect with the statement... All we need in order to provide a theory with an observational basis are statements satisfying this pragmatic property ... Their meaning they obtain from the theory which they belong (1965b, pp. 198-199. Also see 1965b, pp. 204ff.).

In addition to the first problem with the logical empiricist and Popperian test-model (A): that facts, even those facts based upon observable experience, used to test theories are not simply given, but are the result of interpretations), Feyerabend argued that there is a *second* related problem (B): Theories are not just compared with facts. *They are also compared against each other. They compete.*

---

320 See (1965b), especially pp. 202-208. Also see e.g. (1963b), (1963d), (1965a); rp (1981a), pp. 97-103, and (1965c); rp (1981a), pp. 104-131.

321 Feyerabend claimed that Popper, Carnap and Neurath defended the causal 'or pragmatic' theory of observations in the 1930s with excellent arguments, but later abandoned it. See e.g. (1962a), pp. 34ff.; rp (1981a), pp. 49ff., (1965b), p. 152, (1965c), p. 243; rp (1981a), p. 125. Feyerabend's claim that Carnap subscribed to the pragmatic theory of observation has been challenged in Oberdan (1990). For a discussion, see Preston (1997), pp. 45ff.

Ad (B) In Feyerabend's view, this follows from the first point. That is, because specific theories are used in discovering and expressing the facts which may be used to test theories, it could well happen that the facts relevant for testing one particular theory T may be properly described with the help of another competitor theory T′. A theory T may be indirectly tested against another theory T′, as T′ may be used to discover and to state the facts relevant to testing the theory T.

A consequence of Feyerabend's view on the nature of experience as a product of theoretical interpretation, and his subsequent rejection of the assumption of the relative autonomy of facts, is that an empirical epistemology requires a new model of theory testing. Instead of a *monistic test-model* in which a single theory is confronted by 'given' facts, Feyerabend proposed *a pluralistic test-model* according to which:

> the methodological unit to which we must refer when discussing questions of test and empirical content is constituted by a whole set of partly overlapping, factually adequate, but mutually inconsistent theories (1963a; rp 1999a, p. 92, italics in original).[322]

In light of his views on the nature of experience, observation, meaning and theoretical knowledge, it is clear why Feyerabend claimed that the nature of scientific advance cannot be properly captured with a test-model according to which a single theory is compared to the facts. In this section, I have merely *explicated* Feyerabend's pluralistic model on theory testing, and examined the ideas about the nature of experience, perception, meaning, and theoretical knowledge that he used to support it. The next question is: How exactly did Feyerabend argue against monistic test-models, and for his pluralistic test-model?

Feyerabend used a two-pronged argumentative strategy to attack the monistic test-models promoted by classical empiricists, the logical positivists, the logical empiricists, Popper, and Kuhn (all of which envisage a single theory confronted with evidence). He claimed that these monistic accounts do not correctly *describe* actual scientific advance, and that they would inhibit scientific advance if accepted and enforced *normatively*.[323]

According to Feyerabend, scientific advance can only properly be described and understood with a pluralistic test-model according to which theories indirectly compete against each other by confronting the facts as they are stated by competing theories. This pluralistic test-model is at the basis of Feyerabend's methodological arguments for pluralism, as well as his pluralistic conception of scientific method.

322 Kuhn's views are similar, see Hoyningen-Huene (1993), pp. 236ff.

323 See (1962a), pp. 32ff. For Popper, see (1981a), pp. 47ff., fn. 6. For a discussion, see Preston (1997), pp. 124ff.

8.2. Incommensurability as an argument for pluralism

There is a widespread modern myth that incommensurability should be understood as implying the rational incomparability of scientific theories. For example: "Within philosophy of science, theories which, in a radical sense, cannot be compared are said to be incommensurable" (Logue 1995, 397).[324] However, Feyerabend and Kuhn repeatedly rejected such claims. For example, Feyerabend explicitly stated that he had "never inferred 'incomparability' from incommensurability (1978, p. 68).[325] Even so, the myth that they used incommensurability to imply incomparability continues to persist.[326] For example:

> In his original discussion of [incommensurability] in 'Explanation, Reduction, and Empiricism', Feyerabend took incommensurability to consist in absence of logical relations due to semantic variance of the terms used by theories, *resulting in the inability to directly compare the content of theories*" (Sankey and Hoyningen-Huene 2001, ix, italics inserted).

However, the italicized portion of this claim is extremely misleading, as Feyerabend did *not* claim that incommensurability precludes the direct comparison of the content of theories anywhere in (1962a). Rather, as we have seen in chapter 5, he claimed that a particular class of models of theory comparison (formal models) are undermined by the fact that the concepts of certain pairs of theories are incommensurable.

In this section, I argue that far from arguing that incommensurability implies incommensurability, Feyerabend actually argued that incommensurable alternatives offer a *better* means for comparing theories. Incommensurable theories offer a better means for testing the merits of existing theories than the mere development of commensurable alternatives (1962a, p. 66). Feyerabend even illustrated this point with an historical example of a crucial experiment between incommensurable theories. He argued that even though the central concepts of statistical and Phenomenological Thermodynamics were incommensurable, it was still possible to stage a *crucial experiment,* based on a single phenomenon (Brownian Motion) between them. This, as we shall see, is the centerpiece of his meth-

---

324 For a long list of author's who have treated incommensurability as implying incomeparability, see Hoyningen-Huene (1993) 218, fn. 118.

325 Also see (1961b), pp. 160ff.; (1962a), p. 92; (1965b), pp. 214-217; (1965c), p. 23; (1970b), pp. 219ff.; (1970d), pp. 247-249; especially (1970c), pp. 88ff.; and (1977), p. 365, fn. 2. For Kuhn, see e.g. Kuhn (1970), p. 267; Kuhn (1976), pp. 190-191; Kuhn (1979), p. 416; and Kuhn (1983), pp. 670-671, and the excellent discussion in Hoyningen-Huene (1993), pp. 218ff.

326 That incommensurability does not, or was not intended to, imply incomparability, has also been pointed out by Brown (1983); Carrier (2001); Giedymin (1971); Hoyningen-Huene (1993), pp. 218ff.; Moberg (1979); Preston (1997a), 115ff.; and Weber (2002).

odological argument for pluralism. The argument turns on the claim that this test could not have been conducted without the development of an alternative (1962a, p. 66). In short, Feyerabend used the notion of incommensurable concepts as an argument for promoting pluralism, as it increases the testability and empirical content of our theories, not to argue that the content of theories cannot be compared (e.g. 1965c, p. 233; rp 1981a, p. 116).

Feyerabend argued that the greater the difference between an alternative theory and the standard point of view, the better the alternative can *test* the standard point of view:

> Now it seems to me that the more general our knowledge gets the more important it will be to carry out tests in the manner indicated, i.e., not by comparing a single theory with experience, but by staging crucial experiments between theories which, although in accordance with all the known facts, are mutually inconsistent [conceptually; i.e. incommensurable] and give widely different answers in unexplored domains . . .the methodological unit to which we refer when discussing questions of test and empirical content consist of a whole set of partially overlapping, factually adequate, but mutually inconsistent [conceptually, not logically] theories. To the extent to which utilization of such a set provides additional tests which, for empirical reasons, could not have been carried out in a direct manner, the use of a set of this kind is demanded by empiricism. For the basic principle of empiricism is, after all, to increase the empirical content of whatever knowledge we claim to possess (1962a, pp. 66; rp 1981a, pp. 71ff. Cf. e.g. 1965c; p. 226; rp 1981a, p. 109, and 1969e, pp. 62-63; rp 1981a, pp. 159-160).[327]

Feyerabend argued that theories that use incommensurable concepts are *better* suited for the purpose of mutual criticism and improvement than theories that merely logically contradict each other.[328]

Although Feyerabend increasingly argued that incommensurability precludes any judgment of relative verisimilitude or any judgment of approximate truth,[329] he insisted that incommensurable theories can be critically compared:

> It *is* possible to use incommensurable theories for the purpose of mutual criticism (1965c, p. 236, rp 1981a, p. 117, italics in original).

According to Feyerabend, means of critically comparing incommensurable theories "are readily available" (1965c, p. 232; rp 1981a, p. 115). He often discussed several different methods of comparing the relative merits of

---

327 Here, we find Feyerabend simply assuming this methodological principle, not arguing for it, or developing in the context of a new position. Feyerabend is entitled to assume this principle in its context, because it is accepted by the empiricists that he criticized with it.

328 Also see e.g. (1961b), p. 83, (1965b), pp. 216-217, (1965c), pp. 226ff.; rp (1981a), pp. 109ff., and (1967a), pp. 40ff.

329 See e.g. (1970c), pp. 218-219; rp (1981b), pp. 151-152, (1981c) p. 63, especially fn. 2, rp (1981b), p. 238, especially fn. 17, (1977b), p. 365, and (1978a), p. 68.

incommensurable theories,[330] and he explicitly stated that he "never inferred 'incomparability' from incommensurability.

For example, in order to overcome the incommensurability between two theories T and T′, one possibility is to construct a more general theory T′′ describing a common background that provides test statements acceptable to both theories T and T′. Feyerabend argued that:

> the possibility of introducing new critical tests in this way is a further reason why the invention of more general theories should not be made dependent on the refutation of less general theories (1965b, p. 217).

This, then, is another reason for the promotion of alternatives.

The incommensurability between two theories T and T′ can also be overcome through an internal examination of both theories:

> If the concepts are different [incommensurable], how is it possible to have crucial experiments? We require that someone who understands the theory will, under certain circumstances, say that he is wrong. If a proponent of one theory talks to that of another, sooner or later they will discover that they are speaking different languages. But this does not matter because one can learn the other's language, and when he has learned it he will use those terms and then admit that he is wrong. So refutation is possible . . . (1970e, p. 247).

A decision on the relative merits between theories composed of incommensurable concepts is possible by learning the languages of both. Learning how to understand the incommensurable concepts allows one to see the comparative merits of both theories in their own terms, and then to make an informed decision on which to accept. However, the main support for Feyerabend's prescriptive recommendation to use theories containing incommensurable concepts for the purpose of mutual criticism came in the form of the examples, such as Brownian Motion, that he presented, not the abstract suggestions that he made.[331]

Given that Feyerabend argued that using incommensurable theories is a better means for improving our theories than simply confronting one theory with the facts, it is more than just a little ironic that so much of the secondary literature on incommensurability has claimed, and continues to claim, that incommensurability implies incomparability. For example, an entry in the *Oxford Companion of Philosophy* (1995) reads: "Incommensurability. Within philosophy of science, theories which, in a radical sense, *cannot be compared* are said to be incommensurable, a term first given wide currency by Kuhn" (Honderich 1995, p. 397, italics inserted). Devitt con-

330 See e.g. (1961c), pp. 160ff., (1962a), p. 92; rp (1981a), p. 93, (1965b), pp. 214-217, (1965c), p. 233; rp (1981a), p. 116, (1970c), pp. 219ff.; rp (1981b), pp. 152ff., (1970e), pp. 247-249), especially (1970d), pp. 88ff., and (1977b), p. 365, fn. 2.

331 See (1962a), pp. 52-62 and pp. 74-81; rp (1981a), pp. 62-69 and pp. 76-82. For a discussion, see McEnvoy (1975).

tinues to define incommensurability (and he specifically refers to Feyerabend): "Incommensurability: Terms in rival comprehensive theories in a domain differ sufficiently in meaning, especially in reference, to make the theories incomparable" (Devitt 2001, p. 143).[332]

The charge that according to Feyerabend, the content of incommensurable theories cannot be compared was influentially launched by (Achinstein 1964, p. 499, Shapere 1966, pp. 57-58 and pp. 67-68, and especially Scheffler 1967, pp. 16-17). Many other authors have subsequently interpreted incommensurability as implying the impossibility of comparing the content of two theories.[333] Of course, Feyerabend tried to correct the misimpression that incommensurability was intended to imply incomparability on a number of occasions. For example:

> Next comes Laudan's discussion of *incommensurability*. As he tells it, 'Kuhn, Hanson and Feyerabend began to despair about the possibility of any objective yardstick for comparing different theories and suggested that theories were incommensurable and thus not open to objective comparison'. This suggests that we wanted to compare theories, were misguided by some feature of science into believing that a comparative evaluation is impossible and joylessly published this disagreeable consequence. A look at our work reveals an entirely different story. What we 'discovered' and tried to show was that scientific discourse *which contains detailed and highly sophisticated discussions concerning the comparative advantages of paradigms* obeys laws and standards that have only little to do with the naive models philosophers of science have designed for that purpose. *There is* comparative, even 'objective' comparison – but it is a much more complex and delicate procedure than is assumed by rationalists (1981c, p. 63, italics in original, also see fn. 1, and 1981b, pp. 238, fn. 17).

Not everybody has made the common mistake of equating incommensurability with incomparability, or assuming that the latter follows from the former.[334] For example, Couvalis has discussed several methods that Feyerabend suggested for comparing incommensurable theories (Couvalis 1989, pp. 95ff.). Harold Brown also exposed the misunderstanding that incommensurability, as espoused by Kuhn and Feyerabend, was

332 See also Devitt (1984), and Devitt (1997).

333 See e.g. Agassi (1985), p. 51, Balzer (1978), pp. 313-314, Devitt (1979), p. 29, Kitcher (1978), p. 529, Kitcher (1992), p. 95, Kordig (1971), pp. 473-473, Lakatos (1970), p. 179, fn. 1, Levin (1979), p. 407, Newton-Smith (1981), pp. 9-10, Putnam (1981), p. 118, Scheffler (1967), pp. 16-17, more recently Shapere (1999), p. 733, and Szumilewicz (1977). For more sources, see Hoyningen-Huene (1993), p. 218, fn. 118. Hoyningen-Huene argued that Kuhn never meant to imply blanket incomparability either, see Hoyningen-Huene (1993), pp. 218ff. For Kuhn's developing views, which are based on restricting incommensurability to parts of a theory, i.e. to only some of the concepts of two theories, see e.g. Kuhn (1983), pp. 670ff..

334 Other commentators who have not mistaken incommensurability to imply incomparability include Carrier (2001), Giedymin (1971), Hoyningen-Huene (1993), pp. 218ff., Moberg (1979), Preston (1997), pp. 115ff., and Weber (2002).

intended to imply incomparability (Brown 1983). He discussed the ways in which incommensurable theories can be compared, and he outlined the reasons that some well-entrenched philosophical positions have difficulty accepting the idea that scientific progress involves radical reconceptualization.[335]

The reason that Feyerabend was interpreted as having claimed that incommensurability implies incomparability is that in (1962a), Feyerabend wrote: "Now incommensurable theories may not possess any comparable consequences, observational or otherwise. Hence, there may not exist any possibility of finding a characterization of the observations which are supposed to *confirm* two incommensurable theories" (italics inserted). This Is supposed to support the claim that Feyerabend took incommensurability to result in the inability to directly compare the content of theories (Sankey and Hoyningen-Huene 2001, p. ix). However, here Feyerabend claimed that there are no comparable *consequences* of incommensurable theories, not that the *content* of incommensurable theories cannot be directly compared.

While it could be argued that the latter follows from the former on some conceptions of direct theory comparison, the latter is not what Feyerabend claimed, and it is not what Feyerabend meant, and Feyerabend actually explicitly criticized the kind of conceptions of theory comparison needed to make the latter follow from the former. From the claim that incommensurable theories share no comparable consequence, it follows that the content of incommensurable theories cannot be directly compared, *only given* a particular conception of theory comparison according to which the content of two theories is directly compared by comparing their observable consequences, and it is exactly this conception of theory comparison that Feyerabend later rejected as *naïve*, or overly simplistic.

Feyerabend claimed that although the very same physical measurement, or the same experiment, i.e. the same actual physical activity, might produce a numerical result that can be used to *confirm* both of two incommensurable theories, in order to use the result to confirm a theory, it must be still interpreted on the basis of a theory (1962a, p. 93). Given that the two theories, delineated in incommensurable concepts, will use *qualitatively* incompatible concepts to interpret *quantitatively* identical results, they will interpret the same (quantitative) *sentences* as different *statements*:

335 Carrier correctly argues that "all pairs of incommensurable theories necessarily possess at least one realm of phenomena which they jointly address and which provides the basis for their empirical comparison." He also recognizes that Papineau (1979), pp. 137-138 and Laudan (1977), pp. 142-144 reach a similar conclusion. See Carrier (2001), p. 84. In a similar vein, Farrell (2000a) defended Feyerabend's claim that an alternative incomemensurable theory can increase the empirical content of its rival.

> Hence, there may not exist any possibility of finding a *characterization* of the observations which are supposed to *confirm* two incommensurable theories (1981a, p. 93, italics inserted).

I emphasized 'confirm' to highlight the fact that Feyerabend was not discussing the possibility of a crucial experiment, in which one theory is confirmed and the other *disconfirmed*, but rather, was making a point concerning Nagelean reduction, which takes one theory to be reducible to the other, *because* both theories can be confirmed by the same sentence. However, the 'characterization' or interpretation will be different depending on the theories used to make it. Feyerabend's point is that although he agreed with Nagel that one sentence can confirm both theories (i.e. the sentence, 'The temperature of the milk in my glass is 10° Celsius' can be used to confirm both phenomenological and statistical predictions about the temperature of the milk in my glass), this single *sentence* will be interpreted as two mutually exclusive *statements*, on the basis of incommensurable concepts, such as temperature according to Phenomenological Thermodynamics and temperature according to Statistical Thermodynamics. Thus, Feyerabend's use of the term "comparable" in (1962a) should not be interpreted as supporting the claim that according to Feyerabend incommensurability *precludes* the direct comparison of the contents of two theories. He was talking about instances that confirm two theories, not about comparing their contents via a crucial experiment. He simply never directly made the particular claim that incommensurable theories cannot be compared in "Explanation, Reduction and Empiricism", though he has often been interpreted as having done so.

However, on the standard conception of theory comparison, a crucial test is staged between two theories when they make different predictions as to the result of the same experiment, and it turns out that the prediction of one theory is on the basis of the *same evidence* as the prediction of the other is refuted. For example, the proponents of both of two theories A and B agree that experiment Z is a good test of their theories. Theory A predicts that the result of experiment Z will be 2 (plus or minus 0.5), and theory B predicts that the result of experiment Z will be 7 (plus or minus 0.2). If the experiment repeatedly produces a result of 6.9, then a crucial experiment has been staged. The key is that the same piece of evidence confirms one theory and refutes the other (see e.g. Carrier 2001, p. 82).

In the passages cited by Sankey and Hoyningen-Huene, which has so often been used to support interpreting Feyerabend as having argued that incommensurability implies incomparability, Feyerabend did not directly discuss the possibility or impossibility of such crucial experiments. Feyerabend did claim that although two incommensurable theories may predict the same observational result of the same measurement or experiment within their shared domains, the meaning of the terms used to state the result according to the one theory will differ from those of the other. In (1962a), Feyerabend explained this point with regard to each of his three examples of scientific theoretical advance involving incommensurable concepts (1970c, p. 221; rp 1981b, p. 154). In each case, the two theories containing incommensurable concepts agree on the some observable results that can be used to test the theories within their shared domains. For example, when Feyerabend developed the relationship between impetus theory and Newtonian Physics, Feyerabend pointed out that according to impetus theory, a moving object situated in empty space, which is not under the influence of gravity or friction, is still under the influence of impetus (its inner principle of motion). According to both the impetus theory and Newton's theory, *exactly the same motion will result*, and this is why it *seems* reasonable to suggest momentum as an analogue for impetus: "the measure of this magnitude (viz. mv) is *identical* with the measure that has been suggested for the impetus" (1962a, p. 56, rp 1981a, p. 65, italics inserted). However, according to Feyerabend, even though they will quantitatively indistinguishable in their shared domains, impetus cannot be equated, or reduced to by way of logical relations, to momentum, because the concepts are qualitatively incompatible. For example, whereas impetus is supposed to cause the objects motion, momentum is supposed to be the result of the objects motion. Moreover, the inertial motion of Classical Mechanics is supposed to occur without any cause, while impetus is supposed to be the cause. These are the kinds of qualitative differences Feyerabend was alluding to. This was the basis of Feyerabend's claim about the incomparability of *confirming* observable consequences of incommensurable theories. As he put the point in 1970:

> From the fact that the same measuring instrument and the same measuring procedures are there, it doesn't follow that considerable conceptual changes have not taken place (1970e, p. 241).

Feyerabend described the case of Brownian Motion as an example of a "*crucial experiment*" between theories he had explicitly called incommensurable (1962a, p. 66ff.; rp 1981a, p. 71ff, italics inserted). Moreover, he argued that that:

> the more general our knowledge gets the more important it will be to carry out tests in the manner indicated, i.e., not by comparing a single theory with experience, but by staging crucial experiments between theories which, although in accordance with all the known facts, are *mutually inconsistent*[336] and *give widely different answers in unexplored domains* (1962a, p. 66, 1981a, p. 72, italics inserted).

336 That is, *qualitatively* inconsistent, i.e., incommensurable, as the "manner indicated" was illustrated with the example of Phenomenological Thermodynamics and Statistical

On Feyerabend's view, it is the widely different answers in unexplored domains (e.g. small particles suspended in fluids, or velocities close to the speed of light, etc.) that provide ample opportunity to stage crucial experiment. Thus, it is extremely odd that Sankey and Hoyningen-Huene, in an introduction to a book on incommensurability, should claim that *originally* (i.e. in his 1962a), according to Feyerabend, the content of incommensurable theories cannot be directly compared, which implies not only that Feyerabend made this claim (which, as we have seen, he did not), but also that according to Feyerabend, it is *impossible* to stage a crucial experiment between incommensurable theories.[337] To be sure, in later publications, Feyerabend did make claims like:

> With the discussion of incommensurability, I come to a point of Kuhn's philosophy which I wholeheartedly accept. I am referring to his assertion that succeeding paradigms can be evaluated only with difficulty and that they may be altogether incomparable, at least as far as the more familiar standards of comparison are concerned (they may be readily comparable in other respects) (1970c, p. 219; rp 1981b, p. 152).

Feyerabend also emphasized that theories will be "commensurable in some interpretations, *incomparable* in others" (1970c, p. 220; rp 1981b, p. 153, italics inserted). And later Feyerabend claimed that when incommensurable theories are concerned, "A comparison of content and a judgment of verisimilitude cannot be made" (1970c, p. 222 and pp. 227-228; rp 1981b, p. 154, and p. 160). Sankey and Hoyningen-Huene could have used these later statements to support their claim about incommensurability implying the incomparability of content, but note that even here Feyerabend restricted his claims about incommensurability implying the incomparability of the content of theories to some rather unsophisticated models of such comparisons, not to *any* kind of comparison of the content of two theories. In the same article, Feyerabend explained how crucial experiments can, and have, been staged between incommensurable theories (1970c, p. 226; rp 1981b, pp. 158-159). Feyerabend proceeded to explain the usual conception of theory comparison, and to explain why incommensurability exposes the shortcomings of such a model, not the impossibility of comparing incommensurable theories (1970c, p. 220; rp 1981b, p. 153).

With that said, however, Feyerabend did *sometimes* claim that incommensurability, if it is extreme enough, could result in a situation in which

---

Thermodynamics, which was the second source of examples of incommensurable concepts in scientific theories (i.e. the concepts of temperature and entropy).

337 See e.g. (1961c), pp. 160ff., (1962a), p. 92; rp (1981a), p. 93, (1965b), pp. 214-217, (1965c), p. 233; rp (1981a), p. 116, (1970c), pp. 219ff.; rp (1981b), pp. 152ff., (1970e), pp. 247-249, especially (1970d), pp. 88ff., and (1977b), p. 365, fn. 2.

two radically different theories do not share a single statement, and this would *preclude* crucial experiments in which *the same* evidence refutes one theory and confirms another:

> To express it more radically, each theory will possess its own experience, and there will be no overlap between these experiences. Clearly, a crucial experiment is now impossible (1965b, p. 214).

On the other hand, Feyerabend also sometimes claimed that a crucial experiment can be staged between two incommensurable theories in which *we can say* that the *same evidence* refutes one theory and confirms another:

> [two incommensurable theories] may be able to produce the local grammar of sentences which are directly connected with observational procedures. In this case the utterance of one of the sentences in question in accordance with the rules of local grammar, or the utterance of a local statement, as we shall say, can be connected with two "theoretical" statements, one of T, and one of T′ respectively . . . *We may now say* that the empirical content of T′ > the empirical content of T, if for every associated statement of T there is an associated statement provided by T′, but not vice versa. And *we may also say* that T′ has been confirmed *by the very same evidence* that refutes T if there is a local statement S whose associated statement in T′ confirms T′ while its associated statement in T refutes T (1965c, p. 233; rp 1981a, p. 116, italics inserted).

This does not appear to be a case of his having changed his views temporarily, as both publications are dated 1965. Rather, it appears to be a case of Feyerabend having difficulty expressing his views consistently, which partially explains why they have caused so much confusion.

Feyerabend did not claim that it *is* the same evidence that refutes one theory and confirms another. He used the phrases "we can say" and "we may also say" (moreover, he did not specify if these 'sayings' are sentences or statements, a distinction he made in this very passage, and which can be used to explain the apparent tension in his views). Feyerabend was *reluctant* to say that it *is* the same *evidence* that can be used to refute one theory and confirm another given his reservations about scientific realist metaphysics, although he saw no difficulty with having the results of a single experiment confirm one theory and refute another. The apparent tension in Feyerabend's views, however, is easily resolved.

The reason that the *same* evidence cannot be used to stage a crucial experiment between competing incommensurable theories derives from Feyerabend's distinction between sentences as causal events and statements as interpretations. On the one hand, on Feyerabend's view, one and the same physical result of an experiment can be interpreted, on the basis of competing incommensurable theories, as two different sets of evidence, but this does not cause any difficulty in judging which of the two theories that the one result can be interpreted to confirm, and which it can be interpreted to falsify. Thus, a crucial experiment between incommensurable theories can be staged, and one result can confirm one theory while it refutes the other. On the other hand, the point in denying that it is the *same* evidence that can be used to stage a crucial experiment between two incommensurable theories is that *evidence* is only *evidence* as it is interpreted on the basis of some theory. Therefore, the *same* evidence cannot be used to stage a crucial experiment between two incommensurable theories by confirming one theory and disconfirming the other.

To illustrate the point with Brownian Motion, according to Feyerabend, Brownian Motion refutes Phenomenological Thermodynamics *as it is interpreted on the basis of Statistical Thermodynamics*. However, before Statistical Thermodynamics was developed, Brownian Motion did not, and could not, refute Phenomenological Thermodynamics (see below). The important point here is that *one* actual, physical instance of a Brownian particle can be used to stage a crucial experiment between two incommensurable theories, but the actual physical motion of the Brownian particle is not, by itself, evidence. *Interpretations* of the actual instance of Brownian Motion are *evidence*. *Statements* about observations of Brownian Motion are used as evidence to make arguments against theories. When incommensurable theories compete, there are two sets of evidence, based on two mutually exclusive interpretations of the one actual, physical phenomenon. This is because the phenomenon is only evidence as it is interpreted differently by the two incommensurable theories. For example, one instance of a physical phenomena (Brownian Motion), can give rise to one physical observation statement meeting all four pragmatic requirements (recall Feyerabend's semantic views presented in chapter 2), "*That* is Brownian Motion"', which can result in two sets of evidence, based on two different interpretation of the one sentence. For example, the sentence can be interpreted as the statement, 'The particle exhibits strange, unexplained behavior'. This interpretation does not conflict with Phenomenological Thermodynamics. The sentence can also be interpreted as the statement, 'The particle exhibits continual motion because of the impact of the surrounding molecules'. This interpretation is inconsistent with Phenomenological Thermodynamics, but consistent with Statistical Thermodynamics.

An increase in the testability and empirical content of both of the theories delineated in incommensurable concepts is not the only positive consequence of using theories composed of incommensurable concepts for the purpose of mutual criticism. Feyerabend also argued that using incommensurable theories for the purpose of mutual criticism improves our understanding of both of them:

> The use of *a set of theories* with the propertied indicated above [i.e. incommensurable] will also *improve our understanding of each* . . . by making it clear what is denied by the theory accepted in the end. Thus, it seems to me that our understanding of Newton's somewhat obscure notion of absolute space and of its merits greatly improved when we *compare* it with the relational ideas of Berkeley, Huyghens, Leibnitz, and Mach, and when we consider the failure of these latter ideas to give a satisfactory account of the phenomenon of inertial forces. Also the study of general relativity will lead to a deeper understanding of this notion than could be obtained from a study of the *Principia* alone (1962a, p. 67; rp 1981a, p. 73, italics inserted).

To conclude, Feyerabend did not argue that incommensurability implies incomparability, and Feyerabend did not claim that the content of incommensurable theories cannot be directly compared. Instead, he argued that the development of theories consisting of incommensurable concepts promotes pluralism and scientific advance by providing additional means to test our views, which go beyond simply confronting accepted points of view with the facts, or beyond comparing commensurable theories. Feyerabend also made a number of positive suggestions about how scientists should and do proceed when confronted with competing theories delineated with sets of incommensurable concepts, and how crucial experiments can and have been staged between them. Feyerabend also claimed that comparing incommensurable concepts increases our understanding of them. Feyerabend did not originally consider the existence of incommensurable concepts to be a threat to theory comparison, or to directly comparing the content of theories. Later, his views were attacked on the basis of arguments that claimed that his notion of the existence of incommensurable concepts rules out the possibility of directly comparing the content of theories. Feyerabend's response was not to admit that the existence of incommensurable concepts poses difficulties for directly comparing the content of theories, but rather, the existence of incommensurable concepts poses difficulties for overly simplistic models of how to directly compare the content of two theories.

8.3. Feyerabend's methodological argument for pluralism

Feyerabend pluralistic test-model provides him with a general methodological argument for pluralism: *Promoting alternatives can boost the severity of theory-testing.* Feyerabend offered many slightly different version of his argument. It is the heart of his philosophy, and it was developed out of the lessons Feyerabend drew from the impotence of Ehrenhaft's experiments.[338] Feyerabend argued that as increasing testability is not only desirable methodologically, but also an essential feature of scientific advance (according to both the empiricists and Popper), promoting alternatives is also an essential feature of the empirical method (*according to their own accounts*):[339]

> the invention of alternatives in addition to the view that stands in the center of discussion constitutes an essential part of the empirical method (1963a; rp 1999a, p. 93, italics in original)

This is the basis of Feyerabend's general methodological argument for pluralism:

> The function of such concrete alternatives is, however, this: they provide means of criticizing the accepted theory in a manner that goes beyond the criticism provided by a comparison of that theory with the "facts". However closely a theory seems to reflect the facts, however universal its use, and however necessary its existence seems to be to those speaking the corresponding idiom, its factual adequacy can be asserted only *after* it has been confronted with *alternatives whose invention and detailed development must therefore proceed any final assertion of practical success and factual adequacy*. This, then, is the methodological justification of a plurality of theories: such a plurality allows for a much sharper criticism of accepted ideas than does the comparison with a domain of facts that is supposed to be given independently of theoretical considerations (1965b, p. 150, similarly, e.g. 1969e, p. 63; rp 1981a, p. 160).

According to Feyerabend, theoretical pluralism is methodologically *preferable* to theoretical monism because it provides *superior* or 'sharper' means for testing our beliefs – and sharper tests better promote progress in our understanding of the natural world.

338 In fact, almost all of his major publications, and even most of the minor ones, contain some form of a methodological argument for pluralism. See e.g. (1957c); rp (1981a), pp. 207-218, (1958a); rp (1981a), pp. 17-36, (1958b), (1958d); pp. 49ff.; rp (1981a), pp. 236ff.; (1960a); rp (1981a), pp. 37-43, (1995b), pp. 365-367, written 1960/61, (1960e), pp. 321ff.; rp (1981a), pp. 219ff., (1960f), (1961b), p. 81, (1961d), p. 279, (1961f), pp. 399-400, (1961h), pp. 47ff., (1962a); rp (1981a), pp. 44-96, (1962b), (1963a); rp (1999a), pp. 92ff., (1963b), (1963d); rp (1999a), (1964c), pp. 349ff.; rp (1981a), p. 203ff., (1964d); rp (1981a), pp. 161-175, (1965a); rp (1981a), pp. 97-103, (1965b), pp. 148ff., (1965c), pp. 223ff., see especially pp. 254ff., fn. 14; rp (1981a), pp. 104ff. and pp. 107-108, fn. 14, (1966c), pp. 246ff.; rp (1981b), pp. 61ff., (1967a), p. 35ff., (1967b), (1967c), p. 117, (1967e), (1968a), p. 132, (1969a); rp (1999a), pp. 107ff., (1969e), pp. 57ff.; rp (1981a), pp. 157ff., (1970c), pp. 216ff.; rp (1981b), pp. 149ff., (1970b), pp. 276ff., (1970e), pp. 234ff., (1970d), pp. 26ff., and many times thereafter.

339 Here, in "How to be a good empiricist: A plea for tolerance in matters epistemological", we can see that Feyerabend was *using* the principle of testability in an *ad hominem* fashion to make an argument aimed at forcing empiricists, who accept the principle of testability, to also accept pluralism and the promotion of alternative points of views, and thereby become more tolerant.

Feyerabend's methodological argument for pluralism also exposes the role of speculative (or metaphysical) thinking in the empirical sciences (as well as philosophy):

> The function of unusual *metaphysical* ideas is defined accordingly: they play a decisive role in the criticism and the development of what is generally believed and "highly confirmed", and they must therefore be present at *any* stage of the development of our knowledge. A science that is free from metaphysics is on the way to becoming a dogmatic metaphysical system (1965b, p. 150, italics inserted).[340]

It is also the basis for Feyerabend's main methodological maxim (which he applied to his own philosophy):

> Invent and elaborate theories which are inconsistent with the accepted point of view, even if the latter should happen to be highly confirmed and generally accepted (1965c, pp. 223-224; rp 1981a, pp. 105-106).

The key to establishing that pluralism is *better than* monism, because it improves the testability of theories, is to show that general theories can always find ways of avoiding difficulties when they do not face competition. There a number of strategies for establishing this point. First, one could argue for by emphasizing the problem of under-determination, or by using some variation of what is now often called the Duhem-Quine thesis.[341] As Einstein put the point, "it is often, perhaps even always, possible to adhere to a general theoretical foundation by securing the adaptation of the theory to the facts by means of artificial additional assumptions" (Einstein 1949a, p. 23). Faced with recalcitrant experimental data, we always have a choice. We can blame the theories, or we can blame the experiment. In this way, theories can avoid squarely confronting empirical difficulties. The existence of competitors, on the other hand, changes the situation. Competitors may explain or even predict the recalcitrant data, making it more difficult to blame the experiment as opposed to the theory. A competing theoretical alternative can even be used to reinterpret the observations, exposing exactly what and where the problem with the older theories may be. Sometimes Feyerabend took this route:

> Indirect refutations of the kind described [i.e. based on alternatives] occur much more frequently than is usually admitted. It does not seem to be an exaggeration to assume that all *decisive* refutations make use of alternatives. In any case, a refutation that is based on a successful alternative is much *stronger* than is a refutation resulting from the direct comparison of theory and "facts." . . . The reason why a refutation is stronger is easily seen. The direct case is "open," in the sense that a different explanation of the apparent failure of the theory (of the inconsistency between the theory and certain singular statements) might seem to be possible.

340 By metaphysical, Feyerabend means speculative, i.e. not solidly supported by observations.

341 See Duhem (1954), Quine (1980), and for a discussion, see Laudan and Leplin (1991). Also see e.g. (1961c), p. 168, and e.g. (1967a), p. 40.

> The presence of an alternative makes this attitude much more difficult, if not impossible, for we possess now not only the *appearance* of failure (viz., the inconsistency) but also an explanation, on the basis of a successful theory, of why failure *actually occurred* (1965b, fn. 122, pp. 249-250).[342]

However, Feyerabend more often used a different strategy to argue that without alternatives, theories can easily be protected from refutations. He often emphasized the power and depth that general theories have on how we understand the world:

> [T]he influence, upon our thinking, of a comprehensive scientific theory, or of some other general point of view goes much deeper than is admitted by those who would regard it as a convenient scheme for ordering facts only. According to this first idea scientific theories are ways of looking at the world; and their adoption affects our general beliefs and expectations, and thereby our experiences and our conception of reality (1962a, p. 29; rp 1981a, p. 45).

According to Feyerabend, the power that comprehensive scientific theories have limits our capabilities to test them directly:[343]

> A single theory in terms of which everything can be explained –as was the case with Newton's theory of celestial mechanics—is almost like a metaphysical system. *It is impossible to free oneself from it; one can hardly imagine that it might be false; and one has no means of testing its correctness* (1963c, pp. 96-97, italics inserted).

For any number of reasons, a situation can arise in which it appears as if there could only be one interpretation of each particular experience:

> The impression that every fact suggests one and only one interpretation and that therefore our views are 'determined' by the facts, this . . . situation arises whenever a fairly general point of view was held long enough to influence our expectations, our language and thereby our perceptions, and when during that period no alternative picture was seriously considered. We may prolong such a situation either by explaining away diverse facts with the help of *ad hoc* hypotheses . . . or by reducing more successful alternatives to 'instruments of prediction' which, being devoid of any descriptive meaning, cannot clash . . . with any experience; or by devising a criterion of significance according to which such alternatives are meaningless. The important thing is that such a procedure can always be carried out (1958a, p. 167; rp 1981a, p. 34).

According to Feyerabend, myths are constructed such that they are confirmed by all possible experiences (1961i; rp 1999a, pp. 50-77). Protecting a theory from potential refutations can turn a scientific theory into a dogmatic system, or a myth, which is no longer open to fundamental criti-

342 See also (1961c), pp. 155ff., where Feyerabend discusses the import of the Duhem's underdetermination thesis. Also see (1961h), p. 249.

343 Also see e.g. (1964b, p. 251). This also holds for scientific realism.

cisms.[344] According to Feyerabend, a myth supports itself empirically by providing interpretations of observations in its favor:

> Far from being a figment of the imagination that is clearly opposed to what is known to be the real world a myth is . . . a system supported by numerous and very direct and forceful experiences, by experiences moreover, which seem to be far more compelling than the highly sophisticated experimental results upon which modern science bases its picture of the world (1961i, p. 23; rp 1999a, p. 58. Cf. e.g. 1963e, p. 321, 1965b, especially fn. 24, p. 229, and fn. 122, p. 250).

According to Feyerabend, every theory faces and must handle many difficulties, and it is possible to protect or hide them through *ad hoc* and other means. Moreover, sometimes the principle of tenacity is useful and justifiable, although one usually only knows whether it was justifiable with hindsight. This leads to the following predicament:

> refutations are final only as long as ingenious and nontrivial alternative explanations of the evidence are missing. There can be no argument to the effect that the absence of such an alternative at a particular time will make the refutation final. It is very important to point out, in this connection, that the fate of many theories strongly depends upon the belief of their defenders that it will be possible, at some future time, to incorporate into them all the apparently refuting instances (1961c, p. 158).

Here, already in 1961, Feyerabend illustrated his point with the following examples:

> Take, for example, the early belief that both the fixed stars and the planets obey the same laws of circular motion. The obvious irregularity of the motion of the planets was strong refuting evidence against this belief. Yet it was hoped that this apparently refuting evidence would somehow be explainable on the basis of the idea of circular motion, and the attempts to do so finally turned out to be successful. Similar considerations apply to Copernican astronomy (the refuting evidence here was provided by the Aristotelian theory of motion), to the kinetic theory, and the like. In all these cases the conquest of refuting instances is carried out by vastly increasing the domain of alternative explanations (1961c, p. 158).

Here, we can see why alternatives are important.[345] They influence the status of potential refuting instance. On the one hand, without alternatives, refuting instances can appear to be absolute or final (which they are not, according to Feyerabend). On the other hand, without alternatives, potential refuting instances can be set aside, or explained away, and consequently not taken seriously as potentially refuting instances.

Feyerabend's argument for the role of alternatives in staging crucial experiments can be interpreted in a weak version and in a strong version.

344 Preston calls this the "Myth predicament". See Preston (1997), pp. 78ff. for a discussion. For a detailed explanation of how this can occur, see especially (1961i), especially pp. 18ff.; rp (1999a), pp. 56ff., and also e.g. (1965b), pp. 146ff.

345 See e.g. (1961i); rp (1999a), pp. 50-77 and (1965b), p. 152.

According to the weak version, an alternative T′ *facilitates* staging crucial experiments by showing that and how some recalcitrant evidence E refutes theory T, when otherwise the proponents of T would simply explain E away as a *Dreckeffekt.* According to Feyerabend, as in the case with Ehrenhaft's effects, and Brownian Motion to name just two of the many examples Feyerabend discussed, scientists tend to explain away recalcitrant evidence,[346] even in cases in which the evidence is familiar and can be clearly observed with the unaided senses.[347] Alternatives that explain the recalcitrant evidence can prevent this from happening.

According to the strong version, alternatives theories are *necessary* to stage a crucial experiment, because without an alternative T′, the recalcitrant evidence E could never have been used to refute T. This is because of limitations on the possibility of directly and precisely measuring E in order to establish that it actually does refute T without the alternative theory. We will return to this argument in the next section.

In (1965b), Feyerabend illustrated his general methodological argument for pluralism with an episode from the history of science, and several episodes from contemporary science. First, Feyerabend argued "the history of the Scientific Revolution is an excellent example of the necessity of a theoretical pluralism for the purpose of progress" (1965b, p. 237, ft. 22). He claimed that alternatives were needed in order to reinterpret observations that were believed to support Aristotelianism in order to show that they actually conflicted with it:

> What is the role of experience in the scientific revolution of the sixteenth and seventeen centuries? The reply must be: the transition from Aristotelianism to the new physics is *not* a transition to a radical empiricism . . . Quite the contrary, we have here the attempt to use a new point of view for the criticism of both the older *doctrine* and of the older *observations* supporting this doctrine. Considering the omnipresence of the older doctrine—even the most common occurrence was presented in its terms—these supporting observations were of no particular interest. Their support was a very cheap affair. There existed an interdependence between observation and theory that made even empirical arguments circular (1965b, p. 152, italics in original).

By arguing that a correct historical *description* of scientific advance makes use of a pluralistic test-model, Feyerabend was also supporting his *normative* methodological argument for pluralism. This is because the historical episodes were examples that are accepted to be instances of progress by all of the concerned parties.

346 See e.g. (1970d), pp. 36ff. Also see (1963a); rp (1999a), p. 92. Another favorite example of Feyerabend's was the discrepancies in the observation of perihelion of Mercury; again see e.g. (1970d), pp. 40ff.

347 For a longer list, see (1970d), pp. 36ff.

Feyerabend also argued that a number of contemporary areas of research have suffered, and continue to suffer, for a lack of pluralism. For example, Feyerabend repeatedly argued that Copenhagen interpretation of Quantum Mechanics in general, and some interpretations of Bohr's principle of complementarity in specific, are in danger of becoming immune from refutation unless a healthy pluralism is maintained.[348] Here and elsewhere, Feyerabend suggested that progress in Quantum Theory has been jeopardized by failure to consider and develop alternatives:

> It may well be that the refutation of the quantum-mechanical uncertainties presupposes just such an incorporation of the present theory into a wider context which is no longer in accordance with the idea of complementarity and which therefore suggests new and decisive experiments. And it may also be that the insistence, on the part of the majority of contemporary physicists, on the consistency condition[349] will, if successful, forever protect those uncertainties from refutation. This is how modern empiricism may finally lead to a situation where a certain point of view petrifies into dogma by being, in the name of experience, completely removed from any conceivable criticism (1963a; rp 1999a, p. 94).

According to Feyerabend, another contemporary area of research that suffers from a lack of healthy pluralistic competition is cognitive science. More specifically, Feyerabend argued that observations purportedly supporting a mentalistic theory of psychological states could be reinterpreted to support a competitor: materialistic theory, if such a materialistic theory were given a chance:

> That a thought cannot be a material process is established, so it is believed, *by observations*. It is by observation that we discover the difference between the one and the other and refute materialism . . . Introspection *does* indicate, in a most decisive fashion, that my present thought of Aldebaran is not localized, where as Aldebaran is localized: that this thought has no color, where as Aldebaran has a very definite color . . . Is this character of the introspective result proof to the effect that thoughts cannot be material? The answer is no, and the argument is the truism that what *appears* to be different does not on that account need *be* different. Is not the seen table very different from the felt table? . . .And if despite this difference of appearance we are allowed to make an identification, postulating an object in the outer world . . . , then why should the observed difference between a thought and an impression of a brain process prevent us from making another identification, postulating this time an object in the inner . . . world. . . It is, of course, quite possible that such a postulate will run into trouble and be refuted by independent tests . . . The point is that the *prima facia* observed difference between thoughts and the appearance of brain process does *not* constitute such trouble. It is also correct that a language based upon the assumption that the

348 See e.g. (1963a); rp (1999a), p.100, (1965b), e.g. fn. 124, pp. 250-251, and fn. 125, p. 251. Also see e.g. (1956c), p. 262, (1958b), and (1958a), p. 150; rp (1981a), p. 22.

349 Only such theories are then admissible in a given domain which either contain the theories already used in this domain, or which are at least consistent with them inside the domain.

> identification has already been carried out would differ significantly from ordinary English. But this fact can be used as an argument against the identification only after it has been shown that the new language is *inferior* to ordinary English. And such disproof should be based upon the fully developed materialistic idiom and not on the bits and pieces of materialese that are available to the philosophers of today (1965b, p. 189).

Feyerabend illustrated his two-pronged argumentative strategy against monistic models of theory testing, and for his pluralistic test-model (in support of his general methodological argument for pluralism) with an example of a progressive development in the history of science. Moreover, Feyerabend illustrated why and how the competition of alternative theories is methodologically preferable to theoretical monism: *Pluralism helps promote progress.* Pluralism helps promote progress because it offers stronger and sharper possibilities to criticize existing points of view than merely directly testing a theory against what are perceived to be the facts. This argument simply *assumes* the principles of empiricism and of testability; and from that basis, argues that criticism based on alternatives is stronger and sharper, because it offers the possibility of reassessing what is otherwise taken for granted, and of reinterpreting what are otherwise unquestioned (and thus dogmatic) interpretations of observations. Criticism based on alternatives is also stronger and sharper, because it undercuts *ad hoc* maneuvers (as well as other methods of protecting an otherwise well-established point of view from potential problems), and thereby facilitates staging *crucial experiments.* Lastly, Feyerabend also suggested what kind of alternatives are most useful for the purpose of criticizing established points of view: "Alternatives will be the more efficient the more radically they differ from the point of view to be investigated" (1965b, p. 214). He argued that a plurality of competing *incommensurable theories* is best suited to promote progress.

It is important to realize that Feyerabend did not *develop* this idea merely through abstract philosophical considerations such as the Duhem-Quine thesis.[350] *He had experienced such situations first hand*, and these experiences had taught him a lesson that deeply impacted his views on scientific method and scientific knowledge and its development. The lesson was that physical theories can be insulated even from convincing empirical demonstrations to the contrary; and this is the real reason that alternatives are needed. According to Feyerabend, alternatives can be used to persuade the proponents of competitor theories to take potential refutations seriously. If there is no alternative available, the recalcitrant evidence will be explained away. However, if a competitor theory can suggest *how* some

350 Although Feyerabend does refer to "the conclusions of Duhem's argument" (1961c, p. 158).

recalcitrant evidence may, (or even why some evidence actually does), refute the other theory, then it will (help) undermine any *ad hoc* attempts to explain the recalcitrant evidence away.

> Confronted with the experiments the latter [Rosenfeld and Pryce] occasionally acted almost as some of Galileo's opponents must have acted when confronted with the telescope. They pointed out that no conclusions could be drawn from them as they were the result of the interaction of many complex phenomena. In short, they were *Dreckeffekt* (a word that was heard very frequently in the arguments) (1947/1967, appendix 2, p. 5).

According to Feyerabend, alternatives can even be used to stage crucial experiments between incommensurable theories.[351] Feyerabend repeatedly argued that crucial (but still fallible, and thus never absolutely final) experiments are possible through the promotion and competition of alternative theoretical perspectives.[352] A crucial experiment between T and T′ can be staged, when T′ shows *that* some recalcitrant evidence E actually does refute T, and exposes exactly *why* E refutes T. If there is also some independent evidence in support of T′, then T′ has refuted T, and a crucial experiment has been staged.

8.4. Entropy and anomaly: The case of Brownian Motion

Feyerabend's *general methodological argument for pluralism* hinged on the claim that established theories can avoid confronting refuting evidence by *ad hoc* and other means. Alternatives are needed to force established points of view to face up to potential anomalies, not to treat them as *Dreckeffekt.* In this way, alternatives provide the possibility of stronger criticisms; and consequently, pluralism is preferable to monism methodologically. However, Feyerabend also refined this *general argument for pluralism* in a much more ambitious fashion. He offered what I will call an *in principle argument for pluralism.*

The argument is more ambitious, because it suggests that not only is pluralism preferable to monism in promoting progress, but the promotion

351 For Feyerabend on crucial experiments, see e.g. (1961c), pp. 160ff., (1962a), pp. 65-66; rp (1981a), pp. 71-72, (1965b), pp. 216-217, (1965c), p. 232ff.; rp (1981), pp. 115ff., and (1970e), p. 247.

352 See e.g. (1962a), pp. 65-66; rp (1981a), pp. 71-72.

of an alternative theory T′ may always, *in principle*, be necessary in order to discover that a refuting instances of a theory T actually refutes T. Note that the claim is not that the refuting instance cannot be discovered without the alternative theory, but that the phenomena cannot be used to refute the theory without the alternative theory. In other words, Feyerabend argued that in the empirical sciences, we have a situation in which we cannot know whether or not our best theories could be refuted by *known* phenomena, without first developing alternatives to them. This is because it may not be possible to prove that the anomalous character of some phenomena P, for a theory T, actually refutes T on the basis of T alone, i.e. without first developing an alternative theory T′ that can be used to prove that P does refute T.

Feyerabend's argument turns on the claim that some potential refuting instances can *only be established as refuting instances* by developing alternative theories.

> Facts and theories are much more intimately connected than is admitted by the autonomy principle [facts exist and are available independently of whether or not one considers alternatives to the theory to be tested]. Not only is the description of every single fact dependent on *some* theory (which may, of course, be very different from the theory to be tested), but *there also exists facts that cannot be unearthed except with the help of alternatives to the theory to be tested and that become unavailable as soon as such alternatives are excluded* (1963a, p. 22; rp 1999a, pp. 91-92, Feyerabend 1965b, p. 175, italics inserted).

Feyerabend illustrated this more ambitious argument for pluralism by developing the case of Brownian Motion, which became an often-repeated centerpiece of his promotion of pluralism. Feyerabend's argument concerning Brownian Motion first appeared in print in 1960, although there he did not actually use the term 'Brownian Motion' or Brownian particle'. He discussed a "new type of behavior" of "dust-particles", but his basic point was the same (1960e, p. 325, and pp. 328ff.; rp 1981a, p. 223, and pp. 226ff.). Feyerabend borrowed the example from Bohm (1957), although he rarely mentions *this* source in any of the many places he discusses or develops the point.[353]

353 According to Preston, it only first appears in Feyerabend's 1962, see Preston (1997), p. 126. The next appearance of the argument is in (1995b), written 1960/61, pp. 364-365. The argument was developed, reformulated, stated or discussed in (1962a), pp. 65ff.; rp (1981a), p. 71ff., (1963a); rp (1999a), pp. 92-93, (1964b), pp. 253-254, (1964c), pp. 351-353; rp (1981a), pp. 205-206, (1965b), pp. 175ff. and pp. 249-250, fn. 121 and fn. 122, (1965c), p. 224, and p. 249, fn. 25; rp (1981a), p. 106 and p. 113, fn. 25, (1966c), pp. 246-247; rp (1981b), pp. 61-62, (1968a), pp. 132ff., (1969e), pp. 58-59; rp (1981a), pp. 157ff., heavily altered, (1970c), p. 208; rp (1981b), p. 141, (1975a), pp. 38ff., see especially, p. 43, fn. 9, (1981a), pp. 143-145, (1988), pp. 27ff., and (1993a), pp. 27ff. In 1975, he cited D. Bohm and Bub (1966), pp. 456ff. for the idea that alternatives are necessary to stage crucial tests

Brownian Motion refers to the constant but irregular motion of tiny particles suspended in water or distributed in a gas. It is named after Robert Brown, who in 1827, argued that this well-known phenomena could not be biological, but must be physical. He based this claim on the observations that the phenomena occurs in water droplets trapped in rock samples millions of years old, and that almost any kind of material, organic or inorganic, also exhibits the constant but irregular motion in water if it is broken down into fine enough particles.[354] The constant, but irregular, motion of tiny particles is an oddity, because it persists even without the addition of energy.

Feyerabend argued that an investigation of the observational consequences of the Phenomenological Thermodynamics *by itself* (that is, *without the use of an alternative account of heat*) could not have led to *the discovery that the Brownian particle* is a *perpetuum mobile* of the second kind (1966b), and consequently that it *refutes the second law of Phenomenological Thermodynamics.* According to Feyerabend, a direct refutation of the second law is "excluded by physical laws" (1966c, p. 247; rp 1981b, p. 62). Note that Feyerabend's claim is *not* that the observations that refute a theory are discovered with the help of an alternative. Quite to the contrary, Feyerabend argued that the recalcitrant observations are often known well in advance of the refutation.[355] Specifically, observations of the Brownian particle were known well before the invention of the theories that used these observations to refute an accepted point of view (1975a, p. 43, fn. 9). His point was that although the recalcitrant phenomena were widely known, they still could not refute the theory by themselves. He argued that the discovery that Brownian Motion *refutes* the second law of Thermodynamics was impossible without the invention of an alternative.[356] *This* discovery was impossible without the invention of an alternative, because it would have required:

> (a) measurement of the exact *motion* of the particle in order to ascertain the change in its kinetic energy plus the energy spent on overcoming the resistance of the fluid; and (b) it would have required precise measurements of temperature

---

and the example of the Brownian particle to illustrate the point. See (1975a), p. 43, fn. 9. Needless to say, Feyerabend has used the point and the example before 1966.

354 Feyerabend bases his history of the discovery of Brownian Motion and how it was used to refute Phenomenological Thermodynamics on Nye (1972), Einstein's 1926, ***Investigations On the Theory of Brownian Movement***, and Perrin (1920), see (1965b), fn. 121, p. 249. But also see Preston (1997), pp. 126-130, as well as Nye (1972) and Maiocchi (1990).

355 See (1970d) for a list of familiar observations that conflict with once established, or even still established, theories.

356 For this conception of novelty; i.e., new interpretations of known observations on the basis of new theoretical points of view, Feyerabend suggests seeing Zahar (1973). See (1975a), p. 43, fn. 9.

> and heat transfer in the surrounding medium in order to establish that any loss occurring there was indeed compensated by the increase in the energy of the moving particle and the work done against the fluid (1975a, p. 40, 1988, p. 28, 1993a, pp. 27-28, italics in original).

However, because *these* measurements are physically impossible to make, it is impossible to use Brownian Motion to make a *direct* refutation of the phenomenological theory:

> Such measurements are beyond experimental possibilities: neither the heat transfer nor the path of the particle can be measured with the desired precision. Hence, a 'direct' refutation of the second law that would consider only the phenomenological theory and the 'facts' of the Brownian motion, is impossible. It is impossible because of the structure of the world in which we live and because of the laws that are valid in this world (1975a, p. 40, 1988, p. 28, 1993a, pp. 27-28, italics in original).

The reason that direct measurement of the motion is impossible is that the measuring instruments interfere with the system.

According to Feyerabend, Brownian Motion was actually shown to be a refuting instances of Phenomenological Thermodynamics on the basis of measurements that became possible only after the development of an incommensurable competitor (Kinetic Theory), and Einstein's utilization of it to make calculations of the statistical properties of Brownian Motion (1965b, p. 176). The point is not that new physical measurement techniques became available, which allowed experimental physicists to avoid the measurement difficulties mentioned above. The point is that on the basis of Kinetic Theory, predictions about the proportionality of the time and the mean square displacement of the particle can be made, which were confirmed by observations. The physical activities (i.e. observations) that confirmed these predictions (plotting the Brownian Motion) were nothing special. They were available before the development of Kinetic Theory. However, without the predictions based on Kinetic Theory, they have no significance. They did not refute the second-law of Thermodynamics. Thus, by developing the Kinetic Theory and Statistical Mechanics, Einstein claimed that:

> [he] discovered that there would have to be a movement of suspended macroscopic particles open to observation, without knowing that observations concerning the Brownian motion were already long familiar (Einstein 1949a, p. 47).

The first successfully testable quantitative predictions of Brownian Motion were produced independently by Einstein and Smoluchowski, and Jean Perrin is acknowledged as having experimentally confirmed Einstein's predictions about diffusion on the basis of the Kinetic Theory. Thus, only after a competitor to the phenomenological theory was developed could it be proved experimentally that the otherwise well-known phenomena of Brownian Motion actually refute the phenomenological theory.[357]

Feyerabend generalized his point as follows:

> assume that a theory T has a consequence C and that the actual state of affairs in the world is correctly described by C′, where C and C′ are experimentally indistinguishable. Assume furthermore that C′, but not C, triggers, or causes, a macroscopic process M that can be observed very easily and is perhaps well-known. In this case there exist observations, viz., the observations of M, which are sufficient for refuting T, although there is no possibility whatever to find this out on the basis of T and observations alone. What is needed to discover the limitations of T implied by the existence of M is another theory, T′, which implies C′, connects C′ with M, can be independently confirmed, and promises to be a satisfactory substitute for T where this theory can still be said to be correct. Such a theory will have to be inconsistent with T, and it will have to be introduced not because T has been found to be in need of revision, but in order to discover whether T *is* in need of revision (1965b, p. 176, italics in original).

Moreover, Feyerabend suggested that the case of Brownian Motion is by no means an exception to the rule: "Both the relevance and the refuting character of many very decisive facts can be established only with the help of other theories" (1969e, p. 59; rp 1981a, p. 157). However, while he often called such a situation "typical",[358] he very rarely mentioned, let alone developed in any detail, any other examples of direct refutations which are *impossible* to make because of physical limitations on the possibility of making measurements:

> It seems to me that this example is typical of the relation between fairly general theories, or points of view, and the 'facts'. Both the relevance and the refuting character of decisive facts can be established only with the help of other theories which, though factual adequate, are not in agreement with the view to be tested. This being the case, the invention and articulation of alternatives may have to precede the production of refuting facts (1975a, p. 41).

He did, however, sometimes mention some other cases,[359] which are often overlooked in discussions about Feyerabend's argument:[360] The motion of

357 The cause of the motion of tiny particles has been shown to be the result of the impact of the surrounding molecules.

358 Also see (1964c), pp. 351-353; rp (1981a), pp. 205-206, (1963a); rp (1989a), p. 113; rp (1999a), p. 93, (1988), p. 28, (1993a), p. 28. But also see (1981a), p. 143, fn. 13.

359 Interestingly enough, and relevant to any systematic reconstruction of the origins of this argument for pluralism, is the fact that *Feyerabend discussed some of these other cases briefly already in his original 1951 dissertation*, (pp. 89-91) (though not to make this point). It ought to be mentioned that along similar lines, Feyerabend also suggests that the refutation of the Quantum Mechanical uncertainties *may be* awaiting the development of an alternative point of view. See (1963a); rp (1989a), p. 114; rp (1999a), p. 94, and also see (1975a), pp. 41ff.

360 For critical discussions both for and against Feyerabend's argument, see Couvalis (1989), Farrell (2000a), Laymon (1977), Laudan (1989), Preston (1997) especially chapter 7, Worrall (1978), Worrall (1991).

an electron within the shell of an atom, and the idea of observer-independence in the quantum theory of measurement. Feyerabend also claimed that

> A refutation of the basic formulation [of the quantum theory] would . . . have to prove *that there is no conceivable adjustment to the Hamiltonian, or of the operators used,* that could make the theory conform to a given fact (1975a, p. 42, fn. 9, italics in original).

According to Feyerabend, this is *impossible* because on some interpretations the Quantum Theory,

> is an open theory in the sense that apparent inadequacies can be accounted for in an *ad hoc* manner, by adding suitable operators, or elements in the Hamiltonian, rather than by recasting the entire structure (1975a, p. 42, fn. 9).

These examples are supposed to have something in common. It is supposed to be *physically impossible* to make the measurements that are required to *refute* a theory T directly; that is, without first developing a competitor T′. Only after a competitor T′ has been developed, is it possible to make the measurement required to show that the phenomena actually does refute the other theory.

While Feyerabend has been criticized for relying on the development of only a single example, to make his point, he did in fact briefly (two pages) develop a second example: The case of the discovery of observer dependence. Unfortunately, it has been entirely ignored in the secondary literature on Feyerabend's argument,[361] perhaps because of its unusual location.[362]

According to Feyerabend: The idea of observer-independence is part of Newtonian Physics.[363] Difficulties with this idea appeared in electrodynamics, where second order effects of motion were absent. For a long time it was assumed that a solution could be found within the framework of Classical Physics. In other words, it was assumed that the difficulties arose due to a misapplication of the classical point of view, not to the classical point of view itself. In this way, Lorentz had provided what appeared to be a satisfactory solution to the difficulties. However, Einstein suspected that the problem was with the classical point of view itself. He constructed theories that are inconsistent with the idea of observer independence (special theory of relativity and general theory of relativity), which were adopted because of their tremendous empirical success:

361 See e.g. Couvalis (1988), Laymon (1977), and Preston (1997), pp. 126ff.

362 Feyerabend developed this example, as parallel to the case of Brownian Motion, in (1969e), pp. 60-61; rp (1981a), pp. 158-159.

363 For the following, see (1969e), pp. 60-61; rp (1981a), pp. 158-159.

> But one must note that this empirical success was not *preceded* by the complete failure, on empirical grounds, of the classical point of view *taken by itself* (1969e, p. 61; rp 1981a, p. 158, italics in original).

According to Feyerabend:

> This failure was rather a consequence of the success of the new theory and *could not have been demonstrated without this theory* (1969e, p. 61; rp 1981a, p. 158, italics inserted, original italics removed).

As in the case with Phenomenological Thermodynamics, Feyerabend emphasized that he was not claiming that the theory (Classical Physics in this case) was not confronted with empirical difficulties, "Of course, the old theory was in difficulty, Every theory is always in some difficulty or other" (1969e, pp. 61-62; rp 1981a, p. 158). The relevant question, in Feyerabend point of view,

> is whether the difficulty of classical physics was a *fundamental* difficulty, or whether it was accessible to treatment with the help of additional *classical* hypotheses (1969e, p. 62; rp 1981a, p. 158, italics in original).

According to Feyerabend,

> it is clear that the answer to this question cannot consist in the collection of further *facts*. No amount of facts however carefully collected, can show that there exist no classical hypotheses explaining a certain difficult situation (1969e, p. 62; rp 1981a, p. 158, italics in original).

Feyerabend did not explain why this is impossible, but it appears to based on the impossibility of proving a general statement with only a finite number of instances. In any case, Feyerabend argued that:

> The *only* way such an idea [that the problem is not with an application of the classical point of view, but with the point of view itself] can be supported consists in the development of an alternative theory (an alternative to any classical point of view, that is) which 'turns the difficulty into a principle' [Feyerabend is citing Einstein], elaborates its consequences, and compares the resulting point of view with the classical ideas. And this is how Einstein proceeded (1969e, p. 62; rp 1981a, p. 158, italics inserted, original italics removed).

On this basis he concluded that:

> Observer dependence was therefore not refuted by the production of observations inconsistent with it, or by the realization that a language using it was impractical. It was refuted by the construction, and confirmation, of a theory which worked with observer *dependence, and it could not have been refuted in any other way* (1969e, p. 62; rp 1981a, p. 158, italics in original).

Thus, Feyerabend did argue a second example in which the development of an alternative theory was purportedly needed in order to refute its predecessor.

In any case, Feyerabend tried to make the most out of the examples he provided. He implied that in light of even just one case (Brownian

Motion), we should proceed under the assumption that any theory might have anomalies whose refuting character can only be proved by developing alternatives to it. As the only way to find out if this is the case is to go ahead and develop alternative theories, it is always reasonable to promote alternatives:

> *Any* theory T under consideration . . . may be inconsistent with facts which are accessible only indirectly, with the help of an alternative T′. . . it is surely reasonable to demand that the class of refuting instances of a given theory be made as large as possible and that especially those facts which belong to the empirical content of the theory, which refute the theory but which cannot be distinguished from similar, but confirming facts, be separated from the latter and be thus made visible. However, this means that given P(n), it is reasonable to use not only (x)Px, *but as many alternatives as possible* (1964c, pp. 352-353; rp 1981a, p. 206, *italics in original*).

In this way, Feyerabend refined his general argument for pluralism into an *in principle argument for pluralism.* Feyerabend argued that in principle it may be the case that any of even our best theories may have refuting instances that can only be established *as definitively refuting* by developing alternatives to them. Consequently, a pluralism that promotes the constructive criticism of diverse points of view is always methodologically preferable to a monism in which a single theory is tested against the purported facts. *Pluralism is preferable because it better promotes progress.*

8.5. Feyerabend's pluralistic conception of knowledge

On the basis of his arguments for the role of alternatives in promoting scientific advance, Feyerabend concluded that:

> Unanimity of opinion may be fitting for a church, for the frightened victims of some (ancient, or modern) myth, of for the weak and willing followers of some tyrant; variety of opinion is a feature necessary for objective knowledge; and a method that encourages variety is also the only method that is compatible with a humanitarian outlook (1963a; rp 1989a, p. 116; rp 1999a, p. 97) italics in the original).[364]

Promoting alternative points of view is also part and parcel of his pluralist conception of the growth of knowledge according to which "*There is no idea, however ancient and absurd that is not capable of improving our knowledge. The whole history of thought is absorbed into science and is used for improving every single theory*" (1975a, p. 47, italics in original, also see p. 11).

---

364 See similarly, e.g. (1965b), (1970d), (1975a), p. 46, (1988), and (1993a).

Feyerabend did not share the standard view of scientific progress as *convergent* on an ideal point of view.[365] Rather, he conceived of science and its dynamic as *divergent*:

> Knowledge so conceived is not a process that converges toward an ideal view; it is an ever increasing *ocean of alternatives*, each of them forcing the others into greater articulation, all of them contributing, via this process of competition, to the development of our mental faculties (1965c, pp. 224-225; rp 1981a, p. 107).

Feyerabend's pluralistic conception of scientific knowledge, and his corresponding conception of philosophy, is here revealed. Knowledge is not a series of self-consistent theories that converges toward an ideal view, and the advancement of knowledge is not a gradual approach to the truth. Rather, knowledge is an ever-increasing ocean of alternatives, and the history of science is an inseparable part of science itself. In Feyerabend's view, science is aimed at developing our mental capacities through a process of advancing a plurality of competing views that force each other into greater articulation.

365 For some background on this view, see Laudan criticism of convergent realism in Laudan (1984a).

Chapter 9
Feyerabend's philosophical pluralism (1950s-1990s)

Feyerabend was an open-minded, self-critical thinker, who often explicitly changed his opinions, and sometimes even his whole philosophical outlook. He did not think that he had discovered some truth, and then spend the rest of his career trying to convince others of the importance of his insight (apparently an occupational hazard in philosophy). Rather, he was acutely aware of the highly *speculative* nature of the epistemological and metaphysical issues prevalent in the philosophy of science, and accordingly, he was not particularly tied to any specific positions. Yet to Feyerabend, the speculative nature of the issues at stake did not recommend caution. Quite the contrary, Feyerabend was a champion of bold conjectures, creative connections, and the big picture; and he praised others in exactly these terms.[366]

9.1. Commitments and perspectives

Because of the speculative nature of the issues involved in understanding the nature of science and its dynamic (such as questions about the meaning of scientific terms, our relationship to an external world, and how we have knowledge about it, the nature of progress in the sciences over centuries, etc.), philosophical proposals are always open to revision in light of new empirical results (e.g. 1954c). Moreover, from Feyerabend's philosophical perspective, epistemological interpretations of scientific knowledge must conform to both the latest scientific advances, as well as to the available historical evidence.

An excellent example illustrating the speculative character of the issues involved, the relevance of both the latest empirical results and the historical evidence, as well as the appropriate attitude, via a lack of binding commitment with which such speculative issues should be treated, can be

366 See e.g. Feyerabend's praise of Sellars, (1965c), pp. 246-247; rp (1981a), p. 128.

drawn from Feyerabend's contribution to the debates about the theory-ladeness of observation.

In 1958, Hanson coined the phrase 'theory-laden observation' (Hanson 1958). Building on Wittgenstein's distinction between seeing and seeing as (Wittgenstein 1958, XI, 193^{e}-214^{e}), Hanson argued that in order to perceive something as a particular kind of thing, one must already have the concept of that kind of thing at one's disposal. For example, in order to perceive an animal as a duck-bill platypus, one need already know that there are such things as duck-bill platypuses. Otherwise, one sees just some strange animal. More generally, in order for optical sensations to become observations, they must be interpreted according to some concepts and theories. Just how dependent observations can be on theoretical knowledge becomes apparent in cases I which physicists interpret streaks in a cloud chamber as the tracks of sub-atomic particles.

As we have seen, Feyerabend often discussed the significance of then contemporary developments in psychology of perception (Gestalt psychologists who emphasized the extent to which beliefs and expectations effect the organization of our visual perceptions) for various views about the role and status of experience in science.[367] Following Hanson's lead, Feyerabend also argued that such empirical results in the psychology of perception undermine the view that perceptions provide a neutral basis for deciding between competing theories.[368] Observation reports are not made in a theory-neutral language, and this has consequences for simplistic models of theory comparison, according to which theories are evaluated in light of neutral or given facts.

Of course, such insights about the role of theoretical interpretation in making scientific observations, and the complications that can arise for theory testing, were by no means completely new. They can be seen as specifications of Kant's general claim that intuition without concepts is blind (Kant 1781, p. 92-93 A51/B75). More concretely, the debate about theory-ladeness of observation reports, which began in the late 1950s due largely to Hanson, Wittgenstein and Feyerabend, repeats much of the same ground covered in discussions that took place more than a half-century earlier. For example, in Duhem's *The Aim and Structure of Physical Theory*, Duhem argued against Poincaré, who had claimed that "A scientific fact is nothing but a brute fact stated in a convenient language" (Poincaré cited by Duhem 1954, p. 149). Duhem argued that an experiment in physics is not simply an observation of phenomena. It is also an

367 Feyerabend discussed the significance of Gestalt psychology for theory-testing in the guise of "bundles of perception" in his (1963e), pp. 320ff. Also see (1967a), p. 48.

368 See for example, (1958a); rp (1981a), pp. 17-36, (1962a); rp (1981a), pp. 44-96, (1963a); rp (1999a), pp. 50-77, and (1965b), especially fn. 8, pp. 219-224.

interpretation of phenomena on the basis of the theories brought to the experiment (Duhem 1954, pp. 144ff.).

Feyerabend's views on the theory-ladeness of observation have been, and continue to be, the subject of much critical discussion,[369] and Feyerabend tried to clarify his views on the theory-ladeness of observation several times. According to Preston, Feyerabend once accepted the view that "observation is 'theory-laden' in the sense that it is shaped by prior knowledge of the scene involved", but that by 1965 Feyerabend came to think that *this view* is empirically refuted (Preston cited Hanson 1958, p. 19, Preston 1997, pp. 40-41). Here is how Preston supported his claim:

> The view that we see things as we believe them to be has . . . been held without experimental support. I strongly sympathize with this later view, but I must now regretfully admit that it is incorrect. Experiments have shown that not every belief leaves its trace in the perceptual world and that some fundamental ideas may be held without any effect upon perception (1981a, p. 128).

Preston followed Shapere in arguing that Feyerabend was not against theory-independent observation, but against the idea that there is a theory-independent observation *language*. While Feyerabend did indeed argue that observation *statements* are not merely theory-laden, but fully theoretical, Preston errs in claiming that Feyerabend rejected the view that observation is 'theory-laden', in the sense that it is shaped by prior knowledge of the scene involved. Quite to the contrary, Feyerabend was keenly aware that prior knowledge does in fact shape our visual perceptions (and not just the statements we use to describe those perceptions): "Considering that perceptions are influenced by belief in theories" (1962a, p. 95; rp 1981a, p. 93). To take just two more of many examples, Feyerabend wrote:

> It may also be a little unreasonable to assume that phenomena will be unaffected by one's views about their relation to the world . . . (1975a, p. 134, fn. 44).

Or:

> We must also remember the many conflicting views which were held about the surface of the moon, even at Galileo's time, and which may have influenced what observers saw (1975a, p. 130).

Feyerabend cited examples of observation reports of rain and clouds on the moon that can be found in texts by Kepler and da Vinci (Feyerabend referred his readers to chapter 4 of Ronchi's *Optics: The Science of Vision*) as examples of how beliefs can shape perceptions.

---

369 See e.g. Bogen and Woodward (1992), Butts (1966), Chalmers (1985), Churchland (2000), Hardcastle (1994), Hooker (1973), Hull (1972), Kordig (1971), Preston (1997), pp. 40ff., Quine (1993), Shea (1971), and Shapere (1966).

Preston claims that Feyerabend changed his mind concerning some aspects of the theory-ladeness of observation is correct. But the issues that Feyerabend changed his mind about are not whether or not beliefs shape perceptions; but rather (1) whether *every* change in belief can result in a change in perception, and whether *all* perceptions are open to revision in light of changes in beliefs, and (2) more concretely, whether or not the perceptions of the moon by earlier observers were actually *less definite* than ours today.

Ad (1) Feyerabend once supported these *extreme* views, but subsequently moderated them in light of empirical research on the perception of causality. The empirical findings amount to the apparent discovery (Michotte, The *Perception of Causality*, 1963) that in some situations, we perceive (visually) causal relations that conflict with our common-sense views, while in other situations, real causal relations appear to be impossible to perceive visually (1965c, p. 261, fn. 51; rp 1981a, p. 128, fn. 52). To make the points more transparent, consider the following examples: It was once common knowledge that the earth does not move. It has become common knowledge (supposedly) that the sun only *appears* to rise because of the rotation of the earth. However, it still appears to us as if the sun rises. In other words, not *every* change in beliefs effects perceptions. On the other hand, although we know that train tracks are parallel, because of our physical make-up, when we look at them running into the distance from between them, they appear to converge; i.e. not *all* perceptions are open to revision in light of beliefs. On the basis of such evidence, Feyerabend concluded that not *every* conceptual change has consequences for what we perceive, and not *all* perceptions are revisable in light of conceptual change. This is a far cry from Preston's claim that Feyerabend denied that prior knowledge of a scene shapes what is observed.

Ad (2) Struck by some seemingly strange descriptions of the moon that can be found in the history of science (such as those by Anaximander, Xenophanes, Heraclitus, and Berossos), which are clearly incompatible with what one would assume everyone can plainly see, Feyerabend *speculated* that the perceptions of astronomical objects such as the moon, planets, and stars made by earlier scientists may have actually been quite different than those of contemporary individuals (1975a, p. 132). He made this speculation under the influence of Köhler, Anscombe, Wittgenstein, Hanson and others (notably Toulmin). He proposed the thesis that visual impressions of astronomical objects back then might have been less distinct than ours today (1975a, pp. 133-134). A second hypothesis that could explain the strange descriptions of the moon is that the earlier scientists who made them simply disregarded what they saw. For a while, Feyerabend vacillated between these two possible explanations:

> I was for some time inclined towards [the first hypothesis], but it now seems to me that it is ruled out both by physiological (psychological) and by historical information (1975a, p. 133).

The historical information in favor of the second hypothesis (and against the first) amounts to descriptions from the history of science that are *not* indefinite. They match our contemporary perceptions (1975a, p. 133). As for the physiological information, Feyerabend referred back to his earlier claims about the perception of causality discussed above (1975a, p. 133, fn. 38). On the basis of these considerations, Feyerabend rejected the first hypothesis, which he had previously supported, in favor of the second hypothesis:

> We may infer, then, that the occasional disregard for the stability of the face [of the moon] was due not to a lack of clear impressions, but to some widely held views about the unreliability of the sense (1975a, p. 133).

Feyerabend's support for the second hypothesis, that the earlier scientists disregarded their distinct perceptions, because they did not trust them (and not that their perceptions were actually different), was also not set in stone. It too was a speculation open to revision in light of further evidence or enlightening considerations. *An important element of Feyerabend's philosophical perspective is that the value of such speculations are judged on the basis of their potential fertility, not their likely truth.*[370] Feyerabend even explicitly says as much in a note:

> The hypothesis in the text is developed in one particular direction [that they simply disregarded their observations because they did not trust their visual reports, and not their impressions were actually different] not so much because I am convinced that it is true but in order to indicate possible avenues of research and in order to give a clear impression of the complexity of the situation at the time of Galileo (1975a, p. 134, fn. 44).

Similarly, Feyerabend emphasized the speculative nature and his vacillating stance on hypotheses about the potential impact of the telescope on unaided astronomical perception: "I have been considering it [that telescopic observations and acquaintance with new telescopic reports changed not only what was seen through the telescope, but also what was seen with the naked eye] with varying degrees of enthusiasm, for the last two or three years . . . It seems to me, however, that the hypothesis is confronted by many difficulties and must, perhaps, be given up" (1975a, p. 132).

More generally speaking, Feyerabend emphasized that the impact of information on the laws of perceptual organization "cries out for further research" (1965c, p. 247; rp 1981a, p. 128). His fluctuating support for

---

370 Compare Einstein's remarks about his preferences of a deterministic Relativity Theory as a research program as potentially more fruitful than the ontological indeterminism of some interpretations of Quantum Theory. See Einstein (1949a).

various speculative hypotheses concerning the theory-ladeness of observation illustrate his lack of binding commitment to the theses he proposed, and his insistence that hard empirical evidence, not arm chair philosophy, will decide such matters. Feyerabend often made this explicitly clear. For example:

> Even worse, it is assumed that the relation between mental events (inside the human observer) and physical events (outside the human observer) can be decided by philosophical argument, by the abstract considerations of possibilities, without any help from the exigencies of empirical research (1965b, p. 147).

Although I have focused on issues within the debate on the theory-ladeness of observation, the examples are intended to illustrate a more general characteristic of Feyerabend's attitude toward the speculative issues in the philosophy of science, and his lack of binding commitment to proposals about them. In his autobiography, Feyerabend recalled an episode that illustrated just how easily his opinions could be changed.

> Joske was (and perhaps still is) a pushy fellow with high moral standards; he criticized me right and left. I argued with him on some occasions, gave in on others, but lacking an iconoclast world view I floundered about as I had done during the Nazi years. One example, in 1955 Arthur Pap asked me to translate his textbook on analytic philosophy into German. I began working on it. "You're a criminal," said Joske. "It's a bad book." Like a dummy I wrote to Arthur: "I am not going to continue with the translation—it's a bad book." Now the problem was not that I was following Joske despite a contrary opinion of my own; no, the problem was that my perception of things was unstable and could be changed very easily. I had thought that the book was all right; after Joske's criticism I no longer liked it (1995a, pp. 96-97).

Feyerabend did not only change his views about relatively minor hypotheses, such as the example from the debate about the theory-ladeness of observation, or whether or not Pap's book was good. He also made dramatic shifts in his basic philosophical outlook. An example of a major change in philosophical perspective is Feyerabend's fluctuating attitude toward rationalism and relativism. Feyerabend argued from within a rationalist tradition about rationalist theses. He even sometimes attacked relativist views (e.g. 1960e, pp. 334-335; rp 1981a, p. 232; as well as 1964b, pp. 245-246, and p. 250). In the 1970s, however, he began to attack rationalist views and develop and support relativist views. The first hints of Feyerabend's move toward relativism come in the final section of (1970c; rp 1981b, pp. 131-161), where he considers the *possible* relativist and subjectivist implications of incommensurability and pluralism. His relativist views culminate in (1987), where he explicitly endorsed a limited form of relativism. In the 1990s, these relativist views were subsequently tampered.[371] One of the reasons Feyerabend gave for rejecting relativism illustrates his rhetorical flair, and general attitude of lacking any serious commitment toward the views he sometimes espoused: "If on almost every university toilet door there are relativistic theses, then it's time to distance oneself from relativism."[372] Displaying his potential for self-criticism, he also referred to his own earlier relativist advice to keep hands off traditions as an 'idiocy' (1991b, p. 509).

Feyerabend's slide away from rationalism toward relativism is part of a more general shift in his attitude toward science and knowledge, which also illustrates a basic change in outlook. In the publications that he produced in the 1950s and throughout the 1960s, as far as I can tell, Feyerabend never once challenged the assumption that science produces the best form of knowledge. Quite the contrary, Feyerabend supported the view that scientific knowledge should dictate to our other less scientific views. This is most clear when he argued that concepts from everyday language, which have been shown by scientific advance to be mistaken or misleading, should be replaced by scientific concepts (1960f; rp 1991a, p. 45).

As with many things, Feyerabend pushed this Enlightenment view to its extremes, applying it even to controversial developments in Quantum Physics. Feyerabend argued that according to some interpretations, wave Mechanics does not permit the existence of well-defined objects; while according to commonsense (as refined by Classical Physics), there are such objects.[373] Feyerabend's faith in the superiority of scientific knowledge over commonsense determined the outcome: He "invites us to reject common sense and to announce the discovery: objective reality has been found to be a metaphysical mistake" (1981a, p. xii).[374]

Feyerabend's basic faith in the superiority of scientific knowledge over commonsense views comes under attack in his later work, where he does infamously challenge the assumption that science produces the best form of knowledge. He claimed that usually the "excellence of science is *assumed*, it is not *argued for*" (1978a, p. 73, italics in original), and asked, "*What's so great about science?*" (1978a, p. 73, italics in original). In these later

371 See for example (1991a), (1991b) (the manuscripts of were finished in 1989/1990, and 1989 respectively, also see (1993a), p. 268, (1993a), pp. 268ff., (1994a), and (1995a), pp. 151-152.

372 Feyerabend cited by Hoyningen-Huene in (2000b).

373 See e.g. (1961e); rp (1981a), pp. 248-252, and especially the text to fns. 26 and 27.

374 Feyerabend took this stance in (1958a), pp. 160ff.; rp (1981a), pp. 29ff., (1962a), pp. 69ff.; rp (1981a), p. 74ff., (1969e), pp. 43ff.; rp (1981a) pp. 146ff., and (1964d); rp (1981a), pp. 176ff. Also see (1978a), p. 70. Demonstrating his pluralistic philosophical method, and marking a major change of attitude toward Bohr, Feyerabend also argued exactly the other side of this issue in (1970a).

works, the superiority of scientific knowledge is no longer an unquestioned assumption.[375] It is under attack. Eventually, it is outright rejected.

According to Grazia Borrini-Feyerabend's preface to (1999b), Feyerabend:

> constantly revisited and challenged his own previous work. (Authors often write the same book several times under different titles. It has been said that Paul wrote three different books under the same title—the three published editions of Against Method.) [Borrini-Feyerabend refers to Hacking's Feyerabend obituary (Hacking 1994)]. In the last decade of his life, for instance, Paul was not at all pleased with *Science in a Free Society*, which he did not want to see reprinted. Even the detailed treatment of relativism found in *Farewell to Reason* is further extended and, shall I say, overcome by the writings collected in this volume [*Conquest of Abundance*] (1999b, p. xi).

Perhaps more significantly, Feyerabend *revitalized* his 1960s views about the subordination of facts to norms. As we have seen, in his early 1960 publications, Feyerabend often argued from the perspective of normative epistemology. For example:

> which attitude shall we adopt and which kind of life shall we lead? . . . is the fundamental problem of all epistemology. . . [w]e are confronted with a real *decision*, that is, a real choice with a situation which has to be resolved on the basis of our demands and preferences, and which cannot be resolved by proof. It is easy to see that these demands and these preferences concern the welfare of human beings and are therefore ethical demands: epistemology, or the structure of the knowledge we accept, is grounded upon an ethical decision (1961i, pp. 55-56; rp 1999a, p. 71, italics in original).

Or similarly:

> The fact that almost any philosophical doctrine may find realization either in a *cosmology* . . . and/or in a *theory of man* . . . makes it very clear that the procedure leading to the adoption of a philosophical position cannot be *proof* . . . but must be a *decision* on the basis of preferences . . . Philosophers have habitually judged the situation in a very different manner. For them, only *one* of the many existing positions was true and therefore, possible. This attitude, of course, considerably restricts the domain of responsible choice . . . the problem of responsible choice enters even the most abstract philosophical matters and . . . ethics is, therefore, the basis of everything else (1965b, p. 219, fn. 5, italics in original).

Moreover, as we have seen, toward the end of the 1960s, Feyerabend began to favor the historical approach to the philosophy of science, which demands a historical grounding, not a normative one, for epistemological claims. However, in the 1990s, Feyerabend returned to a stance that views "Ethics as a Measure of Scientific Truth" (1992; rp 1999b, pp. 242ff.).

To summarize, Feyerabend lacked a binding commitment to the views that he espoused. Feyerabend's contribution to the debates surrounding

375 This begins as early as (1975a), pp. 295ff. Also see (1988), pp. 231ff.

the theory-ladeness of observation well illustrates that he approached the speculative issues in the philosophy of science with an open-mind, and was willing to try out different sides of the same issues. His speculations were made in the spirit of an effort to make positive proposals that might help further discussions and open the possibilities for further research, not dogmatic pronouncements that he had discovered some truth. In Feyerabend's view, hopefully, such issues would eventually be handled empirically. Not only did he change his views on specific issues, he also changed his basic perspective on the nature of scientific knowledge and epistemology a number of times, and in several fundamental respects, over a career that lasted almost half a century. The twists and turns in Feyerabend's career also well illustrate how even his most basic philosophical commitments fluctuated wildly. He was never convinced of any single point of view.

9.2. Pluralism and opportunism

There is more to Feyerabend's philosophical pluralism than the fact that he was no dogmatist convinced of some truth – tied to a single point of view. Feyerabend was like a chameleon, adapting to the philosophical landscape as it underwent dramatic (perhaps even revolutionary) changes over four decades. As his career progressed, Feyerabend increasingly emphasized his pluralistic philosophical method. He was more and more explicit about his strategy of temporarily adopted conflicting philosophical standpoints, and using and developing ideas that are inconsistent with each other for the purpose of criticizing existing points of view. It is generally accepted that Feyerabend practiced such a pluralistic philosophical method after the publication of *Against Method.* Although he was much less vocal about it, he also practiced a pluralistic philosophical method in his pre-Against Method publications. The detailed study of Feyerabend's pre-1970s philosophical considerations that make up the bulk of this investigation also contain inconsistent ideas *due to the fact* that Feyerabend temporarily adopted conflicting points of view and argued both sides of the same issues. This is part and parcel of the pluralistic philosophical method, complemented and sanctioned by his pluralistic conception of knowledge. The main difference between his pre- and post-*Against Method* publications is not a change in his pluralistic conception of knowledge or in his pluralistic philosophical method (as has often been suggested, see e.g. Preston 1997). In the introduction to two volumes of his collected papers, Feyerabend cheerfully admitted that the ideas that they contain are not consistent (1981a, p. xiv). He claimed that this reflects his belief that "good arguments can be found for the opposite side of any issue" (1981a, p. xiv). However, as we have seen, Feyerabend did not always characterize his own earlier work correctly, and such latter statements about his earlier work have been interpreted as misleading 'glamorization' (Preston 1997, p. 7). Thus, in order to substantiate Feyerabend's latter claim about his earlier work, it is necessary to provide some examples of Feyerabend arguing both sides of the same issue, and temporarily adopting conflicting points of view in his pre-1970 publications.

To begin with, as we have seen, while for the most part Feyerabend was extremely critical of analytical philosophy, he also sometimes practiced it. At least two of his publications contain examples of Wittgensteinian analyses aimed at dissolving apparent philosophical paradoxes.[376] In both cases, Feyerabend set out to clarify the meanings of the terms used to create the apparent paradox, in order to expose the fallacious assumption behind the apparent paradox. In 1956, Feyerabend published "A Note on the Paradox of Analysis" (1956a), which analyses the meanings of the terms that give rise to the apparent paradox, and dissolved it by uncovering the fallacious assumption that gave rise to it.[377] In "Das Problem der Existenz theoretischer Entitäten" (1960f; rp 1999d), Feyerabend provided a second example of a Wittgensteinian analysis of an apparent philosophical paradox. In section 2, "Derivation of an Apparent Paradox", Feyerabend *explicitly* adopted the philosophical method of analysis. He analyzed the meaning of the terms 'observable' and 'theoretical', in order to demonstrate how they create the apparent paradox. He *explicitly* insisted that, "it [the problem of the existence of theoretical entities] is a problem which *cannot* be solved by scientific investigation" (1960f; rp 1999d, p. 21, see similarly p. 17). His linguistic analysis of the meanings of the terms that give rise to the apparent paradox exposed two possible explications of those meanings, and thus, two corresponding formulations of the problem of the existence of theoretical entities. He proceeded to expose the fallacious assumptions behind each of the formulations, and then concluded that:

> the problem of the existence of theoretical entities collapses in upon itself whether one uses the first or the second explanation. This is because in the first case there are no theoretical concepts, and therefore no corresponding problem. And in the second case, *every* concepts is problematic, because every concept is theoretical. No matter how one cuts the cake, it seems impossible to give a rea-

376 For an excellent statement of what I mean by a Wittgensteinian analysis aimed at dissolving an apparent paradox, see Waismann (1965), chapters 1 and 2, pp. 3-43.

377 For a detailed discussion, see Oberheim and Hoyningen-Huene (2000), especially section 2, pp. 368ff.

> sonable sense to the problem of the existence of theoretical entities (1960f; rp 1999d, p. 22, italics in original).

This use of the method of *analytical philosophy* sharply contrasts with Feyerabend's more usual scientific approach. The fact that Feyerabend sometimes adopted *both* philosophical methods supports my pluralistic interpretation of Feyerabend's philosophy, and it undermines Preston's more standard unificationist approach to Feyerabend's ideas as restricted to 'scientific philosophy'. According to Preston, philosophers must *choose* between 'scientific' and 'analytical' philosophy, and Feyerabend chose the former, and "denied that analytical philosophy has any value" (Preston 1997, p. 10). However, as we have seen, while Feyerabend did usually criticize the inherent conceptual conservativism of analytical philosophy, he also *temporarily* adopted its philosophical method when it suited his purposes. Feyerabend's complaint against analytical philosophy was that it does not attempt to change the meanings of the terms it analyses. However, this did not prevent Feyerabend from analyzing the meanings of terms, or even to use such analysis in order to collapse apparent philosophical paradoxes in the spirit of Wittgenstein's later philosophy. He did, not, however *restrict* himself to analysis that leaves the meanings of the terms unchanged. Instead, Feyerabend adopted whatever form of philosophy most suited his immediate purpose, mostly practicing 'scientific' philosophy, but occasionally practicing 'analytical' philosophy as well. On Feyerabend's pluralistic conception of philosophy, philosophers do not need to choose between scientific and analytical philosophy *once and for all*, as Preston appears to suggest. They must merely clearly distinguish between occasions when there are using these radically different philosophical methods. Philosophers, according to Feyerabend's pluralistic conception of philosophy *and how he practiced it in the 1960s*, can *temporarily* adopt whatever method best suits the issue at hand.

A second example of Feyerabend adopting conflicting philosophical perspectives in his pre-1970 arguments is his temporary adoption of both descriptive and normative approaches to the philosophy of science. As we have seen, Feyerabend often attacked methodological principles on the grounds that they do not correctly describe actual scientific advance, as well as on the grounds that they are *undesirable.* In the following example, Feyerabend not only temporarily adopted each of these two conflicting approaches to the philosophy of science, but he also argued both sides of exactly the same issue in doing so. Thus, the example supports my claim that his pre-1970 philosophical ideas do not lend themselves to a neat, unificationist, reconstruction. Rather, Feyerabend practiced pluralism in philosophy in his pre-*Against Method* publications.

The two examples just given are instances of Feyerabend adopting conflicting, competing philosophical methods (scientific versus analytic, normative versus descriptive). Here is an example of how this pluralism plays out. In the early 1960s, Feyerabend made various conflicting claims about the logical empiricist model of explanation. For example, in his 1961 comment on Baker's "The Role of Simplicity in Explanation", Feyerabend began:

> Professor Baker has added to the now quite numerous attacks against logical empiricism by criticizing its theory of explanation. Although I am firmly convinced that the main theses of logical empiricism are wrong *I still think that its model of explanation is correct and should be defended* (1961d, p. 278, italics inserted).

This statement is somewhat striking. In 1961, just one year before Feyerabend famously criticized the logical empiricist model of explanation, Feyerabend *defended* the very same model, even though he was convinced that the main theses of logical empiricism were incorrect. On the basis of (1961d) by itself, one should conclude that Feyerabend thought that the logical empiricist model of explanation was correct and should be defended.

To make matters worse (in terms of consistency), in the 1961 discussion, Feyerabend defended the logical empiricists model of explanation against the same sort of attack that he himself launched just one year later. Baker had attacked the model on the grounds that it does not conform to actual scientific practice. Feyerabend responded that:

> As against Professor Baker's attacks I have to say the following . . . the fact that in medicine (or, for that matter, in alchemy or in sociology) explanations do not often lead to predictions cannot be regarded as a sign that the model is not universally applicable . . . After all, this model is not supposed to be a description of actual scientific practice; it is an ideal to which scientific practice is supposed to conform if it is to be satisfactory. Hence, if a difference is observed between the model and scientific practice, this is regarded to be a criticism of scientific practice, and not of the model. . . what we need are standards for criticism which will help us to improve and to extend our knowledge of the world by improving upon scientific procedure and eliminating unsatisfactory developments (1961d, p. 279).

This is of course a clear instance of Feyerabend supporting normative epistemology, in which Feyerabend defended the logical empiricist model of explanation against descriptive criticisms. However, in his (1962a) Feyerabend launched a long criticism of the logical empiricist model of explanation exactly on the grounds that it "does not give a correct account of actual scientific practice" (1962a, p. 60).

To put the point directly, in 1961 Feyerabend rebuked Baker for attacking the logical empiricist model of explanation on the grounds that it does not conform to actual scientific practice, and in 1962, Feyerabend criticized the very same model of explanation on the grounds that it does not conform to actual scientific practice. This is a clear example of Feyerabend arguing both sides of the same issue – and one directly relevant to his notion of incommensurable concepts, which he introduced as part of his attack on this very same model of explanation.

Notice too, that in arguing both for and against the logical empiricist model of explanation concerning its relation to actual scientific practice, Feyerabend also switched between two competing philosophical schools, which have two different agendas. In 1961, he defended the logical empiricist model of explanation against Baker's criticism by adopting the viewpoint of *normative epistemology*. According to normative epistemology, the principle task of the philosophy of science is to promote *the right* forms of scientific reasoning. In the tradition of Descartes and Locke, the epistemology begins by differentiating between good and bad forms of reasoning, and then *recommends* methods of improving how to reason.[378] In the philosophy of science since the 1950s, normative epistemology has increasingly been overtaken by an historical movement in the philosophy of science that became popular largely in the wake of Kuhn's *The Structure of Scientific Revolutions* (1962). (Although it may be enjoying something of a come back with the rise of Science Studies programs.) According to the contemporary *historical school* in the philosophy of science, the primary epistemological task does *not* involve attempts to make normative recommendations to scientists about how to do good science. Adherents to the historical school often consider this overly presumptuous. For them, the primary epistemological task is to describe or reconstruct scientific rationality, or to describe correctly the conditions for having or attaining scientific knowledge. It is *primarily* a descriptive task, not a normative one. It does have normative elements and consequences, as it still attempts to identify and distinguish good science from bad science, and even just the choice of historical examples to reconstruct involves normative judgments and has normative implications. However, the historical approach is first and foremost a descriptive task, which also has normative implications. The methodological principles that historical philosophers propose are judged and criticized on the basis of how well they describe actual scientific practice.

On the other hand, the normative approach *is* first and foremost normative, and its methodological recommendations should not be criticized on the grounds that they do not conform to actual scientific practice. This is because they are supposed to be *ideals* to which scientists *should* aspire, regardless of whether or not they ever achieve these ideals. Such methodological proposals are the basis of criticism of actual scien-

378 See e.g. Locke (1690) and Descartes ([1701] 1967).

tific practice aimed at improving actual scientific practice, not attempts to describe what scientists actually do.

Because Feyerabend criticized the logical empiricist model of explanation on the grounds that it does not provide a correct description of actual scientific practice at length (not merely as an aside, or a note), it *appears* as if Feyerabend switched away from the perspective of normative epistemology to that of the historical school in the early 1960s. This appearance, however, is misleading. This is easily seen by a closer look at the original 1962 version (but not at the 1981 reprint).[379] In the original version, Feyerabend mentioned and briefly repeated the basis of his criticism of Baker's descriptive attack on the logical empiricist model of explanation (see 1962a pp. 60ff., especially fn. 70). Here Feyerabend again emphasized the need to treat the model as a *normative ideal* of explanation, and to defend it from such descriptive attacks. As a normative ideal, the model is not supposed to capture actual scientific practice. It is supposed to be a standard with which scientific practice can be criticized and improved.

In his (1962a), Feyerabend *both* attacked the logical empiricist model of explanation on descriptive grounds, and he also *defended* the model from such attacks. The attack is staged by interpreting the model as a *historical* claim, and the defense is staged by interpreting the model as a *normative* ideal immune form such descriptive criticisms. Feyerabend adopted the historical perspective to launch his descriptive criticism, and he adopted the normative perspective to defend the model from such criticisms. Feyerabend even then proceeded to criticize the model as a normative ideal. He argued that it would lead to consequences undesirable to the very empiricists promoting it. In (1962a), his praise for the normative perspective must be understood in that context. It does not represent Feyerabend choosing one perspective over another *once and for all*. It does not even represent Feyerabend slowly developing his philosophy from the one perspective to the other. It is merely part of setting the stage for the second prong of a sophisticated two-prong argumentative strategy, in which he attacked the model of explanation on the historical grounds that *as a description* it is inaccurate, and *as a normative ideal* it is undesirable.

Preston has argued that although Feyerabend is often perceived to have been part of the movement toward the historical school in the philosophy of science, together with Kuhn, in the early1960s, at that time Feyerabend actual belonged to the normative school, and only *switched* later, (Preston 1997, p. 17). While there is some truth in this claim (in that Feyerabend gradually placed more and more emphasis on historical con-

379 In the 1981 reprint, one and half pages *on this point* were deleted, see (1962a), pp. 60-62 and compare to (1981a), p. 69.

siderations from the 1950s through the 1980s, where as in the 1990s, he returned to a normative perspective), things are not as black and white as Preston makes them appear. Early on, Feyerabend did repeatedly argue that a purely descriptive approach to the philosophy of science should be subordinate to a normative approach, which is a more important task. Moreover, the normative Popperian influence on Feyerabend's early 1960s publications is often overlooked, especially given the fact that Feyerabend removed most of the traces of it from the 1981 reprints. Even so, the evidence suggests that in the early 1960s, while Feyerabend did *mostly* practice normative epistemology, and did insist that it was superior to purely descriptive approaches, he also used descriptive arguments, that explicitly did *not* belong to normative epistemology, which treated methodological principles as descriptions, and criticized them on the basis that they do not conform to the historical record (See e.g. 1962a, pp. 46ff.; rp 1981a, pp. 57ff.). Such arguments belong to the historical tradition in the philosophy of science, not the normative school.

Furthermore, in the early 1960s, Feyerabend repeatedly emphasized, as we have seen, that the philosophy of science should have a stronger basis in actual scientific practice, as it has historically developed, and even his normative proposals were often grounded on historical examples in addition to methodological principles. Thus, in my view, the 'switch' was not from a strictly normative approach, to a strictly descriptive approach (as Preston presented it), but from a *pluralistic* normative and descriptive approach, in which the descriptive elements were subordinate to the normative elements, to a pluralistic approach in which the normative elements were subordinate to the descriptive elements (and then back again in the 1990s). Feyerabend did not stop making recommendations about how to do good science after 1967, and he did not stop using normatively based arguments after 1967. After all, even in the 1993 version of *Against Method* (and all of the earlier versions), Feyerabend still attacked the universality of methodological rules on both descriptive and normative grounds. Moreover, in the 1990s, he argues that ethics ought to be the measure of truth (1992).

In summary, in 1961, Feyerabend *praised* the logical empiricist model of explanation, and *defended* it (from the standpoint of normative epistemology), from a *descriptive* criticism of it. In 1962, he *criticized* the very same model of explanation (from a historical perspective), on the basis that it does not conform to actual scientific practice. Thus, Feyerabend argued both sides of the same issue by *temporarily* adopting conflicting philosophical perspectives. In 1962, he also proceeded to criticize and attack the very same model from the standpoint of normative epistemology. The strategy of defending normatively, attacking descriptively, defending from descriptive attacks, and then attacking normatively is explicit in the original 1962 version of "Explanation, Reduction and Empiricism". It is more difficult to discern in the 1981 version, because Feyerabend eliminated one and half pages in which he explained why he was switching from a descriptive to a normative position in the middle of his paper. This is a clear example of Feyerabend arguing both sides of the same issue by adopting conflicting interpretations of the logical empiricists model of explanation based on radically different conceptions of the task of philosophy. While he clearly favored normative epistemology, he was not *bound* to it, or to any one particular point of view. He adopted conflicting points of view to make immanent criticisms to promote progress – with no allegiance to any philosophical *isms* other than pluralism. He practiced different competing philosophical methods, normative epistemology and historical epistemology, as well as scientific and analytical philosophy, and he argued both sides of the same issue, for and against the logical empiricist model of explanation. In the mid-1960s, when he set out his pluralistic conception of knowledge, Feyerabend even implied that, for him, more than one position on the same issue could be true: "*For them*, only *one* of the many existing positions was true and therefore, possible" (1965b, p. 219, fn. 5, italics inserted).

These examples of Feyerabend arguing for and against the same idea and adopting different competing philosophical perspectives to do so, supports my interpretation of Feyerabend as an *opportunist* ready and willing to take any stance, be it normative or descriptive, pro or contra the logical empiricist model of explanation, in order to promote constructive critical discussion of the issue at hand. Feyerabend often characterized successful scientists in exactly these terms, and he cited Einstein is support of this method (or lack there of):

> The external conditions which are set for [the scientist] by the facts of experience do not permit him to let himself be too much restricted, in the construction of his conceptual world, by the adherence to an epistemological system. He therefore, must appear to the systematic epistemologist as a type of unscrupulous opportunist . . . (Einstein 1949a, cited by Feyerabend 1970d, p. 18, 1975a, p. 18. Also see 1975a, pp. 175, 179, 187, as well as fns. 58, 70 and 213).

Feyerabend was no systematic epistemologist. He was an unscrupulous opportunist. As Feyerabend viewed the philosophy of science as continuous with science itself, it is no wonder that his characterization of successful scientists matches his own style of and approach to philosophy. Feyerabend took his own methodological advice to heart, even in the 1960s, before he had begun to push pluralism to its extremes and to radicalize and rhetoricize his ideas.

At the very latest, by 1970, Feyerabend had made his pluralistic philosophical method openly explicit. For example, at the beginning of section 9 of (1970c, pp. 214ff.; rp 1981b, pp. 157ff.), Feyerabend pointed out that so far in the article he had criticized Kuhn from Lakatos' point of view. Nevertheless, he then proceeded, explicitly, to defend Kuhn against Lakatos:

> This transition from criticism to defense does not mean that I have changed my mind. Nor can it be completely explained by my cynicism vis-à-vis the business of philosophy of science. It is rather connected with the nature of science itself . . . Contrary arguments bring out the different features that it [science] contains, they challenge us to make a decision, they challenge us to either *accept* this many-faced monster [science] and be devoured by it, or else to change it in accordance with our wishes (1970c, p. 215, italics in original; rp 1981b, p. 148).

Here, Feyerabend practiced pluralism in philosophy on the basis of his pluralistic conception of knowledge that he had developed in the early 1960s, when he argued that contrary arguments promote progress.

Another obvious example of Feyerabend arguing both sides of the same issue concerns his much-repeated call to abandon the concepts of Classical Physics in light of the theoretical advances made in Quantum Theory. Recall that in summarizing his views on realism, Feyerabend characterized his position as an invitation to reject common sense in favor of the view that "objective reality has been found to be a metaphysical mistake" (1981a, p. xii).[380] He repeatedly argued against his interpretation of Bohr's complementarity principle on the grounds that classical concepts could and should be replaced by new incommensurable ones. However in a 1970 article entitled "In Defense of Classical Physics", Feyerabend argued exactly the opposite position. He began by ascribing the views that he himself defended many times over to a talk Popper gave:

> Popper criticized 'the doctrine that science is limited to the search for invariants' and proposed 'the conjecture that thought a search for invariants is undoubtedly one of the most important of all scientific enterprises, it does not constitute or determine the limits of rationality or of the scientific enterprise' [Feyerabend is citing Popper from a 1965 International Conference on the Philosophy of Science in London] (1970a, p. 59).

And then, in an about face to many of his earlier publications, he set out to argue that:

> *I am of a different opinion.* I believe that restricting oneself to a search for invariants may be an advantage and that loosening the framework may have some very unfortunate consequences. More especially *I want to argue that the classical framework has been given up prematurely and that one should continue using and exploiting it* (1970a, p. 59, italics inserted).

380 Feyerabend referred to (1958a), (1962a), (1964d) and (1969e).

Feyerabend concluded the article:

> This encourages one to support the classical view also in fields where it has been definitely abandoned, for example in the quantum theory. One must do this, for it is only in this way, by such swimming against the stream (KRP) that a more critical attitude can be obtained both toward classical physics, and toward the more recent theories which claim to be able to replace it. And if it should be true that classical physics represents a high point in the development of our knowledge, then this would also be the only way to prevent us from slipping, and from losing what we have already achieved (1970a, p. 79).

These conservative sentiments clearly fly in the face of views that Feyerabend had published and republished before 1970, *and they were intended to do so*.[381]

By 1970, the subordination of normative epistemology to the historical school in philosophy was complete, but it did not change Feyerabend's pluralistic conception of knowledge, or his pluralistic philosophical method:

> [In 'Problems of Empiricism'] I made the same historical material support a very different and much more orderly (or "rationalistic conclusion") [than in 'Problems of Empiricism, Part II']. This does not mean that I changed my mind. It means that every case—and, therefore, also the present case—had different aspects and that an intelligent person can always win by concentrating on one side of the issue. In other words it means that there is no room, within critical theory of knowledge, for an "approach to the truth" (1970b, p. 325).

In his post-1970 work, Feyerabend's tendency to jump from one view to another is even clearer. He reminded us to "always remember that the demonstrations and rhetoric used do not express any 'deep convictions' of mine. They merely show how easy it is to lead people by the nose in a rational way" (1975a, p. 32), and he characterized his role as an epistemological anarchist, as "like an undercover agent who plays the game of Reason in order to undercut the authority of Reason (Truth, Honesty, Justice, and so on)" (1975a, pp. 32-33). The point is that this is not at all dissimilar to the role he played it the 1960s. In the 1960s, Feyerabend was like an undercover agent who he played the game of empiricism in order to undercut the authority of experience.

---

381 This paper should be seen in light of Feyerabend's rejection of Popper and his philosophy, and it does represent a major shift in Feyerabend's views. Under Popper's influence, Feyerabend had repeatedly attacked Bohr's ideas in the 1950s and 1960s as conservative and dogmatic. After his subordination of normative epistemology to the descriptive approach (due to conversations with von Weizsäcker, Feyerabend began to attack Popper's views on Quantum Mechanics (1968c) and (1969d). Moreover, Bohr, in Feyerabend's views, now becomes a champion of pluralism and anti-dogmatism (no longer a promoter of conceptual conservatism), see and compare the dramatic differences between, (1961e), especially pp. 384ff. to e.g. (1993a), p. 15, fn. 1, and pp. 31-32.

In Feyerabend's post 1970s works, Feyerabend demonstrated his pluralistic conception of knowledge and method by defending outlandish views. Feyerabend's pluralistic conception of knowledge condones the defense of ideas that deviate from the mainstream. His defenses of witchcraft, astrology, voodoo, etc, can best be understood as examples of Feyerabend practicing the kind of pluralistic philosophy he preached.[382] As Feyerabend himself stated, "My aim is not to establish the truth of propositions, my aim is to make my opponent change his mind" (1978a, pp. 142-143), and "I don't have a philosophy. I have opinions" (2000, p. 162. cf 1975a, pp. 32-33, 1988, p. 23, and 1993a, p. 230.) Feyerabend insisted that his epistemological anarchism is not an attempt to ground a new position or start a new movement. It is against all such positions and movements (1975a, p. 33, fn. 4).

If these remarks are taken seriously, then Feyerabend's support for various endeavors, like astrology, witchcraft, and voodoo, does not represent any commitment to and or serious belief in the efficacy of these as forms of knowledge as opposed to western science.[383] Moreover, his use of some particular concepts or distinctions *for the sake of an efficacious argument* did not commit him to their correctness. His calls to arms in defense of these positions, and his use of their concepts and distinctions, are best understood in the context of Feyerabend's views about the necessary role of alternative views for promoting the kind of critical discussion that leads to progress.

In summary, in the 1960s Feyerabend used different competing philosophical perspectives and methods to argue for, and to promote, pluralism. By 1970, Feyerabend began explicitly promoting the kind of philosophical pluralism that he had practiced in the 1960s. In the 1970s, he was like an undercover agent who plays the game of Reason in order to undercut the authority of Reason. In the 1960s, he was like an undercover agent who he played the game of empiricism in order to undercut the authority of experience. Throughout his career, he was an opportunist ready and willing to adopt whatever thesis or method best served his purpose: to promote progress the through constructive criticism.

382 Also see e.g. Llyod (2000).

383 Although he sometimes promoted astrology, Feyerabend thought it was boring (Paul Hoyningen-Huene, personal communication).

9.3. From philosophical pluralism to epistemological anarchism

Most commentators distinguish Feyerabend's early purported methodological position from his later anarchistic views. For example, Preston suggests that when Feyerabend later challenged the universal validity of methodological rules, he 'officially' foreswore his earlier views. However, Preston also acknowledged that the principles of proliferation and tenacity continue to work in Feyerabend's publications well after 1970. In these cases, *but not in the earlier ones*, Preston claimed that they are used in an *ad hominem* fashion (Preston 1997, p. 219, fn. 12), and consequently do not represent Feyerabend's commitment to them.[384] This allowed Preston to uphold his strict separation of Feyerabend's 1960s and 1970s publications. As we have seen, according to Preston, in the 1960s, Feyerabend was committed to these principles as universally valid methodological rules, but in the 1970s, he merely used them in an *ad hominem* fashion.

However, in his autobiography, Feyerabend recalled that already in his 1952 seminar, Popper:

> started with a line that became widely known: 'I am a Professor of Scientific Method – but I have a problem: there is no scientific method.' 'However,' Popper continued, 'there are some simple rules of thumb, and they are quite helpful' (1995a, p. 88).[385]

Preston would have us believe that Feyerabend ignored or repressed this in his pursuit of a single rational model for the acquisition of knowledge based on supposedly *universally* valid methodological principles, and only rediscovered that there are no such universally valid methodological rules (but only rules of thumb) in the 1970s with the advent of his epistemological anarchism (Preston 1997, p. 177).[386]

Whereas Preston sees Feyerabend as having espoused these methodological suggestions as part of a single pluralist methodology, and then as having rejected his former views in favor of pluralism about methodology at the end of the 1960s, this investigation reveals a different pattern. Beneath Feyerabend's increasingly dramatic break with Popper and his sub-

384 This is particularly ironic given that Feyerabend actually argued for these principles both before and after 1970, while Feyerabend simply assumed the other principles in *ad hominem* arguments before 1970 (with the exception of realism, tenacity and proliferation, for which he also often argued before 1970).

385 See also (1981b), p. 21, where Feyerabend acknowledged Popper's critical rationalism as a forerunner of his epistemological anarchism.

386 Preston also admitted that Feyerabend explicitly stated, when summarizing his views, that these principles are said to be principles "for mnemonic reasons only" (also in an endnote, Preston (1997), p. 218, fn. 11.

sequent attacks on critical rationalism, as well as his subordination of normative epistemology to the historical approach to the philosophy of science, the *central* ideas and arguments that Feyerabend developed before the 1970s carry through to his more radical later publications. While the transition from the 1960s to the 1970s involved a radicalization of his ideas, the nature of the changes to his specific ideas is more of a continuous development, extension, and radicalization, than a outright rejection as Preston has presented it.

The distinction between Feyerabend's early and later philosophy is not original to Preston. Another example of an interpretation of Feyerabend's philosophical development that distinguishes sharply between Feyerabend's early and later ideas is McEnvoy (1975). He argues that the quintessential transition from Feyerabend's critical rationalism to his skeptical fallibilism comes at the end of his (1965b) and concerns changes in Feyerabend's claims about the possibility of comparing incommensurable theories. After 1965, according to McEnvoy, Feyerabend gave up on the possibility of *objectively* comparing incommensurable theories in a Popperian manner (McEnvoy 1975, p. 62). McEnvoy's explication of Feyerabend's views is extremely insightful and revealing. However, while he convincingly shows that Feyerabend's (post-1965 publications) do not fit within Popper's critical rationalism, he did not adequately establish that Feyerabend's earlier views do. McEnvoy did highlight many critical rationalist elements in Feyerabend's philosophy before 1965, and McEnvoy, like Preston, points to lip service, such as Feyerabend's acknowledgements to Popper as the starting point for his discussion. However, McEnvoy overlooked that, even on his own account of Feyerabend's developing views, Feyerabend's pre-1965 ideas do not fit within critical rationalism either. According to Feyerabend, "We have to conclude, therefore, that a formal and 'objective' account of explanation cannot be given" (1962a, p. 95; rp 1981a, p 93). Moreover, McEnvoy also overlooked that even after 1965, Feyerabend continued to argue that crucial experiments between incommensurable theories are possible (1970b, 1970d).

In order to see appreciate the extent to which Feyerabend's views on the pluralistic nature of knowledge change between his 1960s views and the views he promoted under the guise of epistemological anarchism in and after 1975, let us compare Feyerabend's own statements concerning his epistemological views. As we have seen, in 1965, Feyerabend claimed that:

> Knowledge so conceived is not a process that converges toward an ideal view; it is an ever increasing *ocean of alternatives*, each of them forcing the others into

greater articulation, all of them contributing, via this process of competition, to the development of our mental faculties (1965c, pp. 224-225; rp 1981a, p. 107).[387]

In 1975, Feyerabend claimed that:

> Knowledge so conceived is not a series of self-consistent theories that converges toward an ideal view; it is not a gradual approach to the truth. It is rather an ever increasing *ocean of mutually incompatible (and perhaps even incommensurable) alternatives,* each single theory, each fairy tale, each myth that is part of the collection forcing the others into greater articulation and all of them contributing, via this process, to the development of our consciousness (1975a. p. 30, italics on original).

Thus, the pluralist conception of scientific knowledge Feyerabend made famous in *Against Method* is basically the same pluralist conception of knowledge from his pre-1970s publications.[388] Moreover, the arguments that Feyerabend used in favor of his pluralistic conception of knowledge also remain the same (e.g. 1975a, pp. 35ff.) The real difference between Feyerabend's pre- and post-1970's views is that he drastically increased his rhetoric and tried to base his points more firmly on historical considerations, and less so on abstract methodological considerations. Moreover, as we have seen, Feyerabend practiced the same type of pluralistic *ad hominem* approach to philosophy in his pre-1970 publications as the one he 'glamorized' in his post-1970s publications. The basic nature of Feyerabend's later philosophy, progress through criticism regardless of allegiance to *isms,* is the same as it was in his pre-1970 publications. It is also as Popperian as it always was, which was never a very deep commitment.[389]

Attempting to distinguish Feyerabend's views into temporarily separate phases that divide his 1960s ideas from those of the 1970s are unsuccessful because Feyerabend's basic pluralistic conception of knowledge did not change.

387 This claim is *not* part of Feyerabend's post-1970s move toward epistemological anarchism. Rather, *it is explicitly part of a summary of the views expressed in his 1960s publications* as for example his (1963a); rp (1999a), pp. 50-77, and (1962a); rp (1981a), pp. 44-96.

388 This point has been emphasized in Salmon, Earman et al., who also claimed that the basic ideas of *Against Method* "are really ideological and rhetorical [presentations] of views already established in earlier papers" (1992, p. 158). It has also been emphasized in Hellman (1979), p. 191, fn. 4.

389 See chapter 3 and also Farrell (2000b).

Conclusions: Feyerabend's Philosophy? Pluralism for Progress

Feyerabend argued against many particular views on the grounds that *that they would constrain the freedom necessary to develop alternative points of view, and thereby hinder scientific advance.* For Feyerabend this was not merely some abstract danger. As we have seen, he repeatedly argued about cases from contemporary scientific and philosophical developments. In Feyerabend's view, the entrenchment of successful theoretical traditions makes fundamental criticism of them extremely difficult. Feyerabend argued that even in the face of observations that conflict with a reigning theoretical tradition (e.g. Ehrenhaft's effects, the perihelion of Mercury, Brownian Motion), scientists tend to trust theory over their observations, and explain the recalcitrant observations away. However, recalcitrant observations may be taken more seriously if there is an alternative available that purports to explain them. Generally, Feyerabend believed that his role as a philosopher in the 1950s and 1960s was to help contribute to scientific advance by attempting to prevent science and philosophy from becoming dogma in the name of positivism and logical empiricism. This was the motivation for Feyerabend's pluralistic view of knowledge and the pluralistic philosophical method that he both preached and practiced in support of it.

This investigation has focused on the basic insights about the nature of experience and its relation to theoretical knowledge from which Feyerabend developed several central aspects of his pluralistic conception of science and its dynamic. It has revealed how those insights were used to develop the notion of incommensurable concepts, and how Feyerabend used the notion of incommensurable concepts to attack many different forms of conceptual conservativism within science, the philosophy of science and philosophy more generally. It has revealed how those insights were used to develop a new pluralistic model of theory competition, his arguments for pluralism, and a pluralistic conception of scientific knowledge and its advance. Such a *pluralistic conception of knowledge* does not conform to the development of a single philosophical position; rather, it *is best supported by an appropriately pluralistic philosophical method.* Feyerabend made this clear *in the 1960s*:

I cannot conclude this essay without giving at least a hint as to how its results [about pluralism in the sciences] *are related to more general questions of philosophy.* The fact that I have frequently discussed scientific theories must not be misunderstood. I have discussed them because I want to restrict myself to the philosophy of science, but because I regard scientific theories as excellent examples of actual knowledge. Philosophy has always had the tendency to proceed from *a stable system and to judge thought by the criteria implicit in such a system.* It is clear that such criteria will *fail* when applied in the attempt to improve matters in a fundamental way. This is why a consideration of the sciences is important, for despite the great amount of *conservatism* that is still contained in this enterprise, we have here *criticism and progress through revolutions that leave no stone unturned and no principle unchanged*" (1965b, p. 217, italics inserted).

In Feyerabend's philosophical work, Feyerabend's pluralistic *ad hominem* approach to philosophical issues was just such an attempt to avoid the shortcomings of proceeding from a stable system, and to judge thought by the criteria implicit in that system. Feyerabend argued that these shortcomings can be avoided by adopting a pluralistic method. This applied both to scientific research *and* to the philosophy of science:

the task of philosophy, or of any enterprise interested in the advance, rather than embalming of knowledge, would be to encourage the development of such new modes of approach, to participate in their improvement rather than to waste time in showing, what is obvious anyway, that they are different from the established way of thinking (1963b, p. 66. See similarly, 1995b, written 1960/61, p. 246).

Moreover, his general methodological argument for pluralism was supposed to apply to scientific research *and* to the philosophy of science:

The function of such concrete alternatives is, however, this: they provide means of criticizing the accepted theory in a manner that goes beyond the criticism provided by a comparison of [one] theory with the "facts" . . . *The results just mentioned . . . have important implications for philosophy proper* (1965b, pp. 150-151, italics inserted).

Feyerabend attempted to criticize epistemological and methodological views by drawing on diverse, and sometimes even inconsistent perspectives and ideas in the philosophy of science, in order to promote fundamental progress that leaves no philosophical position or principle immune from revision. He continued:

Indeed, it is my contention that once the scientific method has been freed from certain dogmatic elements that still reflect its past involvement with the philosophical tradition, it will provide a basis for the discussion and the solution of *all* philosophical problems dealing with matters of fact. After all, *philosophical doctrines*—be they now formulated explicitly, in the form of a philosophical system, or implicitly, in the form of linguistic rules governing a certain language—*are quite comparable to general theories of physics. They propose an ontology*; they defend this ontology either by reference to observations of a general kind or by theoretical arguments; *and they reject alternatives* in much the same way that a radical empiricist would reject them, *by pointing out that the accepted doctrine is not yet in difficulty and will*

> *perhaps never be in difficulty* (1965b, pp. 150-151, original italics removed, italics inserted).

According to Feyerabend, both philosophical doctrines and general theories in physics posit an ontology that purports to identify the basic kinds of things that exist. They both defend their commitment to these posits by suppressing the promotion of alternative points of view, which could potentially identify their shortcomings and correct them. Feyerabend continued:

> The last assertion can easily be believed. The observations referred to figure in the arguments only to the extent to which they have received a theoretical interpretation on the basis of the system or the linguistic rules chosen: and the theoretical arguments consist in going back and forth between the accepted principles and their consequences. *Philosophical arguments* (with the possible exception of dialectical arguments [which change the meanings of the terms they use]) *are therefore invariably circular. They show what is implied in taking for granted a certain point of view, and do not provide the slightest foothold for a possible criticism* (1965b, pp. 150-151, italics inserted).

According to Feyerabend, general philosophical systems support themselves through the use of *circular arguments*; and thereby protect themselves from the kind of fundamental criticism that can only be achieved by promoting alternative points of view. This applies especially to arguments for realism: "I had heard realistic arguments before; as far as I was concerned they were all circular: you made an assumption that contained a realistic core and then proved realism by revealing the core" (1995a, p. 72). For Feyerabend the consequences are clear:

> This being the case, it is quite impossible to evaluate them in the traditional manner. We must choose a point outside the system or the language defended in order to be able to get an idea of what a criticism would look like. *We must use an alternative.* Such a transition to theoretical pluralism would also seem to increase our chances of solving some philosophical problems (1965b, pp. 150-151, italics inserted).

As we can see, Feyerabend explicitly argued for pluralism in philosophy in the name of progress. According to Feyerabend, the myth-like character of general, stable *philosophical* positions, which posit a basic ontology and support it with circular arguments, *necessitates* the development of alternative points of view. Alternatives are necessary, because without them, fundamental criticism of such general points of view is not possible.

Feyerabend's use and development of a plurality of different positions and ideas cohered with his pluralistic conception of knowledge *in the 1960s*. His pluralistic philosophical method both promoted, and was conducive to, the development of different sides of the many diverse, speculative issues in the philosophy that Feyerabend discussed. On this conception, knowledge is not a series of self-consistent theories that converges toward an ideal view, and the advancement of knowledge is not a gradual approach to the truth. In Feyerabend's view, *this holds both for science and philosophy*. The knowledge arrived at through philosophical discussion is an ever increasing ocean of alternatives aimed at forcing ideas into greater articulation, and contributing through the process of competition, to the development of our mental faculties.

This investigation has attempted to delineate the nature of Feyerabend's conception of knowledge and how he argued for it, as well as Feyerabend's pluralistic conception of philosophy and how he practiced it. Feyerabend thought that the tasks of philosophy are to improve our understanding through immanent criticism of existing points of view, to counter conceptual conservativism (especially concerning the empirical foundations of scientific knowledge), and thereby to help promote the critical development of new points of view. Feyerabend's use of inconsistent ideas cohered with, and was sanctioned by, his pluralistic conception of knowledge. Consequently, in a sense, this investigation has developed Feyerabend's ideas into a sort of philosophy. It is not, however, a philosophy in the usual sense of the term. It is not a cohesive, stable set of doctrines or set of principles. Instead, it is a pluralistic philosophical method that calls for pluralism in science and philosophy in the pursuit of progress.

Literature

Achinstein, P. (1964). "On the Meaning of Scientific Terms." *Journal of Philosophy* **61**: 497-509.

Achinstein, P. (1968). *Concepts of Science*. Baltimore, John Hopkins Press.

Achinstein, P. (2000). "Proliferation. Is It a Good Thing?" *The Worst Enemy of Science? Essays in Memory of Paul Feyerabend*. J. Preston, G. Munévar and D. Lamb (eds.). New York, Oxford University Press: 37-46.

Agassi, E. (1976). "Review of Against Method." *Philosophia* **6**: 165-191.

Agassi, E. (1985). "Commensurability, Incommensurability, and Cumulativity in Scientific Knowledge." *Erkenntnis* **22**: 51-77.

Agassi, E. (2002). "Kuhn's Way." *Philosophy of the Social Sciences* **32**: 394-430.

Ajdukiewicz, K. (1974). *The Scientific World-Perspective and Other Essays 1931-1963*. J. Giedymin (Ed.). Dordrecht: D. Reidel.

Almog, J., J. Perry, et al. (eds.) (1989). *Themes from Kaplan*. New York, Oxford University Press.

Babich, B. (2003). "From Fleck's *Denkstil* to Kuhn's paradigm: conceptual schemes and incommensurability. " *International Studies in the Philosophy of Science* **17**: 75-90.

Bacon, F. (1620). *Novum Organum*. La Salle, Open Court.

Balzer, W. (1978). "Incommensurability and Reduction." *Acta Philosophica Fennica* **30**: 313-335.

Bearn, G. (1986). "Nietzsche, Feyerabend, and the Voices of Relativism." *Metaphilosophy* **17**: 135-154.

Bird, A. (2002). "Kuhn's Wrong Turn." *Studies in History and Philosophy of Science* **33**: 443-463.

Bird, A. (2003). "Kuhn, Nominalism, and Empiricism." *Philosophy of Science* **70**: 690-719.

Blanshard, B. (1941). *The Nature of Thought*. New York, Macmillan.

Bogen, J. and J. Woodward (1992). "Observations, Theories and the Evolution of The Human Spirit." *Philosophy of Science* **59**: 590-611.

Bohm, D. (1957). *Causality and Chance in Modern Physics*. London, Routledge.

Bohm, D. and J. Bub, (1966). Reviews of Modern Physics, No. 38.

Bohr, N. (1934). *Atomic Theory and the Description of Nature.* Cambridge, Cambridge University Press. Reprinted in *The Philosophical Writings of Niels Bohr, Vol. I*, Woodbridge, Ox Bow Press, 1987.

Boyd, R. (1981). "Scientific Realism and Naturalistic Epistemology." *PSA 1980, vol. 2.* P. Asquith and R. Giere (eds.). East Lansing, Philosophy of Science Association: 613-662.

Boyd, R. (1983). "On the Current Status of the Issue of Scientific Realism." *Erkenntnis* **19**: 45-90.

Boyd, R. (1984). "The Current Status of Scientific Realism." *Scientific Realism.* J. Leplin (ed.). Berkeley, University of California Press: 41-82.

Boyd, R. (1985). "*Lex Orandi est Lex Credendi. Images of Science.*" *Essays on Realism and Empiricism, With a Reply From Bas C. Van Fraassen.* P. Churchland and C. Hooker (eds.). Chicago, University of Chicago Press: 3-34.

Boyd, R. (1989). "What Realism Implies and What it Does Not." *Dialectica* **43**: 5-29.

Boyd, R. (1990). "Realism, Approximate Truth, and Philosophical Method." *Scientific Theories. Minnesota Studies in the Philosophy of Science Vol. XIV.* C. W. Savage (ed.). Minneapolis, University of Minnesota Press: 355-391.

Boyd, R. (1992). "Constructivism, Realism, and Philosophical Method." *Inference, Explanation and other Frustrations. Essays in the Philosophy of Science.* J. Earman (ed.). Berkeley, University of California Press: 131-198.

Brown, H. (1983). "Incommensurability." *Inquiry* **26**: 3-29.

Brown, H. (2001). "Incommensurability and Reality." *Incommensurability and Related Matters.* P. Hoyningen-Huene and H. Sankey (eds.). Dordrecht, Kluwer: 123-142.

Bunzl, M. (1994). "Discussion: Scientific Abstraction and the Realist Impulse." *Philosophy of Science* **61**: 449-456.

Butts, R. (1966). "Feyerabend and the Pragmatic Theory of Observation." *Philosophy of Science* **33**: 383-394.

Caneva, K. (2000). "Possible Kuhns in the History of Science: Anomalies of Incommensurable Paradigms." *Studies in History and Philosophy of Science* **31**: 87-124.

Cantor, G. (1975a). "A Critique of Shapin's Social Interpretation of the Edinburgh Phrenology Debate." *Annals of Science* **33**: 245-256.

Cantor, G. (1975b). "The Edinburgh Phrenology Debate: 1803-1828." *Annals of Science* **32**.

Carnap, R. (1956). "The Methodological Character of Theoretical Concepts." *Minnesota Studies in the Philosophy of Science Vol. I.* H. Feigl and M. Scriven (eds.). Minneapolis, University of Minnesota Press: 38-76.

Carnap, R. (1966). *The Philosophical Foundations of Physics.* New York, Basic Books.

Carrier, M. (1991). "What is Wrong With the Miracle Argument?" *Studies in the History and Philosophy of Science* **22**: 23-36.

Carrier, M. (1993). "What is Right With the Miracle Argument: Establishing a Taxonomy of Natural Kinds." *Studies in History and Philosophy of Science* **24**: 391-409.

Carrier, M. (2001). "Changing Laws and Shifting Concepts." *Incommensurability and Related Matters*. P. Hoyningen-Huene and H. Sankey (eds.). Kluwer, Dordrecht: 65-90.

Cartwright, N. (1983). *How the Laws of Physics Lie*. Oxford, Clarendon.

Chalmers, A. (1985). "Galileo's Telescopic Observations of Venus and Mars." *British Journal for the Philosophy of Science* **36**: 175-191.

Chalmers, A. (1978). *What is This Thing Called Science?* Indianapolis, Hackett Pub. Co.

Chalmers, A. (1982). *What is This Thing Called Science?* Second Edition. Indianapolis, Hackett Pub. Co.

Churchland, P. (1979). *Scientific Realism and the Plasticity of the Human Mind*. Cambridge, Cambridge University Press.

Churchland, P. (2000). "To Transform the Phenomena." *The Worst Enemy of Science? Essays in Memory of Paul Feyerabend*. J. Preston, G. Munévar and D. Lamb (eds.). New York, Oxford University Press: 148-158.

Churchland, P. and C. Hooker (eds.) (1985). *Images of Science. Essays on Realism and Empiricism*. Chicago, University of Chicago Press.

Clagett, M. (1959). *The Science of Mechanics in the Middle Ages*. Madison, University of Wisconsin Press.

Coffa, J. (1967). "Feyerabend on Explanation and Reduction." *Journal of Philosophy* **64**: 500-508.

Cohen, R. and Schnelle, T. (1986). *Cognition and Fact*. Dordrecht: D. Reidel.

Collins, H. (1985). *Changing Order*. London, SAGE.

Couvalis, G. (1986). "Should Philosophers Become Playwrights." *Inquiry* **29**: 451-457.

Couvalis, G. (1988). "Discussion: Feyerabend and Laymon on Brownian Motion." *Philosophy of Science* **55**: 415-421.

Couvalis, G. (1989). *Feyerabend's Critique of Foundationalism*. Aldershot, Avebury.

Couvalis, G. (1997). *The Philosophy of Science. Science and Objectivity*. London, SAGE.

Couvalis, G. (1999). "Review Symposia: Radical Fallibilism vs. Conceptual Analysis: The significance of Feyerabend's Philosophy of Science." *Metascience* **8**: 206-216.

Craig, E. (ed.) (1999). *Routledge Encyclopaedia of Philosophy Cd-Rom*. London, Routledge.

Creath, R. (1999). "Carnap, Rudolf (1891-1970)." *Routledge Encyclopaedia of Philosophy*. E. Craig (ed.). London, Routledge.

Culp, S. (1995). "Objectivity in Experimental Inquiry: Breaking Data-Technique Circles." *Philosophy of Science* **62**: 430-450.

Curthoys, J. and W. Suchting (1977). "Feyerabend's Discourse against Method: A Marxist Critique." *Inquiry* **20**: 243-371.

Dampier, S. W. C. (1929). *A History of Science and its Relations With Philosophy and Religion.* Cambridge, Cambridge University Press.

Davidson, D. (1984). "On the Very Idea of A Conceptual Scheme." *Inquiries into Truth and Interpretation.* Oxford, Oxford University Press: 183-198.

Dennis, W. (ed.) (1948). *Readings in the History of Psychology.* New York, Appleton-Century-Crofts.

Descartes, R. ([1701] 1967). *Rules for the Direction of the Mind.* London, Cambridge University Press.

Devitt, M. (1979). "Against Incommensurability." *Australasian Journal of Philosophy* **57**: 29-50.

Devitt, M. (1984). *Realism and Truth.* Oxford, Basil Blackwell.

Devitt, M. (1996). *Coming to Our Senses. a Naturalistic Program for Semantic Localism.* Cambridge, Cambridge University Press.

Devitt, M. (1997). *Realism and Truth.* Princeton, Princeton University Press.

Devitt, M. (2001). "Incommensurability and the Priority of Metaphysics." *Incommensurability and Related Matters.* P. Hoyningen-Huene and H. Sankey (eds.). Dordrecht, Kluwer: 143–157.

Devitt, M. and K. Sterelny (1987). *Language and Reality. An introduction to the Philosophy of Language.* Cambridge, MIT Press.

Dilworth, C. (1981). *Scientific Progress.* Dordrecht, D. Reidel.

Donovan, C. (1978). "Eliminative Materialism Reconsidered." *Canadian Journal of Philosophy* **8**: 289-303.

Douven, I. and J. Brakel (1995). "Is Scientific Realism an Empirical Hypothesis?" *Dialectica* **49**: 3-14.

Duerr, H. (ed.) (1995). *Paul Feyerabend. Briefe an einen Freund.* Frankfurt am Main, Suhrkamp.

Duhem, P. ([1908] 1969) *To Save the Phenomena. An Essay on the Idea of Physical Theory from Plato to Galileo.* Chicago, University of Chicago Press.

Duhem, P. ([1906] 1908). *Ziel und Struktur der physikalischen Theorien.* Leipzig, Johann Ambrosius Barth.

Duhem, P. ([1906] 1954). *The Aim and Structure of Physical Theory.* Princeton, Princeton University Press.

Einstein, A. (1949a). "Autobiographical Notes." *Albert Einstein: Philosopher-Scientist.* P. Schilpp (ed.). Illinois, Open Court: 3-95.

Einstein, A. (1949b). "Remarks Concerning the Essays Brought Together in this Co-operative Volume." *Albert Einstein: Philosopher-Scientist.* P. Schilpp (ed.). Illinois, Open Court: 665-668.

Ellis, B. (1985). "What Science Aims to Do." *Images of Science. Essays on Realism and Empiricism, with a Reply from Bas C. van Fraassen.* P. M. Churchland and C. A. Hooker (eds.). Chicago, The University of Chicago Press: 48-74.

Farrell, R. (2000a). "Rival Theories and Empirical Content Revisited." *Studies in History and Philosophy of Science* **31A**: 137-149.

Farrell, R. (2000b). "Will the Popperian Feyerabend Please Step Forward: Pluralistic, Popperian Themes in the Philosophy of Paul Feyerabend." *International Studies in the Philosophy of Science* **14**: 257-266.

Farrell, R. (2002). "Feyerabend's Metaphysics: Process Realism, or Voluntarist-Idealism?" *Journal for General Philosophy of Science* **32**: 251-369.

Farrell, R. (2003). *Feyerabend and Scientific Values. Tightrope-Walking Rationality.* Dordrecht, Kluwer:

Feyerabend, P. and J. Ferber (1947/1967). "Single Magnetic Northpoles and Southpoles and Their Importance for Science – Ten Lectures delivered at the University of Vienna during the summer semester of 1947 by Dr. Felix Ehrenhaft, U.S. Visiting Professor. Compiled with the assistance of Professor Ehrenhaft and Dr. Schedling." With an introduction and 2 appendices by P. Feyerabend. Philosophical Archive, University of Konstanz: PF: 5-33.

Feyerabend, P. (1951). *Zur Theorie der Basissätze.* Universitätsbibliothek Wien.

Feyerabend, P. (1954a). "Wittgenstein und die Philosophie, I." *Wissenschaft und Weltbild: Monatsschrift für alle Gebiete der Forschung* **7**: 210-220.

Feyerabend, P. (1954b). "Wittgenstein und die Philosophie, II." *Ibid*: 283-287.

Feyerabend, P. (1954c). "Physik und Ontologie." *Ibid*: 464-476.

Feyerabend, P. (1954d). "Ludwig Wittgenstein." *Merkur. Deutsche Zeitschrift für europäisches Denken* **8**: 1021-1038.

Feyerabend, P. (1954e). "Determinismus und Quantenmechanik." *Wiener Zeitschrift für Philosophie, Psychologie, Pädagogik* **5**: 89-111.

Feyerabend, P. (1954f). "Review of *Einführung in die symbolische Logik mit besonderer Berücksichtigung ihrer Anwendunge.* By Rudolf Carnap." *Internationale mathematische Nachrichten* **9**: 35-36, 67.

Feyerabend, P. (1955a). "Review of *Philosophical Investigations.* By Ludwig Wittgenstein." *The Philosophical Review* **64**: 449-483.

Feyerabend, P. (1955b). "Carnaps Theorie der Interpretation theoretischer Systeme." *Theoria* **21**: 55-62.

Feyerabend, P. (1956a). "A Note on the Paradox of Analysis." *Philosophical Studies* **7**: 92-96.

Feyerabend, P. (1956b). "Eine Bemerkung zum Neumannschen Beweis." *Zeitschrift für Physik* **145**: 421-423.

Feyerabend, P. (1956c). "Review of *Kant's Metaphysics and Theory of Science*. By Gottfried Martin." *British Journal for the Philosophy of Science* **7**: 260-263.

Feyerabend, P. (1957a). "Review of *Foundations of Quantum-Mechanics: A Study in Continuity and Symmetry*. By A. Landé." *British Journal for the Philosophy of Science* **7**: 354-357.

Feyerabend, P. (1957b). "Zur Quantentheorie der Messung." *Zeitschrift für Physik* **148**: 551-559.

Feyerabend, P. (1957c). "On the Quantum-Theory of Measurement." *Observation and Interpretation: A Symposium of Philosophers and Physicists. Proceedings of the Ninth Symposium of the Colston Research Society*. S. Körner and M. H. L. Pryce (eds.). London, Butterworths Scientific Pub.: 121-130.

Feyerabend, P. (1957d). "Discussion remarks." *Ibid*: 21, 48-49, 52, 53, 112-113, 139, 146-147, 172, 182, 183, 185.

Feyerabend, P. (1957e). "Die analytische Philosophie und das Paradox der Analyse." *Kant-Studien* **49**: 238-244.

Feyerabend, P. (1958a). "An Attempt at a Realistic Interpretation of Experience." *Proceedings of the Aristotelian Society* **58**: 143-170.

Feyerabend, P. (1958b). "Complementarity." *Proceedings of the Aristotelian Society, Suppl. Vol.* **32**: 75-104.

Feyerabend, P. (1958c). "Review of *Mathematical Foundations of Quantum Mechanics*. By John von Neumann." *British Journal for the Philosophy of Science* **8**: 343-347.

Feyerabend, P. (1958d). "Reichenbach's Interpretation of Quantum-Mechanics." *Philosophical Studies* **9**: 49-59.

Feyerabend, P. (1959). "Review of *The Direction of Time*. By Hans Reichenbach"." *British Journal for the Philosophy of Science* **9**: 336-337.

Feyerabend, P. (1960a). "On the Interpretation of Scientific Theories." *Proceedings of the 12th International Congress of Philosophy, Venice, 1958, Vol. 5*. Florence, Sansoni Editore: 151-159.

Feyerabend, P. (1960b). "O Interpretacij Relacyj Nieokreslonosci." *Studia Filozoficzne* **19**: 21-78.

Feyerabend, P. (1960c). "Professor Landé on the Reduction of the Wave Packet." *American Journal of Physics* **28**: 507-508.

Feyerabend, P. (1960d). "Review of *Patterns of Discovery. An Inquiry into the Conceptual Foundations of Science*. By N.R. Hanson." *The Philosophical Review* **69**: 247-252.

Feyerabend, P. (1960e). "Professor Bohm's Philosophy of Nature. Review of *Causality and Chance in Modern Physics*. By David Bohm." *British Journal for the Philosophy of Science* **10**: 321-338.

Feyerabend, P. (1960f). "Das Problem der Existenz theoretischer Entitäten." *Probleme der Wissenschaftstheorie. Festschrift für Viktor Kraft*. E. Topitsch (ed.). Vienna, Springer Verlag: 35-72.

Feyerabend, P. (1960g). "Ludwig Wittgenstein." *Ludwig Wittgenstein, Schriften, Beiheft*. Frankfurt am Main, Suhrkamp: 30-47.

Feyerabend, P. (1961a). "Comments on Hanson's 'Is There a Logic of Scientific Discovery?'". *Current Issues in the Philosophy of Science. Symposia of Scientists and Philosophers. Proceedings of Section L of the American Association for the Advancement of Science, 1959*. H. Feigl and G. Maxwell (eds.). New York, Holt, Rinehart and Winston: 35-39.

Feyerabend, P. (1961b). "Comments on Sellars' 'The Language of Theories'." *Ibid*: 82-83.

Feyerabend, P. (1961c). "Comments on Grünbaum's 'Law and Convention in Physical Theory'." *Ibid*: 155-161.

Feyerabend, P. (1961d). "Comments on Barker's 'The Role of Simplicity in Explanation'." *Ibid*: 278-280.

Feyerabend, P. (1961e). "Niels Bohr's Interpretation of the Quantum Theory." *Ibid*: 371-390.

Feyerabend, P. (1961f). "Rejoinder to Hanson." *Ibid*: 398-400.

Feyerabend, P. (1961g). "Comments on Hill's 'Quantum Physics and Relativity Theory'." *Ibid*: 441-443.

Feyerabend, P. (1961h). "Review of *An Introduction to the Logic of the Sciences*. By Rom Harré." *British Journal for the Philosophy of Science* **12**: 245-250.

Feyerabend, P. (1961i). *Knowledge Without Foundations: Two Lectures Delivered on the Nellie Heldt Lecture Fund*. Oberlin, Oberlin College.

Feyerabend, P. (1962a). "Explanation, Reduction and Empiricism." *Scientific Explanation, Space, and Time. Minnesota Studies in the Philosophy of Science, Vol. III*. H. Feigl and G. Maxwell (eds.). Minneapolis, University of Minnesota Press: 28-97.

Feyerabend, P. (1962b). "Problems of Microphysics." *Frontiers of Science and Philosophy: University of Pittsburgh Series in the Philosophy of Science Vol. I*. R. Colodny (ed.). Pittsburgh, University of Pittsburgh Press: 189-283.

Feyerabend, P. (1963a). "How to Be a Good Empiricist: A Plea for Tolerance in Matters Epistemological." *Philosophy of Science: The Delaware Seminar*. B. Baumrin (ed.). New York, Interscience Press.

Feyerabend, P. (1963b). "Materialism and the Mind-Body Problem." *The Review of Metaphysics* **17**: 49-66.

Feyerabend, P. (1963c). "Professor Hartmann's Philosophy of Nature. Review of *Philosophie der Natur*. By Nicolai Hartmann." *Ratio* **5**: 81-94.

Feyerabend, P. (1963d). "Mental Events and the Brain." *The Journal of Philosophy* **60**: 295-296.

Feyerabend, P. (1963e). "Review of *Erkenntnislehre*. By V. Kraft." *British Journal for the Philosophy of Science* **13**: 319-323.

Feyerabend, P. (1964a). "Review of *The Concept of the Positron. A Philosophical Analysis*. By N. R. Hanson." *The Philosophical Review* **73**: 264-266.

Feyerabend, P. (1964b). "Review of *Scientific Change*. Edited by A. C. Crombie." *British Journal for the Philosophy of Science* **15**: 244-254.

Feyerabend, P. (1964c). "A Note on the Problem of Induction." *The Journal of Philosophy* **61**: 349-353.

Feyerabend, P. (1964d). "Realism and Instrumentalism: Comments on the Logic of Factual Support." *The Critical Approach to Science and Philosophy: In Honor of Karl R. Popper*. M. Bunge (ed.). London, The Free Press of Glencoe: 280-308.

Feyerabend, P. (1965a). "On the "Meaning" of Scientific Terms." *Journal of Philosophy* **62**: 266-274.

Feyerabend, P. (1965b). "Problems of Empiricism." *Beyond the Edge of Certainty. Essays in Contemporary Science and Philosophy*. R. Colodny (ed.). Pittsburgh, CPS Publications in the Philosophy of Science: 145-260.

Feyerabend, P. (1965c). "Reply to Criticism. Comments on Smart, Sellars and Putnam." *Proceedings of the Boston Colloquium for the Philosophy of Science 1962-64: In Honor of Philipp Frank. Boston Studies in the Philosophy of Science, Vol. II*. R. Cohen and M. Wartofsky (eds.). New York, Humanities Press: 223-261.

Feyerabend, P. (1965d). "Review of *Conjectures and Refutations. The Growth of Scientific Knowledge*. By Karl R. Popper." *Isis* **56**: 88.

Feyerabend, P. (1966a). "Herbert Feigl: A Biographical Sketch." *Mind, Matter, and Method: Essays in Philosophy and Science in Honor of Herbert Feigl*. P. Feyerabend and G. Maxwell (eds.). Minneapolis, University of Minnesota Press: 3-13.

Feyerabend, P. (1966b). "On the Possibility of a Perpetuum Mobile of the Second Kind." *Ibid*: 409-412.

Feyerabend, P. (1966c). "Review of *The Structure of Science*. By Ernest Nagel." *British Journal for the Philosophy of Science* **17**: 237-249.

Feyerabend, P. (1966d). "Wittgenstein's Philosophical Investigations." *Wittgenstein, The Philosophical Investigations: A Collection of Critical Essays*. G. Pitcher (ed.). Garden City, Anchor Books, Doubleday & Co: 104-150.

Feyerabend, P. (1967a). "The Mind-Body Problem." *Continuum* **5**: 35-49.

Feyerabend, P. (1967b). "The Theatre as an Instrument of the Criticism of Ideologies: Notes on Ionesco." *Inquiry* **10**: 298-312.

Feyerabend, P. (1967c). "Review of *Law and Psychology in Conflict.* By J. Marshall." *Inquiry* **10**: 114-120.

Feyerabend, P. (1967d). "Boltzmann, Ludwig." *The Encyclopedia of Philosophy Vol. I.* P. Edwards (ed.). New York, Macmillan Pub. Co.: 334-337.

Feyerabend, P. (1967e). "The Mind-Body Problem." *Continuum* **5**: 35-49.

Feyerabend, P. (1967f). "Wittgenstein's Philosophical Investigations." *Ludwig Wittgenstein. The Man and his Philosophy.* K. Fann (ed.). Sussex, Humanities Press: 214-250.

Feyerabend, P. (1968a). "Science, Freedom, and the Good Life." *The Philosophical Forum* **1**: 127-135.

Feyerabend, P. (1968b). "A Note on Two 'Problems' of Induction." *British Journal for the Philosophy of Science* **19**: 251-253.

Feyerabend, P. (1968c). "On a Recent Critique of Complementarity: Part I." *Philosophy of Science* **35**: 309-331.

Feyerabend, P. (1969a). "Outline of a Pluralistic Theory of Knowledge and Action." *Planning for Diversity and Choice.* S. Anderson (ed.). Cambridge, MIT Press: 275-284.

Feyerabend, P. (1969b). "Science without Experience." *Journal for Philosophy* **56**: 791-794.

Feyerabend, P. (1969c). "Materialism and the Mind-Body Problem." *Modern Materialism: Readings on Mind-Body Identity.* J. O'Connor (ed.). New York, Harcourt, Brace, and World Inc.: 82-98.

Feyerabend, P. (1969d). "On a Recent Critique of Complementarity: Part II." *Philosophy of Science* **36**: 82-105.

Feyerabend, P. (1969e). "Linguistic Arguments and Scientific Method." *Telos* **2**: 43-63.

Feyerabend, P. (1970a). "In Defense of Classical Physics." *Studies in History and Philosophy of Science* **1**: 59-85.

Feyerabend, P. (1970b). "Problems of Empiricism, Part II." *The Nature and Function of Scientific Theories. Essays in Contemporary Science and Philosophy, University of Pittsburgh Series in the Philosophy of Science, Vol. 4.* R. Colodny (ed.). Pittsburgh, University of Pittsburgh Press: 275-353.

Feyerabend, P. (1970c). "Consolations for the Specialist." *Criticism and the Growth of Knowledge.* I. Lakatos and A. Musgrave (eds.). Cambridge, Cambridge University Press: 197-230.

Feyerabend, P. (1970d). "Against Method: Outline of an Anarchistic Theory of Knowledge." *Analysis of Theories and Methods of Physics and Psychology. Minnesota Studies in the Philosophy of Science vol. IV.* M. Radner and S. Winokur (eds.). Minneapolis, University of Minnesota Press: 17-130.

Feyerabend, P. (1970e). "Discussion at the Conference on Correspondence Rules." *Ibid*: 220-259.

Feyerabend, P. (1970f). "Materialism and the Mind-Body Problem." *The Mind/Brain Identity Theory. Controversies in Philosophy.* T. Borst (ed.). London, The Macmillan Press: 142-156.

Feyerabend, P. (1970g). "Philosophy of Science: A Subject with a Great Past." *Historical and Philosophical Perspectives of Science. Minnesota Studies in the Philosophy of Science, Vol. 5.* R. H. Stuewer (ed.). Minneapolis, University of Minnesota Press: 172-183.

Feyerabend, P. (1972). "Von der beschränkten Gültigkeit methodologischer Regeln." *Dialog als Methode. Neue Hefte für Philosophie* **2/3**: 124-171.

Feyerabend, P. (1973). "Die Wissenschaftstheorie - eine bisher unbekannte Form des Irrsinns?" *Natur und Geschichte. X. Kongress für Philosophie, Kiel 8.-12. Oktober 1972.* K. Hübner and A. Menne (eds.). Hamburg, Felix Meiner Verlag: 88-124.

Feyerabend, P. (1974a). "Kuhns Struktur wissenschaftlicher Revolutionen - ein Trostbüchlein für Spezialisten?" *Kritik und Erkenntnisfortschritt. Abhandlungen des Internationalen Kolloquiums über die Philosophie der Wissenschaft, London 1965.* I. Lakatos and A. Musgrave (eds.). Brauschweig, Vieweg & Sohn. **4**: 191-222.

Feyerabend, P. (1974b). "Review of *Objective Knowledge. An Evolutionary Approach.* By Karl R. Popper." *Inquiry* **17**: 475-507.

Feyerabend, P. (1975a). *Against Method. Outline of an Anarchistic Theory of Knowledge.* London, New Left Books.

Feyerabend, P. (1975b). "Let's Make More Movies." *The Owl of Minerva.* C. Bontempo and S. Odell (eds.). New York, McGraw Hill: 201-210.

Feyerabend, P. (1976a). *Wider den Methodenzwang. Skizze einer anarchistischen Erkenntnistheorie.* Frankfurt am Main, Suhrkamp.

Feyerabend, P. (1976b). "Über die Methode. Ein Dialog." *Unter dem Pflaster liegt der Strand (3).* H. Duerr (ed.). Berlin, Karin Kramer Verlag: 99-160.

Feyerabend, P. (1976c). "Logic, Literacy, and Professor Gellner." *British Journal for the Philosophy of Science* **27**: 381-391.

Feyerabend, P. (1977a). "Marxist Fairytales From Australia." *Inquiry* **20**: 372-397.

Feyerabend, P. (1977b). "Changing Patterns of Reconstruction." *British Journal for the Philosophy of Science* **28**: 351-382.

Feyerabend, P. (1978a). *Science in a Free Society.* London, New Left Books.

Feyerabend, P. (1978b). *Der wissenschaftstheoretische Realismus und die Autorität der Wissenschaften.* Braunschweig, Vieweg.

Feyerabend, P. (1978c). "From Incompetent Professionalism to Professionalized Incompetence - The Rise of a New Breed of Intellectuals." *Philosophy of the Social Sciences* **8**: 37-53.

Feyerabend, P. (1978d). "The Gong Show -- Popperian Style." *Progress and Rationality in Science, Boston Studies in the Philosophy of Science, Vol. 58.* G. Radnitzky and G. Andersson (eds.). Dordrecht, D. Reidel Pub. Co.: 387-392.

Feyerabend, P. (1979). "Dialogue on Method." *The Structure and Development of Science.* G. Radnitzky and G. Andersson (eds.). Dordrecht, D. Reidel Pub. Co.

Feyerabend, P. (1980). "Democracy, Elitism and Scientific Method." *Inquiry* 23: 3-18.

Feyerabend, P. (1981a). *Realism, rationalism and scientific method. Philosophical papers.* Cambridge, Cambridge University Press.

Feyerabend, P. (1981b). *Problems of Empiricism: Philosophical Papers Volume 2.* Cambridge, Cambridge University Press.

Feyerabend, P. (1981c). "More Clothes from the Emperor's Bargain Basement. A review of *Progress and its Problems*. By Larry Laudan." *British Journal for the Philosophy of Science* **32**: 57-71.

Feyerabend, P. (1981d). "Über die Methode. Ein Dialog." *Voraussetzungen und Grenzen der Wissenschaft. Die Einheit der Gesellschafts-wissenschaften, Band 25.* G. Radnitzky and G. Andersson (eds.). Tübingen, J. C. B. Mohr (Paul Siebeck): 175-253.

Feyerabend, P. (1982). "Academic Ratiofascism. Comments on Tibor Machan's Review." *Philosophy of the Social Sciences* **12**: 191-195.

Feyerabend, P. (1983a). *Wider den Methodenzwang. Skizze einer anarchistischen Erkenntnistheorie.* Frankfurt am Main, Suhrkamp.

Feyerabend, P. (1983b). "Some Observations on Aristotle's Theory of Mathematics and of the Continuum." *Contemporary Perspectives on the History of Philosophy. Midwest Studies in Philosophy, Vol. 8.* P. French, J. Theodore, E. Uehling and H. Wettstein (eds.). Minneapolis, University of Minnesota Press: 67-88.

Feyerabend, P. (1984). "The Lessing Effect in the Philosophy of Science: Comments on Some of my Critics." *New Ideas in Psychology. An International Journal of Innovative Theory in Psychology* **2**: 127-136.

Feyerabend, P. (1986). "Trivializing Knowledge: A review of Popper's Postscript." *Inquiry* **29**: 93-119.

Feyerabend, P. (1987). *Farewell to Reason.* London, Verso.

Feyerabend, P. (1988). *Against Method. Outline of an Anarchistic Theory of Knowledge.* New York, Verso.

Feyerabend, P. (1989a). "How to Be a Good Empiricist: A Plea for Tolerance in Matters Epistemological." *Readings in the Philosophy of Science.* B. Brody and R. Grandy (eds.). Englewood Cliffs, New Jersey, Prentice Hall: 104-123.

Feyerabend, P. (1989b). "Wem nützt die Wissenschaftstheorie?" *Traditionen und Perspektiven der analytischen Philosophie. Festschrift für Rudolf Haller.* W. L. Gombocz, H. Rutte and W. Sauer (eds.). Wien, Hölder-Pichler-Tempsky: 568-574.

Feyerabend, P. (1991a). *Three Dialogues on Knowledge.* Oxford, Blackwell.

Feyerabend, P. (1991b). "Concluding Unphilosophical Conversation." *Beyond Reason: Essays on the Philosophy of Paul Feyerabend. Boston Studies in the Philosophy of Science, Vol. 132.* G. Munévar (ed.). Dordrecht, Kluwer: 487-527.

Feyerabend, P. (1992). "Ethics as a Measure of Scientific Truth." *From the Twilight of Probability: Ethics and Politics.* W. Shea and A. Spadafora (eds.). USA, Science History Pub.: 106-114.

Feyerabend, P. (1993a). *Against Method. Outline of an Anarchistic Theory of Knowledge.* London, Verso.

Feyerabend, P. (1993d). "Not a Philosopher." *Falling in Love with Wisdom: American Philosophers Talk about their Calling.* D. Karnos and R. Shoemaker (eds.). New York, Oxford University Press: 16-17.

Feyerabend, P. (1994a). "Potentially Every Culture is All Cultures." *Common Knowledge* **3**: 16-22.

Feyerabend, P. (1994b). "Concerning an Appeal for Philosophy." *Ibid*: 10-13.

Feyerabend, P. (1995a). *Killing Time. The Autobiography of Paul Feyerabend.* Chicago, University of Chicago Press.

Feyerabend, P. (1995b). "Two Letters of Paul Feyerabend to Thomas S. Kuhn on a Draft of The Structure of Scientific Revolutions." P. Hoyningen-Huene (ed.). *Studies in History and Philosophy of Science* **26**: 353-387.

Feyerabend, P. (1995c). *Paul Feyerabend. Briefe an einen Freund.* H. Duerr (ed.). Frankfurt am Main, Suhrkamp.

Feyerabend, P. (1997). *Briefwechsel. Paul K. Feyerabend und Hans Albert.* Frankfurt am Main, Fischer.

Feyerabend, P. (1999a). *Paul K. Feyerabend: Knowledge, Science and Relativism. Philosophical Papers Volume 3.* Cambridge, Cambridge University Press.

Feyerabend, P. (1999b). *Conquest of Abundance. A Tale of Abstraction versus the Richness of Being.* Chicago, University of Chicago Press.

Feyerabend, P. (1999c). "On the limited validity of methodological rules." *Knowledge, Science and Relativism. Philosophical Papers Volume 3.* J. Preston (ed.). Cambridge, Cambridge University Press: 138-180.

Feyerabend, P. (1999d). "The Problem of The Existence of Theoretical Entities." *Ibid*: 16-49.

Feyerabend, P. (2000). "Last Interview." *The Worst Enemy of Science? Essays in Memory of Paul Feyerabend.* J. Preston, G. Munévar and D. Lamb (eds.). New York, Oxford University Press: 102-114.

Feyerabend, P. (2007). "More Letters by Paul Feyerabend to Thomas S. Kuhn on *Proto-Structure*." P. Hoyningen-Huene (ed.). *Studies in History and Philosophy of Science* (in print).

Feyerabend, P. (forthcoming [1971/1976]). *Einführung in die Naturphilosophie.* H. Heit and E. Oberheim (eds.). Frankfurt, Suhrkamp.

Feyerabend, P. (forthcoming [1993]). *Conflict and Harmony.* E. Oberheim and G. Munévar (eds.).

Feyerabend, P. and G. Maxwell (eds.) (1966). *Mind, Matter and Method: Essays in the Philosophy of Science in Honor of Herbert Feigl.* Minneapolis, University of Minnesota Press.

Fine, A. (1984a). "The Natural Ontological Attitude." *Scientific Realism.* J. Leplin (ed.). Berkeley, University of California Press: 83-107.

Fine, A. (1984b). "And Not Anti-Realism Either." *Nous* **18**: 51-65.

Fine, A. (1986a). *The Shaky Game. Einstein, Realism and the Quantum Theory.* Chicago, University of Chicago Press.

Fine, A. (1986b). "Unnatural Attitudes: Realist and Instrumentalist Attachments to Science." *Mind* **95**: 149-179.

Fine, A. (1991). "Piecemeal Realism." *Philosophical Studies* **61**: 79-96.

Fleck, L. ([1935] 1980). *Entstehung und Entwicklung einer wissenschaftlichen Tatsache.* Frankfurt am Main, Suhrkamp.

Fleck, L. ([1935] 1979). *Genesis and Development of a Scientific Fact.* T. Trenn and R. Merton (eds.). Chicago, University of Chicago Press.

Fleck, L. ([1927] 1986). "Some Specific Features of The Medical Way of Thinking. *Cognition and Fact: Materials on Ludwik Fleck.*" R. Cohen and T. Schnelle (eds.). Boston, D. Reidel: 39-46.

Fleck, L. ([1935] 1986). "Scientific Observation and Perception in General." *Ibid*: 59-78.

Fleck, L. ([1936] 1986). "The Problem of Epistemology." *Ibid*: 79-112.

Fowler, H. and F. Fowler (1934). *The Concise Oxford Dictionary of Current English.* Oxford, Clarendon Press.

Frank, P. (1950). *Relativity, a Richer Truth.* Boston, Beacon Press.

Franklin, A. e. a. (1989). "Can a Theory-Laden Observation Test the Theory?" *British Journal for the Philosophy of Science* **40**: 229-231.

Gaa, J. (1975). "The Replacement of Scientific Theories: Reduction and Explanation." *Philosophy of Science* **42**: 349-372.

Gallistel, C. (1999). "Behaviorism, Methodological and Scientific." *Routledge Encyclopedia of Philosophy Cd-Rom.* E. Craig (ed.). London, Routledge.

Gellner, E. (1975). "Beyond Truth and Falsehood." *The British Journal for the Philosophy of Science* **26**: 331-342.

Ghosh, T. (1998). "Berkeley's Idealism- Internal Realism and the Incommensurability Thesis." *Indian Philosophical Quarterly* **25**: 241-251.

Giedymin, J. (1970). "The Paradox of Meaning Variance." *British Journal for the Philosophy of Science* **21**: 257-268.

Giedymin, J. (1971). "Consolations for the Irrationalist." *British Journal for the Philosophy of Science* **22**: 39-53.

Giedymin, J. (1982). *Science and Convention: Essays on Henri Poincaré's Philosophy of Science and the Conventionalist Tradition.* Oxford, Pergamon Press.

Giere, R. (1999). *Science without Laws.* Chicago, University of Chicago Press.

Giere, R. and A. Richardson (eds.) (1966). *Origins of Logical Empiricism.* Minneapolis, University of Minnesota Press.

Glymour, C. (1982). "Conceptual Scheming or Confessions of a Metaphysical Realist." *Synthese* **51**: 169-180.

Glymour, C. (1992). "Realism and the Nature of Theories." *Introduction to the Philosophy of Science.* Engelwoodcliffs, Prentice Hall: 104-131.

Golinski, J. (1998). *Making Natural Knowledge. Constructivism and the History of Science.* Cambridge, Cambridge University Press.

Grunfeld, J. (1984). "Feyerabend's Irrational Science." *Logical Analysis* **27**: 221-232.

Hacking, I. (1970). "Imre Lakatos's Philosophy of Science." *British Journal for the Philosophy of Science* **30**: 381-410.

Hacking, I. (1975). *Why Does Language Matter to Philosophy?* Cambridge, Cambridge University Press.

Hacking, I. (1983). *Representing and intervening. Introductory topics in the philosophy of natural science.* Cambridge, Cambridge University Press.

Hacking, I. (1984). "Experimentation and Scientific Realism." *Scientific Realism.* J. Leplin (ed.). Berkeley, University of California Press: 154-192.

Hacking, I. (1993). "Working in a New World: The Taxonomic Solution." *World Changes: Thomas Kuhn and the Nature of Science.* P. Horwich (ed.). Cambridge, MIT Press: 275-310.

Hacking, I. (1994). "Paul Feyerabend, Humanist." *Common Knowledge* **3**: 23-28.

Hanson, N. (1958). *Patterns of Discovery. An Inquiry into the Conceptual Foundations of Science.* Cambridge, Cambridge University Press.

Hardcastle, V. (1994). "The Image of Observables." *British Journal for the Philosophy of Science* **45**: 585-597.

Harré, R. (1959). "Notes on P. K. Feyerabend's Criticisms of Positivism." *British Journal for the Philosophy of Science* **10**: 43-47.

Harris, R. (ed.) (2005). *Rhetoric and Incommensurability.* West Lafayette, Parlor Press.

Hattiangadi, J. (1977). "The Crisis in Methodology: Feyerabend." *Philosophy of the Social Sciences* **7**: 289-302.

Hattiangadi, J. (1978). "Playphilosopher." *Philosophy of the Social Sciences* **8**: 59-61.

Hegel, G. ([1827] 1989) *Encyclopädie der philosophischen Wissenschaften im Grundrisse (Gesammelte Werke 19)*. Hamburg, Meiner.

Hellman, G. (1979). "Against Bad Method." *Metaphilosophy* **10**: 190-202.

Hempel, C. (1948). "Studies in the Logic of Explanation." *Philosophy of Science* **15**: 135-175.

Hempel, C. (1952). "Fundamentals of Concept Formation in Physical Science." *International Encyclopedia of Unified Science*. O. Neurath, R. Carnap and C. Morris (eds.). Chicago, Chicago University Press.

Hempel, C. (1958). "The Theoretician's Dilemma: A Study in the Logic of Theory Construction." *Concepts, Theories and the Mind-Body Problem. Minnesota Studies in Philosophy of Science, Vol. II*. H. Feigl, M. Scriven and G. Maxwell (eds.). Minneapolis, University of Minnesota Press.: 37-98.

Hempel, C. (1965). *Aspects of Scientific Explanation and Other Essays in the Philosophy of Science*. New York, The Free Press.

Hempel, C. (1966). *Philosophy of Natural Science*. Englewood Cliffs, New Jersey, Prentice-Hall.

Hempel, C. and P. Oppenheim (1948). "The Logic of Explanation." *Readings in the Philosophy of Science*. H. Feigl and G. Maxwell (eds.). New York, Appleton-Century-Crofts: 319-352.

Hentschel, K. (1985). "On Feyerabend's Version of 'Mach's Theory of Research and its Relation to Einstein." *Studies in History and Philosophy of Science* **16**: 387-394.

Hesse, M. (1977). "Truth and the Growth of Scientific Knowledge." *PSA 1976*. F. Suppe and P. Asquith (eds.). East Lansing, Philosophy of Science Association: 261-280.

Holcomb, H. (1987). "Circularity and Inconsistency in Kuhn's Defense of his Relativism." *The Southern Journal of Philosophy* **25**: 467-480.

Holton, G. (1978). "Subelectrons, presuppositions, and the Millikan-Ehrenhaft dispute." *Historical Studies in the Physical Sciences* **9**: 161-224.

Honderich, T. (ed.) (1995). *The Oxford Companion to Philosophy*. Oxford, Oxford University Press.

Hooker, C. (1973). "Empiricism, Perception and Conceptual Change." *Canadian Journal of Philosophy* **3**: 59-75.

Hovard, R. (1971). "Theoretical Reduction: The Limits and alternatives to Reductive Methods in Scientific Explanation." *Philosophy of the Social Sciences* **1**: 83-100.

Hoyningen-Huene, P. (1993). *Reconstructing Scientific Revolutions. The Philosophy of Science of Thomas S. Kuhn*. Chicago, Chicago University Press.

Hoyningen-Huene, P. (1995). "Proliferationsprinzip, philosophisches." *Philosophie und Wissenschaftstheorie*. J. Mittelstraß (ed.). Stuttgart, Weimar.

Hoyningen-Huene, P. (1999). "Feyerabends Kritik an Kuhns normaler Wissenschaft." *Rationalität, Realismus, Revision*. J. Nida-Rümelin (ed.). Berlin, Walter de Gruyter: 465-470.

Hoyningen-Huene, P. (2000a). "Paul Feyerabend and Thomas Kuhn." *The Worst Enemy of Science? Essays in Memory of Paul Feyerabend*. J. Preston, G. Munévar and D. Lamb (eds.). New York, Oxford University Press: 102-114.

Hoyningen-Huene, P. (2000b). "Paul K. Feyerabend: An Obituary." *Ibid*: 3-15.

Hoyningen-Huene, P. (2001). "Paul Feyerabend und Thomas Kuhn." *Journal for General Philosophy of Science* **33**: 61-83.

Hoyningen-Huene, P. (2002). "Paul Feyerabend - ein postmoderner Philosoph? Ein Portrait." *Information Philosophie* März: 30-37.

Hoyningen-Huene, P., E. Oberheim, et al. (1996). "On Incommensurability." *Studies in History and Philosophy of Science* **27**: 131-141.

Hoyningen-Huene, P. and H. Sankey (eds.) (2001a). *Incommensurability and Related Matters*. Dordrecht, Kluwer.

Hoyningen-Huene, P. and H. Sankey (2001b). "Introduction." *Incommensurability (and related matters)*. P. Hoyningen-Huene and H. Sankey (eds.). Dordrecht, Kluwer: vi-xxxiv.

Hoyningen-Huene, P. and H. Sankey (2001c). "Incommensurability Bibliography." *Ibid*: 303-316.

Hoyningen-Huene, P. and P. Schaber (2003). "Incommensurability of Knowledge: Theories and Values." *Unity of Knowledge in Transdisciplinary Research for Sustainability*, G. Hirsch Hadorn (ed.) Oxford, Eolss Publishers Co.

Hoyningen-Huene, P. (2005). "Three Biographies: Kuhn, Feyerabend and Incommensurability." *Rhetoric and Incommensurability* R. Harris (ed.). West Lafayette, Parlor Press: 150-175.

Hull, R. (1972). "Feyerabend's Attack on Observation Sentences." *Synthese* **23**: 374-399.

Irzik, G. and T. Grünberg (1995). "Carnap and Kuhn: Arch Enemies or Close Allies?" *British Journal for the Philosophy of Science* **46**: 285-309.

Irzik, G. and T. Grünberg (1998). "Whorfian Variations on Kantian Themes: Kuhn's Linguistic Turn." *Studies in History and Philosophy of Science* **29**: 207-221.

Jacobs, S. (2002). "Polanyi's Presagement of the Incommensurability Concept." *Studies in History and Philosophy of Science* **33**: 105-120.

Jacobs, S. (2003). "Two Sources of Michael Polanyi's Prototypal notion of incommensurability: Evans-Pritchard on Azande witchcraft and St Augustine on conversion." *History of the Human Sciences* **16**: 57-76.

Jaki, S. (1969). "Introductory Essay." *To Save the Phenomena* by Pierre Duhem. University of Chicago Press: Chicago: ix-xxvi.

Jakovljevic, D. (1996). "Realismus oder Instrumentalismus?" *Facta Universitatis.* **1**: 187-194.

Jarvie, I. (1999). "Popper, Karl Raimund (1902-94)." *Routledge Encyclopedia of Philosophy Cd-Rom.* E. Craig (ed.). London, Routledge.

Kadvany, J. (1996). "Reason in History: Paul Feyerabend's Autobiography." *Inquiry* **39**: 141-146.

Kant, I. (1781). *Critique of Pure Reason.* London, Macmillan.

Kenshur, O. (1984). "The Rhetoric of Incommensurability." *The Journal of Aesthetics and Art Criticism* **42**: 367-382.

Keuth, H. (2005). The Philosophy of Karl 'Popper. Cambridge, Cambridge University Press.

Kirkham, R. (1992). *Theories of Truth: A Critical Introduction.* Cambridge, MIT Press.

Kitcher, P. (1978). "Theories, Theorists and Theoretical Change." *The Philosophical Review* **87**: 519-547.

Kitcher, P. (1982). "Genes." *British Journal for the Philosophy of Science* **33**: 337-359.

Kitcher, P. (1992). "The Naturalists Return." *The Philosophical Quarterly* **101**: 53-115.

Kitcher, P. (1993). *The Advancement of Science. Science Without Legend, Objectivity Without Illusions.* Oxford, Oxford University Press.

Kitcher, P. (1999). Explanation. *Routledge Encyclopedia of Philosophy Cd-Rom.* E. Craig (ed.). London, Routledge.

Koertge, N. (1972). "For and Against Method." *British Journal for the Philosophy of Science* **23**: 274-290.

Köhler, W. (1920). *Die Physischen Gestalten in Ruhe und in stationärem Zustand, Eine naturphilosophische Untersuchung,* Braunschweig, Friedr. Vieweg & Sohn.

Köhler, W. ([1920] 1938). "Physical Gestalten. " *A Source Book of Gestalt Psychology.* W. Ellis (ed.). London, Kegan: 17-54.

Köhler, W. (1947). *Gestalt Psychology. An Introduction to New Concepts in Modern Psychology.* New York, Mentor Books.

Kordig, C. (1971). "The Comparability of Scientific Theories." *Philosophy of Science* **38**: 467-485.

Kordig, C. (1971). "The Theory Ladeness of Observation." *Review of Metaphysics* **24**: 448-484.

Körner, S. (ed.) (1957). *Observation and Interpretation in the Philosophy of Physics.* New York, Dover.

Kripke, S. (1980). *Naming and Necessity.* Cambridge, Harvard University Press.

Kuhn, T. (1957). *The Copernican Revolution.* Cambridge, Harvard University Press.

Kuhn, T. (1962). *The Structure of Scientific Revolutions.* Chicago, Chicago University Press.

Kuhn, T. ([1962] 1970a). *The Structure of Scientific Revolutions.* Chicago, University of Chicago Press.

Kuhn, T. (1970b). "Reflections on my Critics." *Criticism and the Growth of Knowledge: Proceedings of the International Colloquium the Philosophy of Science, London, 1965,*. I. Lakatos and A. Musgrave (eds.). Cambridge, Cambridge University Press: 231-278.

Kuhn, T. (1970c). "Logic of Discovery or Psychology of Research?" *Ibid*:1-23.

Kuhn, T. (1974). "Second Thoughts on Paradigms." *The Structure of Scientific Theories,*. F. Suppe (ed.). Urbana, University of Illinois Press: 459-482.

Kuhn, T. (1977a). *The Essential Tension: Selected Studies in Scientific Tradition and Change.* Chicago, University of Chicago Press.

Kuhn, T. (1977b). *Die Entstehung des Neuen. Studien zur Struktur der Wissenschaftsgeschichte.* Frankfurt, Suhrkamp.

Kuhn, T. (1979a). "Foreword." *Genesis and Development of a Scientific Fact* by Ludwik Fleck. T. Trenn and R. Merton (eds.). Chicago: University of Chicago Press: vi-xi.

Kuhn, T. (1979b). "Metaphor in Science." *Metaphor and Thought.* A. Ortony (ed.). Cambridge, Cambridge University Press: 409-419.

Kuhn, T. (1981). *What are Scientific Revolutions?* Boston, MIT Press.

Kuhn, T. (1983). "Commensurability, Comparability, Communicability." *PSA 1982. Proceedings of the 1982 Biennial Meeting of the Philosophy of Science Association.* P. D. Asquith and T. Nickles (eds.). East Lansing, Philosophy of Science Association: 669-688.

Kuhn, T. (1989). "Possible Worlds in History of Science. " *Possible Worlds Humanities, and Sciences. Proceedings of Nobel Symposium 65.* T. Allen (ed.), Berlin, de Gruyter: 9-32.

Kuhn, T. (1993). "Afterwords." *World Changes: Thomas Kuhn and the Nature of Science.* P. Horwich (ed.). Cambridge, MIT Press: 311-341.

Kuhn, T. (1999). "Remarks on Incommensurability and Translation." *Incommensurability and Translation.* R. Favretti, G. Sandri and R. Scazzaieri (eds.). Cheltenham, E-dawed Elgar: 33-37.

Kuhn, T. (1991). "The Road Since Structure." *PSA 1990, Proceedings of the 1990 Biennial Meeting of the Philosophy of Science Association.* A. Fine, M. Forbes and L. Wessels (eds.). East Lansing, Philosophy of Science Association **2**: 3-13.

Kuhn, T. (1999). "Remarks on Incommensurability and Translation." *Incommensurability and Translation.* R. Favretti, G. Sandri and R. Scazzaieri (eds.), Cheltenham, U.K.: Edawed Elgar: 33-37.

Kuhn, T. (2000). *The Road Since Structure: Philosophical Essays, 1970-1993, with an Autobiographical Interview.* Chicago, University of Chicago Press.

Kukla, A. (1994). "Scientific Realism, Scientific Practice, and the Natural Ontological Attitude." *British Journal for the Philosophy of Science* **45**: 955-975.

Kukla, A. (1995). "The Two anti-Realisms of Bas van Fraassen." *Studies in the History and Philosophy of Science* **26**: 431-454.

Kuppermann, J. (1975). "Precision in History." *Mind* **84**: 374-389.

Lakatos, I. (1963-1964). "Proofs and Refutations." *British Journal for the Philosophy of Science* **14**: 474-478.

Lakatos, I. (1970). "Falsification and the Methodology of Scientific Research Programmes." *Criticism and the Growth of Knowledge*. I. Lakatos and A. Musgrave (eds.). Cambridge, Cambridge University Press: 91-196.

Larvor, B. (1998). *Lakatos: An Introduction*. London, Routledge.

Laudan, L. (1976). "Two Dogmas of Methodology." *Philosophy of Science* **43**: 585-597.

Laudan, L. (1977). *Progress and its Problems. Towards a Theory of Scientific Growth*. Berkeley, University of California Press.

Laudan, L. (1984a). "A Confutation of Convergent Realism." *Scientific Realism*. J. Leplin (ed.). Berkeley, University of California Press: 218-249.

Laudan, L. (1984b). *Science and Values. The Aims of Science and Their Role in Scientific Debate*. Berkeley, University of California Press.

Laudan, L. (1989). "For Method: Or, Against Feyerabend." *An Intimate Relation*. J. Brown and J. Mittelstrass (eds.). Dordrecht, Kluwer: 299-318.

Laudan, L. and J. Leplin (1991). "Empirical Equivalence and Underdetermination." *Journal of Philosophy* **88**: 449-472.

Laymon, R. (1977). "Feyerabend, Brownian Motion, and the Hiddenness of Refuting Facts." *Philosophy of Science* **44**: 225-247.

Lehrer, K. (1990). *Theory of Knowledge*. Boulder, Westview.

Leplin, J. (ed.) (1984a). *Scientific Realism*. Berkeley, University of California Press.

Leplin, J. (1984b). "Truth and Scientific Progress." *Scientific Realism*. J. Leplin (ed.). Berkeley, University of California Press: 193-217.

Levin, M. E. (1979). "On Theory-Change and Meaning-Change." *Philosophy of Science* **46**: 407-424.

Lipton, P. (1991). *Inference to the Best Explanation*. London, Routledge.

Lipton, P. (1993). "Coordinating science." Review of *Reconstructing Scientific Revolutions* by P. Hoyningen-Huene. *Nature* 364: 770.

Lipton, P. (2001). "Kant on Wheels." *London Review of Books* 23: 30-31.

Llyod, E. (1996). "The Anachronistic Anarchist." *Philosophical Studies* **81**: 247-261.

Llyod, E. (2000). "Feyerabend, Mill, and Pluralism." *The Worst Enemy of Science? Essays in Memory of Paul Feyerabend.* J. Preston and G. Munévar (eds.). New York, Oxford University Press: 115-124.

Locke, J. (1690). "An Essay Concerning Human Understanding." *Philosophical Classics.* W. Kaufmann (ed.). Englewood Cliffs, Prentice-Hall. **2**: 164-202.

Lugg, A. (1977). "Feyerabend's Rationalism." *Canadian Journal of Philosophy* **7**: 755-775.

Lycan, W. (1999). *Philosophy of Language. A contemporary introduction.* London, Routledge.

Machamer, P. (1973). "Feyerabend and Galileo: The Interaction of Theories, and the Reinterpretation of Experience." *Studies in History and Philosophy of Science* **4**: 1-46.

Machamer, P., L. Darden, et al. (2000). "Thinking about Mechanisms." *Philosophy of Science* **67**: 1-25.

Machan, T. (1982). "Anarchosurrealism Revisited. Reply to Feyerabend's Comments." *Philosophy of the Social Sciences* **12**: 197-199.

Mackinnon, E. (1979). "Scientific Realism: The New Debates." *Philosophy of Science* **46**: 501-532.

Maiocchi, R. (1990). "The Case of Brownian Motion." *British Journal for the History of Science* **23**: 257-283.

Martin, M. (1971). "Referential Variance and Scientific Objectivity." *British Journal for the Philosophy of Science* **22**: 17-26.

Maudlin, T. (1995). "The Irrelevance of Incommensurability: Reflections of Torretti's Creative Understanding." *Studies in History and Philosophy of Science* **25**: 1005-1012.

McDowell, J. (1994). *Mind and World.* Boston, Harvard University Press.

McEnvoy, J. (1975). "A 'Revolutionary' Philosophy of Science: Feyerabend and the Degeneration of Critical Rationalism into Sceptical Fallibilism." *Philosophy of Science* **42**: 49-66.

McMullin, E. (1984). "A Case for Scientific Realism." *Scientific Realism.* J. Leplin (ed.). Berkeley, University of California Press: 8-56.

McMullin, E. (1987). "Explanatory Success and the Truth of Theory." *Scientific Inquiry in Philosophical Perspective.* N. Rescher (ed.). Lanham, University Press of America.

McMullin, E. (1991). "Selective Anti-Realism." *Philosophical Studies* **61**: 97-108.

Meynell, H. (1978). "Feyerabend's Method." *Philosophical Quarterly* **28**: 242-252.

Milton, J. (1999). Bacon, Francis (1561-1626). *Routledge Encyclopedia of Philosophy.* London, Routledge.

Moberg, D. (1979). "Are There Rival, Incommensurable Theories?" *Philosophy of Science* **46**: 244-262.

Moore, H. (1972). "Dewey and the Philosophy of Science." *Man World* **5**: 158-168.

Motterlini, M. (ed.) (2000). *For and Against Method.* Chicago, University of Chicago Press.

Munévar, G. (1999). "Review Symposia: Radical Fallibilism vs. Conceptual Analysis: The significance of Feyerabend's Philosophy of Science." *Metascience* **8**: 216-226.

Munévar, G. (2000a). "Preface." *The Worst Enemy of Science? Essays in Memory of Paul Feyerabend.* J. Preston, G. Munévar and D. Lamb (eds.). New York, Oxford University Press: v-vi.

Munévar, G. (2000b). "A *Rehabilitation* of Paul Feyerabend." *Ibid*: 58-79.

Musgrave, A. (1979). "How to Avoid Incommensurability." *Acta Philosophica Fennica* **30**: 336-346.

Musgrave, A. (1982). "Realism Vs. Constructive Empiricism." *Philosophical Quarterly* **32**: 262-271.

Musgrave, A. (1985). "Realism Vs. Constructive Empiricism." *Images of Science. Essays on Realism and Empiricism, with a Reply from Bas C. van Fraassen.* P. Churchland and C. Hooker (eds.). Chicago, The University of Chicago Press: 197-221.

Musgrave, A. (1988). "The Ultimate Argument for Scientific Realism." *Relativism and Realism in Science.* R. Nola (ed.). Dordrecht, Kluwer Academic Press: 229-252.

Musgrave, A. (1989). "NOA's Ark -- Fine for Realism." *Philosophical Quarterly* **39**: 383-398.

Musgrave, A. (1999). *Essays on Realism and Rationalism.* Amsterdam, Adolphi.

Nagel, E. (1949). "The Meaning of Reduction in the Natural Sciences." *Science and Civilization.* R. Stauffer (ed.). Madison, University of Wisconsin Press: 99-145.

Nagel, E. (1961 [1966]). *The Structure of Science. Problems in the Logic of Scientific Explanation.* London, Routledge & Kegan Paul.

Neto, J. (1991). "Feyerabend's Skepticism." *Studies in the History and Philosophy of Science* **22**: 543-555.

Newton-Smith, W. (1981). *The Rationality of Science.* London, Routledge.

Niaz, M. and M. Rodriguez. (2001). "Do We Have to Introduce History and Philosophy of Science or is it already 'Inside' Chemistry?" *Chemistry Education: Research and Practice in Europe* **2**: 159-164.

Nordmann, A. (1990). "Goodbye and Farewell: Siegel vs. Feyerabend." *Inquiry* **33**: 317-331.

Nye, M. (1972). *Molecular Reality. A Perspective on the Scientific Work of Jean Perrin.* London, Elsevier.

Oberdan, T. (1990). "Positivism and the Pragmatic Theory of Observation." *PSA 1990, vol. 1.* A. Fine, M. Forbes and L. Wessels (eds.). East Lansing, Philosophy of Science Association.

Oberheim, E. (1997). "Bibliographie Paul Feyerabends." *Journal for General Philosophy of Science* **28**: 211-234.

Oberheim, E. (1999). "The Works of Paul Feyerabend." *Paul Feyerabend's Knowledge, Science and Relativism. Philosophical Papers Volume 3.* J. Preston (ed.). Cambridge, Cambridge University Press: 227-251.

Oberheim, E. (2005). "On the historical origins of the contemporary notion of incommensurability: Paul Feyerabend's assault on conceptual conservativism." *Studies in the History and Philosophy of Science* **36**: 363-390.

Oberheim, E. (forthcoming). "Feyerabend, Paul K." *Dictionary of Modern American Philosophers.* M. Holme (ed.). Bristol, Thoemmes Press.

Oberheim, E. and P. Hoyningen-Huene (1997). "Incommensurability, Realism and Metaincommensurability." *Theoria* **12**: 447-465.

Oberheim, E. and P. Hoyningen-Huene (1999). "Review Symposia: Radical Fallibilism vs. Conceptual Analysis: The significance of Feyerabend's Philosophy of Science." *Metascience* **8**: 226-233.

Oberheim, E. and P. Hoyningen-Huene (2000). "Essay review of John Preston's *Feyerabend: Philosophy, Science and Society* (Cambridge: Polity, 1997)." *Studies in the History and Philosophy of Science* **31**: 363-375.

Papineau, D. (1979). *Theory and Meaning.* Oxford, Clarendon.

Papineau, D. (1996). "Theory-Dependent Terms." *Philosophy of Science* **63**: 1-20.

Perrin, J. (1920). *Die Atome.* Leipzig.

Polanyi, M. ([1946] 1964). *Science, Faith and Society.* Chicago, University of Chicago Press.

Polanyi, M. (1951). *The Logic of Liberty.* Chicago, University of Chicago Press.

Polanyi, M. (1958). *Personal Knowledge: Towards a post-critical philosophy.* London, Kegan.

Popper, K. (1934). *The Logic of Scientific Discovery.* London, Hutchinson.

Popper, K. (1935). *Logik der Forschung.* Vienna, Springer.

Popper, K. (1948)."Naturgesetze und Theoretische Systeme." *Gesetz und Wirklichkeit.* S. Moser (ed.). Innsbruck, Hochschulverlag: 65-84.

Popper, K. (1954). *The Open Society and Its Enemies.* Princeton.

Popper, K. (1957a). *Die offene Gesellschaft und ihre Feinde, Band I. Der Zauber Platons.* München, Bern. 7th edition 1992, Reworked by J.C.B. Mohr (Paul Siebeck). Tübingen, Francke Verlag.

Popper, K. (1957b). "The Aim of Science." *Ratio* **1**: 24-35.

Popper, K. ([1957] 1972). "The Aim of Science. " *Objective Knowledge. An Evolutionary Approach.* Oxford, Clarendon: 191-205.

Popper, K. (1958). *Die offene Gesellschaft und ihre Feinde, Band II. Falsche Propheten: Hegel, Marx und die Folgen*. München, Bern. 7th edition 1992, Reworked by J.C.B. Mohr (Paul Siebeck). Tübingen, Francke Verlag.

Popper, K. (1959). "New Appendicies." *The Logic of Scientific Discovery*. London, Hutchinson.

Popper, K. (1963). *Conjectures and Refutations. The Growth of Scientific Knowledge*. London, Routledge.

Popper, K. (1968). *The Logic of Scientific Discovery. Second Edition*. London, Hutchinson.

Popper, K. (1972). *Objective Knowledge. An Evolutionary Approach*. Oxford, Oxford University Press.

Popper, K. (1975). "The Rationality of Scientific Revolutions." *Problems of Scientific Revolution*. R. Harré (ed.). Oxford, Clarendon Press: 72-101.

Preston, J. (1992). "Review of *Feyerabend's Critique of Foundationalism*." *International Studies in the Philosophy of Science* **6**: 155-158.

Preston, J. (1997). *Feyerabend. Philosophy, Science and Society*. Oxford, Blackwell.

Preston, J. (1997a). "Feyerabend's Retreat from Realism." *Philosophy of Science* **64**: 421-431.

Preston, J. (1997b). Feyerabend's Polanyian Turns. *Appraisal* **1**: 30-36.

Preston, J. (1998). "Science as Supermarket: 'Post-Modern' Themes in Paul Feyerabend's Later Philosophy of Science." *Studies in History and Philosophy of Science* **29**: 425-447.

Preston, J. (1999). "Review Symposia: Radical Fallibilism vs. Conceptual Analysis: The significance of Feyerabend's Philosophy of Science. Author's Response." *Metascience* **8**: 233-243.

Preston, J. (2002). "Paul Feyerabend." *The Stanford Encyclopedia of Philosophy*. E. Zalta (ed.).

Preyer, G. (1998). "The Received View, Incommensurability and the Comparison of Theories. Beliefs As the Basis of Theorizing." *Protosociology* **12**: 40-58.

Psillos, S. (1994). "Thermodynamics: Resisting the Pessimistic Meta-Induction." *Studies in History and Philosophy of Science* **25**: 159-190.

Psillos, S. (1999). *Scientific Realism: How science tracks the truth*. London, Routledge.

Putnam, H. (1965). "How Not to Talk About Meaning." *Boston Studies in the Philosophy of Science. Volume 2. In Honor of Philipp Frank*. R. Cohen and M. Wartofsky (eds.). New York, Humanities Press.

Putnam, H. (1975a). *Mind, Language, and Reality. Philosophical Papers, Vol. 2*. Cambridge, Cambridge University Press.

Putnam, H. (1975b). "The Meaning of 'Meaning'." *Mind, Language and, Reality. Philosophical Papers, Vol. 2*. H. Putnam (ed.). Cambridge, Cambridge University Press: 215-271.

Putnam, H. (1975c). *Mathematics, Matter and Method. Philosophical Papers Vol. 1*. Cambridge, University Press.

Putnam, H. (1981). *Reason, Truth and History*. Cambridge, Cambridge University Press.

Putnam, H. (1983). *Realism and Reason. Philosophical Papers, Vol. 3*. Cambridge, Cambridge University Press.

Putnam, H. (1984). "What is Realism?" *Scientific Realism*. J. Leplin (ed.). Berkeley, University of California Press: 140-153.

Putnam, H. (1987). *The Many Faces of Realism*. La Salle, Open Court.

Putnam, H. (1990a). *Realism with a Human Face*. Cambridge, Harvard University Press.

Putnam, H. (1990b). "A Defense of Internal Realism." *Realism with a Human Face*. J. Conant (ed.). Cambridge, Harvard University Press: 30-42.

Putnam, H. (1993). "Meaning and Reference." *Meaning and Reference*. A. Moore (ed.). New York, Oxford University Press: 150-161.

Putnam, H. (1994a). "The Question of Realism." *Words and Life*. J. Conant (ed.). Cambridge, Harvard University Press: 295-312.

Putnam, H. (1994b). "Three Kinds of Scientific Realism." *Ibid*: 492-498.

Quine, W. (1993). "In Praise of Observation Sentences." *The Journal of Philosophy* **40**: 107-116.

Quine, W. V. O. (1980). *From a Logical Point of View*. Cambridge, Mass., Harvard University Press.

Recanati, F. (1989). *Direct Reference*. Oxford, Blackwell.

Reiner, R. and P. Pierson (1995). "Hacking's Experimental Realism: An Untenable Middle Ground." *Philosophy of Science* **62**: 60-69.

Rorty, R. (1979). *Philosophy and the Mirror of Nature*. Princeton, Princeton University Press.

Rosenthal-Schneider, I. (1949): "Presuppositions and Anticipations in Einstein's Physics." *Albert Einstein: Philosopher-Scientist. Vol 1*. P. Schilpp (ed.). La Salle, Open Court: 131-146.

Russell, B. (1948). *Human Knowledge: Its Scope and Limits*. London, Allen and Unwin.

Salmon, M. H., J. Earman, et al. (1992). *Introduction to the Philosophy of Science*. Englewood Cliffs, Prentice-Hall.

Salmon, N. (1981). *Reference and essence*. Princeton, Princeton University Press.

Salmon, W. C. (1989). "Four Decades of Scientific Explanation." *Scientific Explanation. Minnesota Studies in the Philosophy of Science, Vol. XIII.* P. Kitcher and W. Salmon (eds.). Minneapolis, University of Minnesota Press: 43-94.

Saltzwedel, J. (1995). "Kopfschmerz der Erkenntnis." *Der Spiegel* **28**: 147-149.

Sankey, H. (1993). "Kuhn's Changing Concept of Incommensurability." *British Journal for the Philosophy of Science* **44**: 759-774.

Sankey, H. (1994). *The Incommensurability Thesis.* Aldershot, Avebury.

Sankey, H. (1997a). *Rationality, Relativism and Incommensurability.* Aldershot, Ashgate.

Sankey, H. (1997b). "Incommensurability: The Current State of Play." *Theoria* **12**: 425-445.

Sankey, H. (1998). "Taxonomic Incommensurability." *International Studies in the Philosophy of Science* **12**: 7-16.

Sankey, H. (2000). "What is Scientific Realism?" *Divinatio: Studia Culturologica Series* **12**: 103-120.

Sankey, H. (2001). "Scientific Realism: An Elaboration and a Defense." *Theoria* **98**: 35-54.

Sankey, H. (2002). "Realism, Method and Truth." *The Problem of Realism.* M. Marsonet (ed.). Ashgate, Aldershot: 64-81.

Sankey, H. and P. Hoyningen-Huene (2001). "Introduction." *Incommensurability and Related Matters.* P. Hoyningen-Huene and H. Sankey (eds.). Dordrecht, Kluwer: vii-xxxiv.

Schaffer, S. (1989). "Glass Works: Newton's Prisms and the Use of Experiment." *The Use of Experiment. Studies in the Natural Sciences.* D. Gooding, T. Pinch and S. Schaffer (eds.). Cambridge, Cambridge University Press: 67-104.

Scheffler, I. (1967). *Science and Subjectivity.* Indianapolis, Hackett.

Schilpp, P. (ed.) (1951). *Albert Einstein: Philosopher-Scientist.* P. Schilpp (ed.). Illinois, Open Court.

Sellars, W. (1961). "The Language of Theories." *Current Issues in the Philosophy of Science.* H. Feigl and G. Maxwell (eds.). New York, Holt, Rinehart, and Winston: 57-77.

Settle, T. (1969). "The Point of Positive Evidence-Reply to Professor Feyerabend." *British Journal for the Philosophy of Science* **20**: 352-355.

Shand, J. (1986). "Grayling, Feyerabend and the Constancy of Sense." *Analysis* **46**: 211-212.

Shapere, D. (1966). "Meaning and Scientific Change." *Mind and Cosmos.* R. Colodny (ed.). Pittsburgh, Pittsburgh University Press: 41-85.

Shapere, D. (1969). "Notes Toward a Post-Positivistic Interpretation of Science, Part 1." *Reason and the Search for Knowledge. Investigations in the Philosophy of Science.* D. Shapere (ed.). Dordrecht, Reidel: 102-119.

Shapere, D. (1984). *Reason and the Search for Knowledge. Investigations in the Philosophy of Science.* Dordrecht, Reidel.

Shapere, D. (1999). "Incommensurability." *Routledge Encyclopedia of Philosophy.* E. Craig (ed.). Routledge: 732-736.

Shapin, S. (1975). "Phrenological Knowledge and the Social Structure of Early Nineteenth Century Edinburgh." *Annals of Science* **32**: 219-243.

Shapiro, A. E. (1996). "The Gradual Acceptance of Newton's Theory of Light and Color, 1672-1727." *Perspectives on Science* **4**: 59-140.

Shea, W. (1971). "Beyond Logical Empiricism." *Dialogue* **10**: 223-242.

Simmons, L. (1994). "Three Kinds of Incommensurability Thesis." *American Philosophical Quarterly* **31**: 119-131.

Smart, J. (1965). "Conflicting views about explanation." *Boston Studies in the Philosophy of Science. Volume 2. In Honor of Philipp Frank.* R. Cohen and M. Wartofsky (eds.). New York, Humanities Press.

Sokal, A. and J. Bricmont (1998). *Fashionable Nonsense.* New York, St. Martin's Press.

Spinner, H. (1973). "Science without Reduction." *Inquiry* **16**: 16-94.

Suchting, W. (1978). "Rising Up from Downunder. Comments on Feyerabend's 'Marxist Fairytales from Australia'." *Inquiry* **21**: 337-361.

Suppe, F. (1972). "What's Wrong With the Received View on the Structure of Scientific Theories?" *Philosophy of Science* **39**: 1-19.

Szumilewicz, I. (1977). "Incommensurability and the Rationality of the Development of Science." *British Journal for the Philosophy of Science* **28**: 345-350.

Theocharis, T. and M. Psimopoulos (1987). "Where Science has Gone Wrong." *Nature* **329**: 595-598.

Tibbets, P. (1976). "Feyerabend on Ideology, Human Happiness, and the Good Life." *Man World* **9**: 362-371.

Torretti, R. (1990). *Creative Understanding: Philosophical Reflections of Physics.* Chicago, Chicago University Press.

Van Fraassen, B. (1980). *The Scientific Image.* Oxford, Clarendon.

Van Fraassen, B. (1984). "To Save the Phenomena." *Scientific Realism.* J. Leplin (ed.). Berkeley, University of California Press: 250-259.

Van Fraassen, B. (1985). "Empiricism in the Philosophy of Science." *Images of Science. Essays on Realism and Empiricism, with a Reply from Bas C. van Fraassen.* P. Churchland and C. Hooker (eds.). Chicago, The University of Chicago Press: 245-308.

Van Fraassen, B. (1997). "Sola Experientia? Feyerabend's Refutation of Classical Empiricism." *Philosophy of Science* **64**: 385-395.

Waismann, F. (1965). *The Principles of Linguistic Philosophy.* New York, St Martin's Press.

Warnock, G. (1999). "Ordinary language philosophy, schools of." *Routledge Encyclopedia of Philosophy*.

Watkins, J. (1966). "Review of Mind and Cosmos edited by R. Colodony." *Philosophy* **41**: 359-362.

Watkins, J. (1978). "Corroboration and the Problem of Content-Comparison." *Progress and Rationality in Science*. G. Radnitzky and G. Andersson (eds.). Dordrecht, Reidel: 339-378.

Watkins, J. (2000). "Feyerabend among Popperians, 1948-1978." *The Worst Enemy of Science? Essays in Memory of Paul Feyerabend*. J. Preston, G. Munévar and D. Lamb (eds.). New York, Oxford University Press: 47-57.

Weber, M. (1993). "The 'Anything Goes' Philosopher: No Rules is a Good Rule for a Scientist to Follow, Paul Feyerabend Tells Marcel Weber." *The Times Higher Education* Supplement 10.

Weber, M. (2002). "Incommensurability and Theory Comparison in Experimental Biology." *Biology and Philosophy* **17**: 155-169.

Weber, M. (2004). *Philosophy of Experimental Biology*. Cambridge, Cambridge University Press.

Whewell, W. (1860) *Philosophy of Discovery*. Chapters Historical and Critical, London.

Wisdom, J. (1974). "The Incommensurability Thesis." *Philosophical Studies* **25**: 299-301.

Wittgenstein, L. (1922). *Tractatus Logico-Philosophicus*. London, Routledge.

Wittgenstein, L. (1958). *Philosophical Investigations*. New York, MacMillan Publishing Company.

Worrall, J. (1978). "Against Too Much Method." *Erkenntnis* **13**: 279-295.

Worrall, J. (1978). "Is the Empirical Content of a Theory Dependent on its Rivals." *The Logical and Epistemology of Scientific Change*. I. Niiniluoto and R. Tuomela (eds.). Amsterdam, North-Holland. **13**: 2-4.

Worrall, J. (1982). "Scientific Realism and Scientific Change." *Philosophical Quarterly* **32**: 201-231.

Worrall, J. (1991). "Feyerabend and the Facts." *Beyond Reason, Essays on the Philosophy of Paul Feyerabend*. G. Munévar (ed.). Dordrecht, Kluwer.

Zahar, E. (1973). "Why did Einstein's Programme supercede Lorentz's?" *British Journal for the Philosophy of Science* **24**: 95-123 and 233-262.

Zahar, E. (1981). "Second Thoughts about Machian Positivism: A Reply to Feyerabend." *British Journal for the Philosophy of Science* **32**: 267-276.

Index

Bei Fragen zur Produktsicherheit wenden Sie sich bitte an:
If you have any questions regarding product safety,
please contact:

Walter de Gruyter GmbH
Genthiner Straße 13
10785 Berlin
productsafety@degruyterbrill.com